T0177343

Multiphase Lattice Boltzmann Methods: Theory and Application

Multiphase Lattice Boltzmann Methods: Theory and Application

Haibo Huang

Department of Modern Mechanics,
University of Science and Technology of China

Michael C. Sukop

Department of Earth and Environment
Florida International University

Xi-Yun Lu

Department of Modern Mechanics
University of Science and Technology of China

WILEY Blackwell

Library of Congress Cataloging-in-Publication Data

Huang, Haibo (Engineering professor)
 Multiphase lattice Boltzmann methods : theory and application / Haibo Huang, Michael C. Sukop, and Xi-Yun Lu.
     pages cm
 Includes bibliographical references and index.
  ISBN 978-1-118-97133-8 (cloth)
  1. Lattice Boltzmann methods. 2. Multiphase flow. 3. Fluid dynamics. 4. Fluid dynamics–Mathematical models. I. Sukop, Michael C. II. Lu, Xi-Yun. III. Title.
  TA357.5.M84H83 2015
  530.13'2–dc23

2015004408

Contents

Preface, xi

About the companion website, xiii

1 Introduction, 1

    1.1 History of the Lattice Boltzmann method, 2

    1.2 The Lattice Boltzmann method, 3

    1.3 Multiphase LBM, 6

        1.3.1 Color-gradient model, 7

        1.3.2 Shan−Chen model, 7

        1.3.3 Free-energy model, 8

        1.3.4 Interface tracking model, 9

    1.4 Comparison of models, 9

    1.5 Units in this book and parameter conversion, 11

    1.6 Appendix: Einstein summation convention, 14

        1.6.1 Kronecker δ function, 15

        1.6.2 Lattice tensors, 15

    1.7 Use of the Fortran code in the book, 16

2 Single-component multiphase Shan−Chen-type model, 18

    2.1 Introduction, 18

        2.1.1 "Equilibrium" velocity in the SC model, 20

        2.1.2 Inter-particle forces in the SC SCMP LBM, 20

    2.2 Typical equations of state, 21

        2.2.1 Parameters in EOS, 27

    2.3 Thermodynamic consistency, 28

        2.3.1 The SCMP LBM EOS, 29

        2.3.2 Incorporating other EOS into the SC model, 31

    2.4 Analytical surface tension, 32

        2.4.1 Inter-particle Force Model A, 32

        2.4.2 Inter-particle Force Model B, 33

    2.5 Contact angle, 34

2.6 Capillary rise, 36

2.7 Parallel flow and relative permeabilities, 39

2.8 Forcing term in the SC model, 40
    2.8.1 Schemes to incorporate the body force, 42
    2.8.2 Scheme overview, 44
    2.8.3 Theoretical analysis, 44
    2.8.4 Numerical results and discussion, 46

2.9 Multirange pseudopotential (Inter-particle Force Model B), 55

2.10 Conclusions, 58

2.11 Appendix A: Analytical solution for layered multiphase flow in a channel, 58

2.12 Appendix B: FORTRAN code to simulate single component multiphase droplet contacting a wall, as shown in Figure 2.7(c), 60

3 Shan and Chen-type multi-component multiphase models, 71

3.1 Multi-component multiphase SC LBM, 71
    3.1.1 Fluid–fluid cohesion and fluid–solid adhesion, 73

3.2 Derivation of the pressure, 73
    3.2.1 Pressure in popular papers (2D), 74
    3.2.2 Pressure in popular papers (3D), 75

3.3 Determining G_c and the surface tension, 76

3.4 Contact angle, 78
    3.4.1 Application of Young's equation to MCMP LBM, 79
    3.4.2 Contact angle measurement, 79
    3.4.3 Verification of proposed equation, 80

3.5 Flow through capillary tubes, 83

3.6 Layered two-phase flow in a 2D channel, 85

3.7 Pressure or velocity boundary conditions, 87
    3.7.1 Boundary conditions for 2D simulations, 87
    3.7.2 Boundary conditions for 3D simulations, 89

3.8 Displacement in a 3D porous medium, 91

4 Rothman–Keller multiphase Lattice Boltzmann model, 94

4.1 Introduction, 94

4.2 RK color-gradient model, 96

4.3 Theoretical analysis (Chapman–Enskog expansion), 99
    4.3.1 Discussion of above formulae, 103

4.4 Layered two-phase flow in a 2D channel, 103
    4.4.1 Cases of two fluids with identical densities, 104
    4.4.2 Cases of two fluids with different densities, 106

4.5 Interfacial tension and isotropy of the RK model, 110
    4.5.1 Interfacial tension, 110
    4.5.2 Isotropy, 110

4.6 Drainage and capillary filling, 111

4.7 MRT RK model, 113

4.8 Contact angle, 114
    4.8.1 Spurious currents, 115

4.9 Tests of inlet/outlet boundary conditions, 117

4.10 Immiscible displacements in porous media, 118

4.11 Appendix A, 121

4.12 Appendix B, 122

5 Free-energy-based multiphase Lattice Boltzmann model, 136

5.1 Swift free-energy based single-component multiphase LBM, 136
    5.1.1 Derivation of the coefficients in the equilibrium
        distribution function, 138

5.2 Chapman–Enskog expansion, 143

5.3 Issue of Galilean invariance, 146

5.4 Phase separation, 149

5.5 Contact angle, 154
    5.5.1 How to specify a desired contact angle, 154
    5.5.2 Numerical verification, 155

5.6 Swift free-energy-based multi-component multiphase LBM, 158

5.7 Appendix, 158

6 Inamuro's multiphase Lattice Boltzmann model, 167

6.1 Introduction, 167
    6.1.1 Inamuro's method, 167
    6.1.2 Comment on the presentation, 169
    6.1.3 Chapman–Enskog expansion analysis, 170
    6.1.4 Cahn–Hilliard equation (equation for order parameter), 173
    6.1.5 Poisson equation, 174

6.2 Droplet collision, 175

6.3 Appendix, 178

7 He–Chen–Zhang multiphase Lattice Boltzmann model, 196

   7.1 Introduction, 196

   7.2 HCZ model, 196

   7.3 Chapman–Enskog analysis, 199
      7.3.1 N–S equations, 199
      7.3.2 CH equation, 202

   7.4 Surface tension and phase separation, 202

   7.5 Layered two-phase flow in a channel, 204

   7.6 Rayleigh–Taylor instability, 205

   7.7 Contact angle, 210

   7.8 Capillary rise, 213

   7.9 Geometric scheme to specify the contact angle and its hysteresis, 215
      7.9.1 Examples of droplet slipping in shear flows, 218

   7.10 Oscillation of an initially ellipsoidal droplet, 219

   7.11 Appendix A, 222

   7.12 Appendix B: 2D code, 223

   7.13 Appendix C: 3D code, 238

8 Axisymmetric multiphase HCZ model, 253

   8.1 Introduction, 253

   8.2 Methods, 253
      8.2.1 Macroscopic governing equations, 253
      8.2.2 Axisymmetric HCZ LBM (Premnath and Abraham 2005a), 255
      8.2.3 MRT version of the axisymmetric LBM (McCracken and Abraham 2005), 256
      8.2.4 Axisymmetric boundary conditions, 258

   8.3 The Laplace law, 258

   8.4 Oscillation of an initially ellipsoidal droplet, 259

   8.5 Cylindrical liquid column break, 263

   8.6 Droplet collision, 265
      8.6.1 Effect of gradient and Laplacian calculation, 267
      8.6.2 Effect of BGK and MRT, 274

   8.7 A revised axisymmetric HCZ model (Huang et al. 2014), 276
      8.7.1 MRT collision, 276
      8.7.2 Calculation of the surface tension, 277
      8.7.3 Mass correction, 278

8.8 Bubble rise, 279

    8.8.1 Numerical validation, 281

    8.8.2 Surface-tension calculation effect, 283

    8.8.3 Terminal bubble shape, 284

    8.8.4 Wake behind the bubble, 284

8.9 Conclusion, 286

8.10 Appendix A: Chapman–Enskog analysis, 288

    8.10.1 Preparation for derivation, 288

    8.10.2 Mass conservation, 289

    8.10.3 Momentum conservation, 289

    8.10.4 CH equation, 291

9 Extensions of the HCZ model for high-density ratio two-phase flows, 292

9.1 Introduction, 292

9.2 Model I (Lee and Lin 2005), 293

    9.2.1 Stress and potential form of intermolecular forcing terms, 293

    9.2.2 Model description, 294

    9.2.3 Implementation, 297

    9.2.4 Directional derivative, 298

    9.2.5 Droplet splashing on a thin liquid film, 299

9.3 Model II (Amaya-Bower and Lee 2010), 301

    9.3.1 Implementation, 302

9.4 Model III (Lee and Liu 2010), 304

9.5 Model IV, 305

9.6 Numerical tests for different models, 306

    9.6.1 A drop inside a box with periodic boundary conditions, 306

    9.6.2 Layered two-phase flows in a channel, 311

    9.6.3 Galilean invariance, 313

9.7 Conclusions, 316

9.8 Appendix A: Analytical solutions for layered two-phase flow in a channel, 317

9.9 Appendix B: 2D code based on Amaya-Bower and Lee (2010), 319

10 Axisymmetric high-density ratio two-phase LBMs (extension of the HCZ model), 334

10.1 Introduction, 334

10.2 The model based on Lee and Lin (2005), 334

    10.2.1 The equilibrium distribution functions I, 336

    10.2.2 The equilibrium distribution functions II, 336

10.2.3 Source terms, 337

10.2.4 Stress and potential form of intermolecular forcing terms, 337

10.2.5 Chapman–Enskog analysis, 338

10.2.6 Implementation, 340

10.2.7 Droplet splashing on a thin liquid film, 342

10.2.8 Head-on droplet collision, 342

10.3 Axisymmetric model based on Lee and Liu (2010), 345

10.3.1 Implementation, 347

10.3.2 Head-on droplet collision, 348

10.3.3 Bubble rise, 353

References, 359

Index, 371

Preface

When we began working with the multiphase Lattice Boltzmann method (LBM), we always asked which of the available models was the best or most suitable for the particular multiphase conditions we investigated and what the advantages and disadvantages of each model were. Over the years we have gained experience with these issues and in this book we share these experiences to hopefully make it easier for researchers to choose an appropriate model.

In this book most of the popular multiphase LBMs are analyzed both theoretically and through numerical simulation. We present many of the mathematical derivations of the models in greater detail than they can be found in the existing literature. Our approach to understanding and classifying the various models is principally based on simulation compared against analytical and observational results and discovery of undesirable terms in the derived macroscopic equations and sometimes their correction. However, we are not numerical analysts, and this limits our ability to fully characterize the numerics of the models. Our hope is that readers can understand some of the advantages and disadvantages of each model quickly and choose the model most suitable for the problems they are interested in.

More LBM books appear with each passing year. As far as we know, this book is the first one addressing all of the popular multiphase models. The book is intended as a reference book on the current state of multiphase LBMs and is targeted at graduate students and researchers who plan to investigate multiphase flows using LBMs.

This book grew out of roughly a decade of teaching and research at the University of Science and Technology of China and Florida International University, USA. We are grateful for the support we received from the National Natural Science Foundation of China (Grant Nos. 11172297 and 11472269), the Program for New Century Excellent Talents in University, Ministry of Education, China (NCET-12-0506), and Alexander von Humboldt Foundation, Germany. We also acknowledge funding from the United States National Science Foundation. This material is based upon work supported by the National Science Foundation under Grant No. 0440253. Any opinions, findings, and conclusions or recommendations are those of the authors and do not necessarily reflect the views of the National Science Foundation. In addition we thank our universities, departments, colleagues, and students for enabling us conduct our research and pursue the creation of this book. Finally, Jianlin Zhao, China University of Petroleum, Eastern China, produced two of the book's figures and Matlab code for the Maxwell construction, while Alejandro Garcia of Florida International

University assisted by producing gfortran versions of the book's code, which are available on the internet.

Haibo Huang, Michael C. Sukop, Xi-Yun Lu
University of Science and Technology of China, Hefei, China
Florida International University, Miami, Florida
November 2014

About the companion website

This book is accompanied by a companion website:

www.wiley.com/go/huang/ boltzmann

The website includes:

- Compaq Fortran and GNU Fortran (GFortran) codes for downloading

CHAPTER 1

Introduction

Multiphase fluid phenomena and flows occur when two or more fluids that do not readily mix (such as air and water) share an interface. Multiphase fluid interactions are nearly ubiquitous in natural and industrial processes. Multiphase phenomena and flows can involve single component multiphase fluids, e.g., water and its own vapor, and multi-component multiphase fluids, e.g., oil/water. Some practical examples of multiphase fluid problems are the recovery and enhanced recovery of petroleum resources from reservoirs, non-aqueous phase liquid contamination of groundwater, soil water behavior, surface wetting phenomena, fuel cell operation, and the movement and evolution of clouds.

Computational fluid dynamics (CFD) has become very important in fluid flow studies. The Lattice Boltzmann method (LBM) has developed very quickly in the last two decades and has become a novel and powerful CFD tool – particularly for multiphase flows. The LBM has some major advantages compared to traditional CFD methods. First, it originates from Boltzmann's kinetic molecular dynamics – a more foundational level than normal continuum approaches. The LBM is able to recover the traditional macroscopic scale continuity and Navier–Stokes (N–S) equations, which are discretized and solved numerically in the common CFD methods. In the LBM, the more fundamental Boltzmann equation is directly discretized. Alternatively, the LBM can be viewed from its discrete-particle, more molecular-dynamics-like lattice gas origins. Second, in the LBM the pressure is usually related to the density through an ideal gas equation of state (for single-phase flow) or through a non-ideal van der Waals-like equation of state for some types of complex multiphase fluids. The pressure fields can be obtained directly once the density field is known. Hence, the Poisson equation – which can be computationally expensive – does not have to be solved in the LBM. The third advantage of the LBM is that the method is easy to parallelize due to the locality of much of the computation. Finally, no-slip boundary condition can be easily handled by simple bounce-back scheme.

The LBM has had great success in studies of single-phase flows, with commercial software known as POWERFLOW (Exa Corporation, https://www.exa.com/), based on the LBM, appearing about ten years ago. In contrast, multiphase LBMs

Multiphase Lattice Boltzmann Methods: Theory and Application, First Edition.
Haibo Huang, Michael C. Sukop and Xi-Yun Lu.
© 2015 John Wiley & Sons, Ltd. Published 2015 by John Wiley & Sons, Ltd.
Companion Website: www.wiley.com/go/huang/boltzmann

are still undergoing development and there are many multiphase Lattice Boltzmann models available.

1.1   History of the Lattice Boltzmann method

LBMs trace their roots to cellular automata, which were originally conceived by Stanislaw Ulam and John von Neumann in the 1940s. Cellular automata consist of a discretization of space on which individual cells exist in a particular state (say 0 or 1), and update their state at each time step according to a rule that takes as input the states of some set of the cell's neighbors. Sukop and Thorne (2006) provide an introduction to cellular automata. Wolfram (1983, 2002) studied simple cellular automata systematically and inspired some of the earliest application to fluids, leading to the first paper to propose a lattice gas cellular automaton (LGCA) for the N–S equations (Frisch et al. 1986). The use of a triangular grid restored some of the symmetry required to properly simulate fluids. Rothman and Zaleski (1997), Wolf-Gladrow (2000), Succi (2001), and Sukop and Thorne (2006) all provide instructive information on this model and the extensions that appeared. All of the LGCA models suffer from inherent defects, however, in particular the lack of Galilean invariance for fast flows and statistical noise (Qian et al. 1992, Wolf-Gladrow 2000). These are explicit particle-based Boolean models that include the random fluctuations that one would expect at a molecular level of gas simulation and hence required extensive averaging to recover the smooth behavior expected at macroscopic scales.

A second major step towards the modern LBM was taken by McNamara and Zanetti (1988), who dispensed with the individual particles of the LGCAs and replaced them with an averaged but still directionally discrete distribution function. This completely eliminated the statistical noise of the LGCA. A major simplification was introduced by Qian et al. (1992): the collision matrix of Higuera et al. (1989) is replaced by a single relaxation time, leading to the Bhatnagar, Gross, and Krook (BGK) model. After that, the LBM developed very quickly. Sukop and Thorne (2006) showed that there were fewer than 20 papers on the topic in 1992; more than 600 were published in 2013.

Later Lallemand and Luo (2000) and Luo (1998) showed that the LBM can be derived from the continuous Boltzmann equation (Boltzmann 1964/1995). Hence, it can be considered as a special discretized form of the Boltzmann equation (Nourgaliev et al. 2003). From the Chapman–Enskog expansion (Wolf-Gladrow 2000), the governing continuity and N–S equations can be recovered from the LBM. Without solving Poisson's equation, the pressure field can be obtained directly from the density distributions.

Today, the use of LBM spans a broad variety of disciplines. For example, an overview of the LBM for material science and engineering can be found in

Raabe (2004). Application of the LBM to biophysics can be found in Boyd et al. (2005) and Sun et al. (2003).

1.2   The Lattice Boltzmann method

The LBM can be derived from the BGK approximation of the Boltzmann equation (He and Luo 1997),

$$\frac{\partial f}{\partial t} + \boldsymbol{\xi} \cdot \nabla f + \mathbf{F} \cdot \nabla_{\boldsymbol{\xi}} f = -\frac{f - f^{eq}}{\tau},$$ (1.1)

where $f(\mathbf{x}, \boldsymbol{\xi}, t)$ is the single-particle distribution function in the phase space $(\mathbf{x}, \boldsymbol{\xi})$, and $f^{eq}(\mathbf{x}, \boldsymbol{\xi})$ is the Maxwell–Boltzmann distribution function. $\mathbf{x}$ is the position vector, $\boldsymbol{\xi}$ is the microscopic velocity, $\mathbf{F}(\mathbf{x},t)$ is a body force, and τ is the relaxation time, which determines the kinematic viscosity.

In the lattice BGK method, a discrete distribution function f_i is introduced to represent the fluid. This distribution function satisfies the following Lattice Boltzmann equation (He and Luo 1997):

$$f_i(\mathbf{x} + \mathbf{e}_i \Delta t, t + \Delta t) = f_i(\mathbf{x}, t) - \frac{1}{\tau}(f_i(\mathbf{x}, t) - f_i^{eq}(\mathbf{x}, t)) + S_i(\mathbf{x}, t),$$ (1.2)

where $f_i(\mathbf{x}, t)$ is the density distribution function related to the discrete velocity direction i and τ is a relaxation time, which is related to the kinematic viscosity by $v = c_s^2(\tau - 0.5)\Delta t$, where c_s is the sound speed. $S_i(\mathbf{x}, t)$ is the source term added into the standard Lattice Boltzmann equation. The equilibrium distribution function $f_i^{eq}(\mathbf{x}, t)$ can be calculated as (Luo 1998)

$$f_i^{eq}(\mathbf{x}, t) = w_i \rho \left[1 + \frac{\mathbf{e}_i \cdot \mathbf{u}}{c_s^2} + \frac{(\mathbf{e}_i \cdot \mathbf{u})^2}{2c_s^4} - \frac{(\mathbf{u})^2}{2c_s^2} \right].$$ (1.3)

In Eqs (1.2) and (1.3) the $\mathbf{e}_i$ are the discrete velocities, as defined below, and w_is are weights, as given in Table 1.1. ρ is the macroscopic density and $\mathbf{u}$ is the macroscopic velocity vector. Discrete velocity models are usually specified as DnQm, where n is the space dimension and m is the number of velocities.

Table 1.1   Overview of the weighting coefficients and sound speeds.

Model	w_i	c_s^2
D2Q7	$\frac{1}{2}$ $(i = 0)$, $\frac{1}{12}$ $(i = 1, \dots, 6)$,	$c^2/4$
D2Q9	$\frac{4}{9}$ $(i = 0)$, $\frac{1}{9}$ $(i = 1, 2, 3, 4)$, $\frac{1}{36}$ $(i = 5, 6, 7, 8)$	$c^2/3$
D3Q15	$\frac{2}{9}$ $(i = 0)$, $\frac{1}{9}$ $(i = 1, \dots, 6)$, $\frac{1}{72}$ $(i = 7, \dots, 14)$	$c^2/3$
D3Q19	$\frac{1}{3}$ $(i = 0)$, $\frac{1}{18}$ $(i = 1, \dots, 6)$, $\frac{1}{36}$ $(i = 7, \dots, 18)$	$c^2/3$

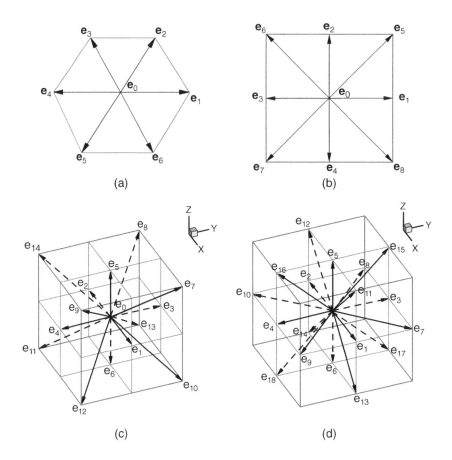

Figure 1.1   Discrete velocity models (a) D2Q7, (b) D2Q9, (c) D3Q15, and (d) D3Q19.

The popular 2D and 3D discrete velocity models are D2Q7, D2Q9, D3Q15, and D3Q19, which are shown in Figure 1.1.

For the D2Q7 model (Frisch et al. 1986), the discrete velocities are

$$[\mathbf{e}_0, \mathbf{e}_1, \mathbf{e}_2, \mathbf{e}_3, \mathbf{e}_4, \mathbf{e}_5, \mathbf{e}_6]$$
$$= c \begin{bmatrix} 0 & 1 & \frac{1}{2} & -\frac{1}{2} & -1 & -\frac{1}{2} & \frac{1}{2} \\ 0 & 0 & \frac{\sqrt{3}}{2} & \frac{\sqrt{3}}{2} & 0 & -\frac{\sqrt{3}}{2} & -\frac{\sqrt{3}}{2} \end{bmatrix}.$$

For the D2Q9 model, the discrete velocities are given by (Qian et al. 1992)

$$[\mathbf{e}_0, \mathbf{e}_1, \mathbf{e}_2, \mathbf{e}_3, \mathbf{e}_4, \mathbf{e}_5, \mathbf{e}_6, \mathbf{e}_7, \mathbf{e}_8]$$
$$= c \begin{bmatrix} 0 & 1 & 0 & -1 & 0 & 1 & -1 & -1 & 1 \\ 0 & 0 & 1 & 0 & -1 & 1 & 1 & -1 & -1 \end{bmatrix}.$$

For the D3Q15 model (Wolf-Gladrow 2000), the velocities are

$$[\mathbf{e}_0, \mathbf{e}_1, \mathbf{e}_2, \mathbf{e}_3, \mathbf{e}_4, \mathbf{e}_5, \mathbf{e}_6, \mathbf{e}_7, \mathbf{e}_8, \mathbf{e}_9, \mathbf{e}_{10}, \mathbf{e}_{11}, \mathbf{e}_{12}, \mathbf{e}_{13}, \mathbf{e}_{14}]$$

$$= c \begin{bmatrix} 0 & 1 & 0 & 0 & -1 & 0 & 0 & 1 & -1 & 1 & 1 & -1 & 1 & -1 & -1 \\ 0 & 0 & 1 & 0 & 0 & -1 & 0 & 1 & 1 & -1 & 1 & -1 & -1 & 1 & -1 \\ 0 & 0 & 0 & 1 & 0 & 0 & -1 & 1 & 1 & 1 & -1 & -1 & -1 & -1 & 1 \end{bmatrix}.$$

For the D3Q19 model (Wolf-Gladrow 2000), they are

$$[\mathbf{e}_0, \mathbf{e}_1, \mathbf{e}_2, \mathbf{e}_3, \mathbf{e}_4, \mathbf{e}_5, \mathbf{e}_6, \mathbf{e}_7, \mathbf{e}_8, \mathbf{e}_9, \mathbf{e}_{10}, \mathbf{e}_{11}, \mathbf{e}_{12}, \mathbf{e}_{13}, \mathbf{e}_{14}, \mathbf{e}_{15}, \mathbf{e}_{16}, \mathbf{e}_{17}, \mathbf{e}_{18}]$$

$$= c \begin{bmatrix} 0 & 1 & -1 & 0 & 0 & 0 & 0 & 1 & -1 & 1 & -1 & 1 & -1 & 1 & -1 & 0 & 0 & 0 & 0 \\ 0 & 0 & 0 & 1 & -1 & 0 & 0 & 1 & 1 & -1 & -1 & 0 & 0 & 0 & 0 & 1 & -1 & 1 & -1 \\ 0 & 0 & 0 & 0 & 0 & 1 & -1 & 0 & 0 & 0 & 0 & 1 & 1 & -1 & -1 & 1 & 1 & -1 & -1 \end{bmatrix}.$$

In the above equations, c is the lattice speed and is defined as $c = \frac{\Delta x}{\Delta t}$. Here, we define 1 lattice unit (Δx) as 1 lu, 1 time step (Δt) as 1 ts, and 1 mass unit as 1 mu. There are other velocity models available, for example the D3Q27 model (He and Luo 1997), but we do not use them in simulations in this book.

In Eq. (1.3) w_is are weighting coefficients that can be derived theoretically (He and Luo 1997). c_s^2 can be derived from

$$c_s^2 \delta_{\alpha\beta} = \sum_i w_i e_{i\alpha} e_{i\beta}, \tag{1.4}$$

where $\delta_{\alpha\beta} = 1$ when $\alpha = \beta$, otherwise $\delta_{\alpha\beta} = 0$ and we use the Einstein summation convention as detailed in the appendix to this chapter. Hence, $c_s^2 = \sum_i w_i e_{ix} e_{ix}$ or $c_s^2 = \sum_i w_i e_{iy} e_{iy}$. As a detailed example, the computation of c_s^2 for the D2Q9 model is given in the following (calculation of each term from $i = 0$ to $i = 8$ is shown):

$$\sum_{i=0}^{8} w_i e_{ix} e_{ix} = 0 + \frac{1}{9}c^2 + 0 + \frac{1}{9}c^2 + 0$$

$$+ \frac{1}{36}c^2 + \frac{1}{36}c^2 + \frac{1}{36}c^2 + \frac{1}{36}c^2 = \frac{1}{3}c^2 = c_s^2, \tag{1.5}$$

while the contribution from $\alpha \neq \beta$ ($x \neq y$) is

$$\sum_{i=0}^{8} w_i e_{ix} e_{iy} = 0 + \frac{1}{9}c \times 0 + \frac{1}{9} 0 \times c + \frac{1}{9}(-c) \times 0 + \frac{1}{9} 0 \times (-c)$$

$$+ \frac{1}{36}c^2 - \frac{1}{36}c^2 + \frac{1}{36}c^2 - \frac{1}{36}c^2 = 0. \tag{1.6}$$

In Eq. (1.3) ρ is the density of the fluid, which can be obtained from:

$$\rho = \sum_i f_i. \tag{1.7}$$

This is simply the sum of the f_i, revealing them as portions of the overall density associated with one of the discrete velocity directions. For $S_i = 0$, the macroscopic fluid velocity is given by

$$\mathbf{u} = \frac{1}{\rho} \sum_i f_i \mathbf{e}_i, \tag{1.8}$$

or in terms of the vector components of $\mathbf{u}$ as

$$u_\alpha = \frac{1}{\rho} \sum_i f_i e_{i\alpha}, \tag{1.9}$$

which means the discrete velocities weighted by the directional densities. Application examples for viscous single-phase flow can be found in Yu et al. (2003), Dünweg and Ladd (2009), Aidun and Clausen (2010), and many others. A discussion on the H theorem in the context of the LBM can be found in Succi et al. (2002).

1.3 Multiphase LBM

Numerous macroscopic numerical methods have been developed for solving the two-phase N–S equations (Scardovelli and Zaleski 1999), such as the front-tracking method, the volume-of-fluid (VOF) method, the level set method, and so on. The first three methods are the most popular ones. However, the front-tracking method is usually not able to simulate interface coalescence or break-up (Liu et al. 2012; Scardovelli and Zaleski 1999). In the VOF and level set methods an interface reconstruction step or interface reinitialization is usually required, which may be non-physical or complex to implement (Liu et al. 2012). In addition, numerical instability may appear when the VOF and level set methods are applied to simulate surface-tension-dominated flows in complex geometries (Scardovelli and Zaleski 1999).

Compared to common CFD methods, the LBM has many advantages (Chen and Doolen 1998). First, it is based on the molecular kinetic theory (Luo 1998). At the macroscopic scale it is able to recover N–S equations. Second, for single-phase flow simulations it usually involves an ideal-gas equation of state. Hence, it is not necessary to solve a Poisson equation for the pressure in the LBM. This saves significant computer central processing unit (CPU) time compared with common CFD methods. Third, it is easy to program and parallelize with much of the computational burden local to a node.

In the last decade LBM has become a numerically robust and efficient technique for simulating both single-phase and multiphase fluids (Chen and Doolen 1998; Guo and Shu 2013; He et al. 1999; Lee and Lin 2005; Rothman and Keller 1988; Shan and Chen 1993; Swift et al. 1995). Compared with conventional methods for multiphase flows, LBM usually automatically maintains sharp interfaces, and explicit interface tracking is not needed (Házi et al. 2002; Inamuro et al. 2004; Sankaranarayanan et al. 2003).

There are several popular multiphase LBM models. The earliest one is the color-gradient model proposed by Gunstensen et al. (1991), which is based on the Rothman–Keller (RK) multiphase lattice gas model (Rothman and Keller 1988). The Shan–Chen (SC) model (Shan and Chen 1993) appeared soon after and is based on incorporation of an attractive or repulsive force, which leads to phase separation. The free-energy (FE) model was proposed by Swift et al. (1995), then the He-Chen-Zhang (HCZ) model (He et al. 1999) was proposed. Coupling with common CFD techniques, some other less popular multiphase LBMs have also been proposed, such as symmetric free-energy-based multi-component LBM (Li and Wagner 2007), the front-tracking LBM (Lallemand et al. 2007), finite-difference LBM for binary fluid (Xu 2005), the total variation diminishing LBM (Teng et al. 2000), the model of Nourgaliev et al. (2002), etc.

We provide an introduction to each of these popular models below and the models are examined in detail in the chapters that follow.

1.3.1 Color-gradient model

In the two-component model, one component is red-colored fluid and the other is blue-colored fluid. Two distribution functions are used to represent the two fluids. In addition to the common collision step in the LBM, there is an extra collision term in the model (Latva-Kokko and Rothman 2005a). There is also a re-coloring step in the model. Grunau et al. (1993) modified the model to handle binary fluids with different density and viscosity ratios. More recently, Ahrenholz et al. (2008) improved the RK model and used a multiple relaxation time (MRT) LBM to handle cases of higher viscosity ratios and lower capillary numbers. One advantage of the RK model is that the surface tension and the ratio of viscosities can be adjusted independently (Ahrenholz et al. 2008). Huang et al. (2013) confirmed that although the RK model is able to correctly simulate density-matched cases, it is usually unable to handle high-density ratio cases. The possible reasons are given in Huang et al. (2013). For cases of density ratio of order $O(10)$, a scheme to improve the RK model is suggested (Huang et al. 2013).

1.3.2 Shan–Chen model

The second type of multiphase LBM model is the SC model (Shan and Chen 1993, Shan and Chen 1994, Sukop and Thorne 2006). In the single-component multiphase (SCMP) SC model, incorporating a forcing term into the corresponding Lattice Boltzmann equation replaces the ideal gas equation of state (EOS) in single-phase LBMs by a non-ideal non-monotonic EOS (Shan and Chen 1993). In the multi-component multiphase (MCMP) SC model, each component is represented by its own distribution function (Shan and Doolen 1995).

The SC SCMP model works well with density ratios of $O(10)$ (Huang et al. 2011a), but the surface tension and the ratio of densities and viscosities cannot be adjusted independently. Some parameters have to be determined through

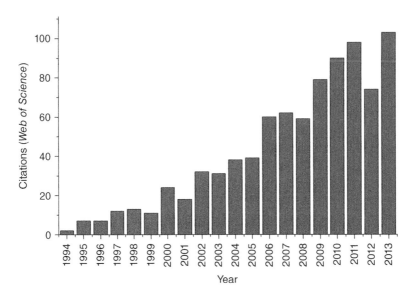

Figure 1.2   Citations in each year for the SC model (Shan and Chen 1993) (data from *Web of Science*).

numerical experiments (Ahrenholz et al. 2008). Falcucci et al. (2010) and Sbragaglia et al. (2007) argue that by adopting a multirange pseudopotential form (including interactions beyond nearest neighbors), the ratio of densities and the surface tension can be adjusted. Shan (2008) outlined a general approach for calculating the pressure tensor in the SC LBM with interactions beyond nearest neighbors. The extension of the interaction beyond the range of nearest neighbors is able to eliminate the spurious currents (Shan 2006, Wagner 2003), which are small-amplitude artificial and unphysical velocity fields near the interface. This finding may expand the possible applications of the SC model. However, some studies (Huang et al. 2011a; Shan et al. 2006) indicate that there is a defect in the original forcing strategy of the SC model and suggest a correct one.

Web of Science citations of the first SC model paper (Shan and Chen 1993) are illustrated in Figure 1.2. From the figure we can see that the citations have increased with time. Study and/or application of the model are very active (approximately 100 citations per year). It is worth mentioning that Shan and Chen (1993) is first in the list of most-cited articles that have been published in *Physical Review E* since 1993. To the end of 2013, the total number of citations is about 860 in the *Web of Science*. Refer to http://pre.aps.org/ for additional information.

1.3.3 Free-energy model
The third type of multiphase LBM model is the FE LBM (Swift et al. 1996, 1995). In this model the thermodynamic issue of the non-monotonic EOS is

Table 1.2 Citations of main articles about Lattice Boltzmann multiphase models.

Model	Article	Total citations through 2013	Citations in 2013
RK	Rothman and Keller (1988)	314	12
	Gunstensen et al. (1991)	468	43
SC	Shan and Chen (1993)	860	103
	Shan and Chen (1994)	398	40
FE	Swift et al. (1995)	468	38
	Swift et al. (1996)	498	50
HCZ	He et al. (1999)	277	38
	Lee and Lin (2005)	146	31

incorporated into the pressure tensor in the N–S equations and the normal equilibrium distribution function is revised (Swift et al. 1995). However, the original FE model (Swift et al. 1995) is not Galilean invariant for the viscous terms in the N–S equation (Luo 1998; Swift et al. 1995). Holdych et al. (1998) improved the model by redefining the stress tensor and Galilean invariance was recovered to $O(u^2)$, which is consistent with the spirit of the LBM.

Inamuro et al. (2004) achieved a high-density ratio through improving Swift's FE model (Swift et al. 1995), but the model has to solve a Poisson equation, which decreased the simplicity of the usual LBM. Zheng et al. (2006) proposed a Galilean-invariant FE LBM model. This model is simpler than that of Inamuro et al., but only valid for density-matched cases (Fakhari and Rahimian 2010).

1.3.4 Interface tracking model

The fourth type of multiphase LBM model is the interface tracking model proposed by He et al. (1999) (the HCZ model). In the HCZ model, two distribution functions and two corresponding LBEs are used. Macroscopically, the Cahn–Hilliard interface tracking equation and the N–S equations can be recovered from the Lattice Boltzmann equations. Based on this model, many models have been developed to access higher density ratios (Amaya-Bower and Lee 2010; Lee and Fischer 2006; Lee and Lin 2005; Lee and Liu 2010) or to enhance numerical stability by extending into an MRT version (McCracken and Abraham 2005).

All the above Lattice Boltzmann multiphase models are under active development. The citation trends are similar to those of the SC model. In Table 1.2, the total citations garnered by each model (based on only the most representative articles) are listed.

1.4 Comparison of models

There are some theoretical analyses (He and Doolen 2002; Luo 1998; Swift et al. 1995) of the RK, SC, and FE models, and a few numerical analyses (Hou et al.

Table 1.3 Comparison of Lattice Boltzmann multiphase models.

Model	Maximum density ratio	Convenient to specify wetting condition?	Efficiency	Accuracy
RK	$O(10)^*$	Yes	Not so efficient	Accurate
SC SCMP	$O(10^2)$	Yes	Very efficient	Less accurate
SC MCMP	$O(1)$	Yes	Very efficient	Less accurate
FE	$O(10)$	No, density gradient required	Not so efficient	Accurate
HCZ	$O(10)$	Yes	Efficient	Accurate
Lee–Lin	$O(10^3)$	No, density gradient required	Efficient	Accurate

* The revised RK (Huang et al. 2013). For the RK model (Latva-Kokko and Rothman 2005a), only density-matched cases can be simulated correctly.

1997; Huang et al. 2011b). Hou et al. (1997) compared the SC and RK models, and focused on drop/bubble simulation. However, there are no quantitative comparisons with other available analytical solutions in that work. Huang et al. (2011b) evaluated the performance of the RK, SC, and FE models for multi-component flow in porous media.

In this book, all of these popular models, the RK model (Grunau et al. 1993; Gunstensen et al. 1991; Rothman and Keller 1988), the SC model (Shan and Chen 1993; Shan and Doolen 1995), the FE model (Inamuro et al. 2004; Swift et al. 1995), and the HCZ model (He et al. 1999; Lee and Lin 2005), will be evaluated in detail. The emphasis is on the strengths and weakness of each model.

The models are compared in Table 1.3. For the RK model (Latva-Kokko and Rothman 2005a), only density-matched cases can be simulated correctly. Huang et al. (2013) extended the RK model to handle higher density ratios. According to our experience, the SC model is very efficient but less accurate. The FE model is as efficient as the RK model. Potentially, there are some similarities between them (Huang et al. 2011b). For the original HCZ model (He et al. 1999), the density ratio is about $O(10)$. Later the HCZ model was extended by Lee and Lin (2005) (Lee–Lin model) to handle high-density ratios. How to specify the wetting condition is an important topic in multiphase flow problems, especially for flows in porous media. In the RK, SC, and HCZ models, a "wall density" can be specified to obtain the desired contact angles (Huang et al. 2014b; Huang and Lu 2009) or a fluid-surface force can be incorporated (Huang et al. 2007; Martys and Chen 1996; Sukop and Thorne 2006). In the FE and Lee–Lin models, the wetting condition can only be implemented by specifying the density gradient on the wall (Liu et al. 2013). Specifying "wall density" or force is more convenient than the density gradient scheme (Liu et al. 2013). For more details readers should refer to the corresponding chapters.

The structure of the book is shown in Figure 1.3 and it is organized as follows. In each chapter, first the model will be introduced briefly. Second, the relevant

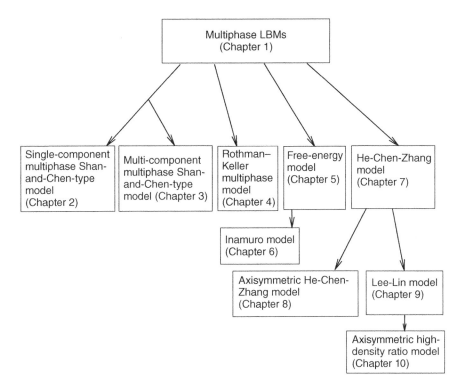

Figure 1.3   Content of the book.

numerical analysis, such as the Chapman–Enskog expansion, and other important formula derivations relevant to the model are presented. Then applications
of the model are given, such as contact angles, bubble rise, and multiphase flows
in porous media.

1.5   Units in this book and parameter conversion

In this book, if the units of a variable are not specified, the units are in lattice
units. In other words, the units are combinations of the basic units listed in
Table 1.4. For example, velocity is given in lu/ts, density in mu/ lu^3, pressure
in mu/(lu ts^2), surface tension in mu/ ts^2, etc.

   Usually, parameter conversion can be performed through non-dimensional
parameters, e.g., Reynolds number Re, Weber number We, or capillary number
Ca. For single-component flows, the parameter conversion is easier than for multiphase flows because often only Re is considered. Hilpert (2011), Sukop and
Thorne (2006) provide a procedure for Re-matching for gravity-driven flow in a
slit. As another example, in a lid-driven cavity flow, the fluid is water at 20° C
($\nu = 10^{-6}$ m^2/s), cavity dimension (characteristic length) is $L = 0.01$ m, and the

Table 1.4   Units in Lattice Boltzmann methods.

	Unit
Mass	mu (mass unit)
Length	lu (lattice unit)
Time	ts (time step)
Temperature	tu (temperature unit)

lid velocity is $U = 0.01$ m/s. The Reynolds number in this flow is

$$Re = \frac{U_{phys}L_{phys}}{v_{phys}} = \frac{0.01 \text{ m/s} \times 0.01 \text{ m}}{10^{-6} \text{ m}^2/\text{s}} = 100 = \frac{U_{LBM}L_{LBM}}{v_{LBM}}. \tag{1.10}$$

One can first choose the LBM length L, pick an LBM kinematic viscosity v, and then compute the LBM lid velocity U. Change parameters if U is too big (> 0.1 lu/ts). Alternatively, if the dimension of the cavity and the velocity in LBM are assumed to be $L = 100$ lu and $U = 0.1$ lu/ts, the kinematic viscosity can be calculated from $v_{LBM} = \frac{U_{LBM}L_{LBM}}{Re} = 0.1$ lu^2/ts. Then the relaxation time can be obtained through $\tau = \frac{v}{c_s^2 \Delta t} + 0.5 = 0.8$, where τ is supposed to be a non-dimensional parameter. The non-dimensional time step is

$$\Delta t^* = \frac{\Delta t_{LBM} U_{LBM}}{L_{LBM}} = \frac{\Delta t_{phys} U_{phys}}{L_{phys}}. \tag{1.11}$$

Hence, in this case, $\Delta t^* = \frac{U_{LBM}\Delta t_{LBM}}{L_{LBM}} = 10^{-3}$. In the physical situation, the time step is Δt_{phys}, which corresponds to the time step in the LBM $\Delta t_{LBM} == \frac{\Delta t^* \times L_{LBM}}{U_{LBM}} = \frac{10^{-3} \times 100 \text{ lu}}{0.1 \text{ lu/ts}} = 1$.

From the above calculation procedure, we can see that if U is chosen to be smaller, say $U = 0.001$ and L is fixed, the calculated $\tau = 0.503$, which is very close to 0.5. That means the non-dimensional time step $\Delta t^* = \frac{U\Delta t}{L} = 10^{-5}$ is smaller. For unsteady flow problems, not only grid-independence studies but also time step-independence studies should be performed in CFD simulations. For unsteady flow problems, a smaller non-dimensional time step (Δt^*) is preferred, which means that a smaller τ should be adopted.

For the LBM, the *CFL* (Courant–Friedrichs–Lewy) number is $CFL = \frac{e_a \Delta t}{\Delta x_a} = 1$, which is fixed to be unity. A 'predictor–corrector' algorithm was introduced in Nourgaliev et al. (2003) to relax constraints imposed by the *CFL* condition.

For the cases of multiphase flows, here an example about oil–water flow in porous media is used to illustrate how to convert the parameters between LBM and reality. The density and viscosity of oil (dyed PCE) and water are shown in Table 1.5 (Pan et al. 2004). The subscripts n and w denote non-wetting and wetting fluid, respectively.

To make the simulation analogous to reality, the following non-dimensional parameters (capillary number *Ca*, dynamic viscosity ratio *M*, and Bond

Table 1.5 Properties of an experimental multiphase system.

NWP (dyed PCE) density ρ_n	1.6×10^3 kg/ m^3
WP (water) density ρ_w	1.0×10^3 kg/ m^3
NWP dynamic viscosity η_n	1.844×10^{-3} Pa·s
WP dynamic viscosity η_w	1.0×10^{-3} Pa·s
Interfacial tension σ	3.623×10^{-2} (kg/ s^2)
Dynamic viscosity ratio M	1.844
Bond number Bo	8.9×10^{-5}

NWP, non-wetting phase; PCS, tetrachloroethylene; WP, wetting phase.

number Bo) should be identical in both the simulations and reality:

$$Ca = \frac{u_w \eta_w}{\sigma},$$

$$M = \frac{\eta_n}{\eta_w},$$

$$Bo = g_0(\rho_n - \rho_w)\frac{R^2}{\sigma}, \tag{1.12}$$

where u_w is the wetting phase Darcy velocity, g_0 is the gravitational constant, and R is the mean pore radius.

Using the SC model, Pan et al. (2004) proposed two ways to achieve the desired dynamic viscosity ratio in the above flow system (Table 1.5). One approach matches both the density and kinematic viscosity ratios (Case A) and the other approach assumes that the densities for both fluids are identical and matches the dynamic viscosity ratio by changing the kinematic viscosity of each fluid (Case B). Obviously, in Case B the body force effect is neglected because the Bond number $Bo = 0$ due to $\rho_n = \rho_w$. As mentioned in Pan et al. (2004) in their work, capillary force dominates and $Bo = 0$ is acceptable.

Here we discuss Case A. If the densities of the water and oil (PCE) are set to be 1 mu/ lu^3 and 1.6 mu/ lu^3, respectively, then the density ratio matches that in reality. The surface tension is related to the interaction force between the two components in the SC model (Chapter 3). After the strength of the fluid/fluid interaction (a parameter in the interaction formula) is set, the surface tension is determined. Suppose the relaxation time of the wetting fluid is unity ($\tau_w = 1$) because the dynamic viscosity ratio is

$$\frac{\eta_{n,phys}}{\eta_{w,phys}} = \frac{\eta_{n,LBM}}{\eta_{w,LBM}} = \frac{\rho_n c_s^2(\tau_n - 0.5)}{\rho_w c_s^2(\tau_w - 0.5)} \tag{1.13}$$

and by rearrangement the relaxation time of the non-wetting fluid is

$$\tau_n = \frac{\eta_n}{\eta_w}\frac{\rho_w}{\rho_n}(\tau_w - 0.5) + 0.5 = \frac{1.844}{1.6}(1 - 0.5) + 0.5 = 1.08. \tag{1.14}$$

Through the definition of Ca in Eq. (1.12), we have

$$Ca = \frac{u_{w,phys}\eta_{w,phys}}{\sigma_{phys}} = \frac{u_{w,LBM}\eta_{w,LBM}}{\sigma_{LBM}}. \tag{1.15}$$

Hence, to match a case with the physical $Ca = 10^{-3}$, if $\sigma_{LBM} = 0.1$ mu/ ts^2 (possibly determined by fitting the Laplace law to drops and bubbles), displacement velocity $u_{w,LBM}$ can be calculated in lattice units as follows:

$$\begin{aligned} u_{w,LBM} &= \frac{\sigma_{LBM}Ca}{\eta_{w,LBM}} = \frac{\sigma_{LBM}Ca}{\rho_w c_s^2(\tau_{w,LBM} - 0.5)\Delta t} \\ &= \frac{0.1 \text{ mu/ts}^2 \times 10^{-3}}{1 \text{ mu/lu}^3 \times \frac{1}{3} \text{ lu}^2/\text{ts}^2 \times (1.0 - 0.5) \times 1 \text{ ts}} = 6 \times 10^{-4} \text{lu/ts}. \tag{1.16} \end{aligned}$$

This $u_{w,LBM}$ can be used to specify the inlet velocity of the displacement to match the case with $Ca = 10^{-3}$. For the gravity effect, if in our LBM simulation $R = 10$ lu, then g_0 in lattice units can be calculated from the definition of Bo:

$$Bo = g_{0,phys}(\rho_{n,phys} - \rho_{w,phys})\frac{R_{phys}^2}{\sigma_{phys}} = g_{0,LBM}(\rho_{n,LBM} - \rho_{w,LBM})\frac{R_{LBM}^2}{\sigma_{LBM}}. \tag{1.17}$$

That is

$$\begin{aligned} g_{0,LBM} &= \frac{\sigma_{LBM}Bo}{R_{LBM}^2(\rho_{n,LBM} - \rho_{w,LBM})} \\ &= \frac{0.1 \text{ mu/ts}^2 \times 8.9 \times 10^{-5}}{100 \text{ lu}^2 \times (1.6 - 1.0) \text{ mu/lu}^3} = 1.48 \times 10^{-7} \text{ lu/ts}^2. \tag{1.18} \end{aligned}$$

For more examples of parameter conversion between LBM simulations and physical reality, please refer to Chapter 2 (Section 2.6), Chapter 7 (Sections 7.6 and 7.7), and Chapter 8 (Sections 8.6 and 8.8).

1.6 Appendix: Einstein summation convention

In this book, the subscripts α, β, and γ denote Cartesian components (for 2D cases, it is x or y coordinates; for 3D cases, it is x, y, or z coordinates).

According to the Einstein summation convention, when an index variable (the subscript) appears twice in a single term it implies summation of that term over all the values of the index. For example, the kinematic energy E in 2D cases is

$$E = \frac{1}{2}m(u_x^2 + u_y^2), \tag{1.19}$$

where m is mass of an object. Obviously, the indices (subscripts) can range over the set x, y for 2D cases. Eq. (1.19) can be reduced by the convention to

$$E = \frac{1}{2}mu_\alpha u_\alpha. \tag{1.20}$$

For 3D cases, the equation (1.20) means $E = \frac{1}{2}m(u_x^2 + u_y^2 + u_z^2)$.

The index that is summed over is a summation index, in this case α. It is also called a dummy index since any symbol can replace α without changing the meaning of the expression, provided that it does not collide with index symbols in the same term (Hazewinkel 1993). Note that dummy indices do not appear in the 'answer', e.g., $I = \int f(\theta)d\theta = \int f(x)dx$, where θ and x are dummy variables.

An index that is not summed over is a free index and should be found in each term of the equation or formula if it appears in any term (Hazewinkel 1993). The connection between the boldface notation and Einstein summation can be illustrated in the following examples. The boldface notation $\mathbf{v} \cdot \mathbf{w}$, $\nabla \cdot \vec{\tau}$, and $\nabla^2 s$ can be expressed as $v_i w_i$, $\partial_j \tau_{ji}$, and $\partial_i \partial_i s = \partial_i^2 s$, respectively. Here $\mathbf{v}$, $\mathbf{w}$ are vectors, $\vec{\tau}$ is a second-order tensor, and s is a scaler.

1.6.1 Kronecker δ function

When the subscripts in $\delta_{\alpha\beta}$ are identical, for example $\alpha = \beta = x$ (or $\alpha = \beta = y$ in 2D cases), then $\delta_{\alpha\beta} = 1$, otherwise $\delta_{\alpha\beta} = 0$. Hence, we have $u_\alpha u_\beta \delta_{\alpha\beta} = u_\alpha u_\alpha = u_x^2 + u_y^2$ for 2D cases.

In the following, two additional examples demonstrate the operation of the δ function. For 2D cases ((x,y) coordinates), $\delta_{\alpha\beta}\delta_{\alpha\beta} = 2$ because according to the summation rule

$$\delta_{\alpha\beta}\delta_{\alpha\beta} = (\delta_{\alpha x}\delta_{\alpha x}) + (\delta_{\alpha y}\delta_{\alpha y}) = (\delta_{xx}\delta_{xx} + \delta_{yx}\delta_{yx}) + (\delta_{xy}\delta_{xy} + \delta_{yy}\delta_{yy})$$

$$= \delta_{xx}\delta_{xx} + \delta_{yy}\delta_{yy} = 2. \tag{1.21}$$

It is easy to derive $\delta_{\alpha\beta}\delta_{\alpha\beta} = 3$ for 3D cases.

For 2D cases ((x,y) coordinates), $\delta_{\alpha\beta}\delta_{\alpha\gamma} = \delta_{\beta\gamma}$ because

$$\delta_{\alpha\beta}\delta_{\alpha\gamma} = \delta_{x\beta}\delta_{x\gamma} + \delta_{y\beta}\delta_{y\gamma}. \tag{1.22}$$

It is obvious that providing $\beta = \gamma$ (no matter $\beta = \gamma = x$ or $\beta = \gamma = y$), $\delta_{\alpha\beta}\delta_{\alpha\gamma} = 1$ (otherwise $\delta_{\alpha\beta}\delta_{\alpha\gamma} = 0$), which exactly means $\delta_{\alpha\beta}\delta_{\alpha\gamma} = \delta_{\beta\gamma}$.

1.6.2 Lattice tensors

For the first- and third-order lattice tensors in the D2Q9 model, because of the symmetry of the velocities in the velocity model we have

$$\sum_{i=0}^{8} e_{i\alpha} = 0 \tag{1.23}$$

and

$$\sum_{i=0}^{8} e_{i\alpha}e_{i\beta}e_{i\gamma} = 0. \tag{1.24}$$

For the second-order lattice tensor in the D2Q9 model we have:

$$\sum_{i=1}^{4} e_{i\alpha} e_{i\beta} = 2c^2 \delta_{\alpha\beta} \quad \text{and}$$

$$\sum_{i=5}^{8} e_{i\alpha} e_{i\beta} = 4c^2 \delta_{\alpha\beta}. \tag{1.25}$$

For the fourth-order lattice tensor in the D2Q9 model, the formula is a little bit more complex:

$$\sum_{i=1}^{4} e_{i\alpha} e_{i\beta} e_{i\gamma} e_{i\delta} = 2c^4 \delta_{\alpha\beta\gamma\delta} \quad \text{and}$$

$$\sum_{i=5}^{8} e_{i\alpha} e_{i\beta} e_{i\gamma} e_{i\delta} = 4c^4 (\delta_{\alpha\beta}\delta_{\gamma\delta} + \delta_{\alpha\gamma}\delta_{\beta\delta} + \delta_{\alpha\delta}\delta_{\beta\gamma}) - 8c^4 \delta_{\alpha\beta\gamma\delta} \tag{1.26}$$

If the weighting factor w_i in the equilibrium distribution function (Eq. (1.3)) is added into the above tensors, we have

$$\sum_{i=0}^{8} w_i e_{i\alpha} = 0, \tag{1.27}$$

$$\sum_{i=0}^{8} w_i e_{i\alpha} e_{i\beta} = \frac{1}{3} c^2 \delta_{\alpha\beta}, \tag{1.28}$$

$$\sum_{i=0}^{8} w_i e_{i\alpha} e_{i\beta} e_{i\gamma} = 0, \tag{1.29}$$

and

$$\sum_{i=0}^{8} w_i e_{i\alpha} e_{i\beta} e_{i\gamma} e_{i\delta} = c_s^4 (\delta_{\alpha\beta}\delta_{\gamma\delta} + \delta_{\alpha\gamma}\delta_{\beta\delta} + \delta_{\alpha\delta}\delta_{\beta\gamma}). \tag{1.30}$$

In the following chapters, in the Chapman–Enskog expansion analysis the above formulae are used extensively.

1.7   Use of the Fortran code in the book

The codes provided with this book were originally written and edited using Compaq Visual Fortran version 6.5.0. The format is strictly based on the fixed format (Fortan77): before each line, there are seven blank spaces. For a long line exceeding the fixed line length of 80 columns, "&", "*" or "%" at the beginning of the next line can be used to extend the content onto that line.

Lines "c===================================" in the book's listing of the code are used to separate the different files only. Please do not include these lines in the separate Fortran files that will be needed to compile the code. Lines "c- - - - - - - - - - - - - - - - - " in the code are used to separate small segments inside a subroutine.

If Compaq Visual Fortran is used to compile the files, choose Win32 Release in the menu Build-> Set Active Configuration. Executable files compiled using Win32 Release are much more efficient than those compiled using the Win32 Debug option. If you encounter an error "forrt1: severe <170> : Program Exception – stack overflow", modify the following option and compile again. In "Project->Setting->Link->Output->Reserve", set the reserve value to be at least 200,000,000. This means you reserve more memory when the executable file runs.

Create a new folder "out" in the working directory before running the code and all output files will be written into the subfolder. The codes given in this book have not been optimized for memory or efficiency.

We also provide instructions designed to make the codes compatible with gfortran when directly copied from the PDF. This is necessary due to the fact that directly copying from the PDF changes much of the spacing and involves characters that are not compilable by gfortran. We suggest not using Adobe reader for directly copying and pasting due to the program not working well. We suggest using other alternative PDF viewer/editors such as Foxit, which do not have these problems.

The code should be compiled using the command "gfortran -ffree-form *.for" in order to compile all the files at once. Each file should be tested individually using gfortran in order to catch errors. All c's used to comment should be changed to !. All continuation characters, including &, *, and %, should be changed to & and placed at the end of the preceding line, not at the beginning of the new line.

When copying directly from a PDF to a text editor, the system uses * as a linebreak separating code. If this occurs, then the previous line must be joined with the next line with the * in between. This is only in the case when * is used as an operator and not as a replacement for &.

A line of - - - - - - - - - ending in - tends to pull the next line up. The next line should be moved back down where it is part of the code.

Thick apostrophes are not readable by the compiler and should be replaced with a standard'.

Finally, we have made a repository of codes in both the Compaq and gfortan forms available on the internet (www.wiley.com/go/huang/boltzmann).

CHAPTER 2

Single-component multiphase Shan–Chen-type model

2.1 Introduction

The SC model appeared in 1993–1994 (Shan and Chen 1994, 1993). With the incorporation of an attractive or repulsive force between 'particles' in the LBM model, phase separation due to the non-ideal gas behavior of a single chemical component (SCMP) or phase separation due to mutual repulsion of one or more different chemical components (MCMP) can be realized. This model is elegant in its simplicity and intuitive connection to classical non-ideal gas equations of state and hydrophobic phase separation. Evaporation, condensation, and cavitation can be simulated with the SCMP model, while diffusion can be simulated as a zero inter-particle force in the MCMP model (Sukop and Thorne 2006). Despite these positive attributes, the original SC model has a number of limitations, for example it is generally limited to low-density ratios between liquid and vapor phases, and the surface tension cannot be specified independently of inter-particle force.

In the last 20 years there have been many applications of the SCMP SC model. The model has been applied to study bubble rise (Sankaranarayanan et al. 2002), cavitation (Falcucci et al. 2013; Sukop and Or 2005), relative permeability in porous media (Chen et al. 2014; Huang and Lu 2009; Martys and Chen 1996), the 3D moving contact line problem (Hyväluoma and Harting 2008), and air entrainment by moving contact lines (Chan et al. 2013). The SC model is also extended to simulate thermal two-phase flows (Biferale et al. 2012; Házi and Márkus 2008; Zhang and Chen 2003), suspensions of solid particles in liquid and/or vapor phases (Joshi and Sun 2009), and flow of soft-glassy systems (Benzi et al. 2009b). More application examples can be found in Chen et al. (2014); Gong and Cheng (2012); Hazi and Markus (2009); Hyväluoma et al. (2007); Joshi and Sun (2010); Sbragaglia et al. (2006); Sukop and Or (2003, 2004). In

Multiphase Lattice Boltzmann Methods: Theory and Application, First Edition.
Haibo Huang, Michael C. Sukop and Xi-Yun Lu.
© 2015 John Wiley & Sons, Ltd. Published 2015 by John Wiley & Sons, Ltd.
Companion Website: www.wiley.com/go/huang/boltzmann

the original SC model, an attractive force between fluid at the computational node and fluid at its nearest neighbors (eight neighbors for the D2Q9 model in the Cartesian mesh) is incorporated into the model, which can lead to non-ideal gas behavior and phase separation. The SC model has been shown to lack thermodynamic consistency (explained in detail in Section 2.3) (Benzi et al. 2006; He and Doolen 2002; Shan and Chen 1994; Swift et al. 1995) and the surface tension in the model cannot be adjusted independently of the density ratio (He and Doolen 2002). However, in many applications this thermodynamic inconsistency is not of primary importance. Recently, Sbragaglia et al. (2007) argued that surface tension can be adjusted for constant density ratios. The proposed method (Falcucci et al. 2010; Sbragaglia et al. 2007) extended the inter-particle interaction up to next-nearest neighbors. Kupershtokh et al. (2009) also proposed a similar strategy. Spurious current near the interface in simulations with the SC model originates from insufficient isotropy of the discrete gradient operator (Shan 2006). Extension of interactions beyond nearest neighbors can eliminate the spurious currents in the SC simulations (Shan 2006).

In the SCMP SC model, by incorporating an attractive forcing term into the corresponding LBE, the ideal gas EOS that characterizes single-phase LBMs is substituted by a non-ideal non-monotonic EOS. An introduction to non-ideal EOS can be found in Sukop and Thorne (2006). Some researchers (He and Doolen 2002, Sankaranarayanan et al. 2002, Yuan and Schaefer 2006) have proposed a simple implementation strategy to incorporate different EOS into the SC LBM to achieve high-density ratios (see Section 2.3.2). An alternative way to incorporate different EOS into the SC LBM is proposed by Zhang et al. (2004) and Zhang and Kwok (2004). Yuan and Schaefer (2006) investigated the magnitude of spurious currents and the coexistence curves for five different EOS. Sofonea et al. (2004) proposed an alternative numerical scheme using flux limiter techniques to reduce spurious currents and ensure improved low-viscosity numerical stability. Huang et al. (2011a) addressed the important issue of surface tension variations in the SC LBM. Through studying different forcing strategies in the SC model, it is found that in terms of matching analytical expectations of surface tension the SC model in combination with the forcing strategy of He et al. (1998) is able to generate a much more accurate surface tension than the other forcing strategies (Huang et al. 2011a).

The SC model is usually unable to handle high-density ratio multiphase simulations due to numerical instability. To improve numerical stability, Yu and Fan (2010) extended the SC model to an MRT version.

In this chapter the SCMP SC model is briefly introduced. The specification of the contact angle in the model is then illustrated. The forcing term in the SC model is discussed in detail in Section 2.8. In addition, the mid-range interaction in the SC model (Sbragaglia et al. 2007) involving a stencil extending beyond nearest neighbors is discussed.

2.1.1 "Equilibrium" velocity in the SC model

This scheme was proposed by Shan and Chen (1993). After the collision step, the momentum of fluid particles is calculated as

$$\rho u_\alpha = \sum_i f_i e_{i\alpha}. \tag{2.1}$$

Suppose a momentum $\mathbf{F}\tau$ is contributed to the fluid particles from internal or external body forces, then the momentum of the fluid particles would reach a new equilibrium state of $\rho \mathbf{u}^{eq}$ after a time $\Delta t = \tau$. From Newton's law of motion, the components of the "equilibrium" velocity $\mathbf{u}^{eq}$ are calculated from

$$u_\alpha^{eq} = u_\alpha + \frac{F_\alpha \tau}{\rho}, \tag{2.2}$$

where we recognize that the density is equal to the mass for a unit volume. According to the SC model, this velocity should be substituted into Eq. (1.3) to calculate f_i^{eq}. In Eq. (2.2), the force acting on the fluid includes the inter-particle force $\mathbf{F}_{int}$ and external force $\mathbf{F}_{ext}$. For simplicity $\mathbf{F}_{ext} = 0$ for the moment.

In the model, however, the actual fluid velocity is not defined by $\mathbf{u}^{eq}$, but rather by $\mathbf{u}^*$, which according to Shan and Doolen (1995) can be calculated by

$$u_\alpha^* = u_\alpha + \frac{F_\alpha \Delta t}{2\rho}. \tag{2.3}$$

Note that in this scheme, the "equilibrium" velocity $\mathbf{u}^{eq}$, which is used in Eq. (1.3), and the "physical" velocity $\mathbf{u}^*$ may not be identical.

2.1.2 Inter-particle forces in the SC SCMP LBM

In the original D2Q9 SC model, the inter-particle force is defined as (Martys and Chen 1996)

$$\mathbf{F}_{int}(\mathbf{x}, t) = -G\psi(\mathbf{x}, t) \sum_i w_i \psi(\mathbf{x} + \mathbf{e}_i \Delta t, t)\mathbf{e}_i, \tag{2.4}$$

where G is a parameter that controls the strength of the inter-particle force and ψ is a mean-field potential. It is seen from Eq. (2.4) that the interaction is only applied to nearest neighbors. In Shan and Chen (1993) ψ is defined as

$$\psi(\rho) = \rho_0[1 - \exp(-\rho/\rho_0)], \tag{2.5}$$

where ρ_0 is a constant. In Shan and Chen (1994) ψ is defined as

$$\psi(\rho) = \psi_0[-\exp(-\rho_0/\rho)], \tag{2.6}$$

where ψ_0 and ρ_0 are arbitrary constants. This second potential yields a model with "...behavior consistent with an isothermal process..." (Shan and Chen 1994). This inter-particle force is subsequently referred to as **Inter-particle Force Model A**.

If interactions with next-nearest neighbors are also involved in the force computation, the value of the surface tension may be adjusted without changing the density ratio (Sbragaglia et al. 2007). This inter-particle force is defined as (Sbragaglia et al. 2007)

$$\mathbf{F}_{int}(\mathbf{x}, t) = -\psi(\mathbf{x}, t) \sum_i w_i [G_1 \psi(\mathbf{x} + \mathbf{e}_i \Delta t, t) + G_2 \psi(\mathbf{x} + 2\mathbf{e}_i \Delta t, t)] \mathbf{e}_i, \qquad (2.7)$$

where G_1 and G_2 are parameters that control the interactions with nearest and the next-nearest neighbors, respectively. It is referred to as **Inter-particle Force Model B**.

For Inter-particle Force Model A (Eq. (2.4)), one finds from Taylor expansion as described in Appendix A in Benzi et al. (2006):

$$\begin{aligned}
F_\alpha &= -G\psi \sum_i w_i \psi(\mathbf{x} + e_{i\alpha} \Delta t) e_{i\alpha} \\
&= -G\psi \Big\{ \sum_i w_i e_{i\alpha} \psi + \Delta t \sum_i w_i e_{i\alpha} e_{i\beta} \partial_\beta \psi \\
&\quad + \tfrac{1}{2} \Delta t^2 \sum_i w_i e_{i\alpha} e_{i\beta} e_{i\gamma} \partial_\beta \partial_\gamma \psi + \tfrac{1}{6} \Delta t^3 \sum_i w_i e_{i\alpha} e_{i\beta} e_{i\gamma} e_{i\delta} \partial_\beta \partial_\gamma \partial_\delta \psi \Big\} + \dots \\
&\approx -\tfrac{G}{2} \Delta t c_s^2 \partial_\alpha \psi^2 - \tfrac{G}{2} \Delta t^3 c_s^4 \psi (\partial_\alpha \nabla^2 \psi).
\end{aligned} \qquad (2.8)$$

where the subscripts α, β, and γ indicate the coordinates x or y for the two-dimensional cases considered here. The inter-particle force can be translated into an excess pressure with respect to that of the ideal gas by

$$-\partial_\alpha p_{\alpha\beta} + \partial_\beta (c_s^2 \rho) = F_\beta. \qquad (2.9)$$

Then the total pressure tensor can be obtained as follows (Benzi et al. 2006):

$$p_{\alpha\beta} = \left[c_s^2 \rho + \tfrac{1}{2} G c_s^2 \psi^2 + \tfrac{1}{2} G c_s^4 (\psi \nabla^2 \psi + \tfrac{1}{2} |\nabla \psi|^2) \right] \delta_{\alpha\beta} - \tfrac{1}{2} G c_s^4 \partial_\alpha \psi \partial_\beta \psi. \qquad (2.10)$$

If the Taylor expansion is only expanded to $O(\Delta t)$ in Eq. (2.8), then from Eq. (2.9) we obtain the EOS in the SCMP SC model for Inter-particle Force Model A:

$$p = c_s^2 \rho + \frac{c_s^2 G}{2} \psi^2(\rho). \qquad (2.11)$$

2.2   Typical equations of state

"Ideal" or "perfect" gas laws characterize the behavior of gases at low density. Laws that give the pressure/density relations of a gas are known as EOS. The ideal gas law is commonly written as

$$PV = nRT \quad \text{or} \quad P = \frac{nRT}{V}, \qquad (2.12)$$

where P is pressure (atm), V is volume (L), n is the number of mols, R is the gas constant (0.0821 atm L mol^{-1} K^{-1}), and T is temperature (K). $V_m = V/n$ is the volume occupied by one mol of substance at a particular temperature and pressure. Using V_m, gas laws can be re-written in a way that does not require the number of mols n,

$$P = \frac{RT}{V_m}. \tag{2.13}$$

The van der Waals EOS was derived in 1873 and led to a Nobel prize for van der Waals in 1910. It accounts for the behaviors of many real gases while retaining conceptual simplicity. It is given by

$$P = \frac{RT}{V_m - b} - a\left(\frac{1}{V_m}\right)^2, \tag{2.14}$$

where a (atm L^2 mol^{-2}) is a parameter that characterized the attraction of gas molecules for one another, and b (L mol^{-1}) is effectively a minimum molar volume such that as V_m approaches b, no further compression is possible and the pressure rises rapidly.

The van der Waals EOS can be presented graphically in $P - V_m$ space, i.e., pressure is plotted against V_m. Once V_m is know, the density corresponding to a particular V_m can be calculated as $\rho = \frac{M}{V_m}$, for example $M = 44.8 \times 10^{-3}$ kg/mol is the molecular weight of the gas CO_2. That means V_m is proportional to $\frac{1}{\rho}$ for a specific chemical material (e.g., CO_2). Figure 2.1 shows the $P - V_m$ plot for CO_2. The van der Waals EOS with parameters a and b taken from Atkins (1978) was used to plot the curves at various temperatures. For different temperatures CO_2 may show subcritical, critical, and supercritical behaviors. At high temperature (373 K) CO_2 is supercritical and no distinct liquid and vapor phases can be discerned. A critical temperature is reached at $T = 304$ K; below this temperature phase separation into liquid and vapor is possible, and different molar volumes V_m (i.e., different densities because $\rho = \frac{M}{V_m}$) of the substance may coexist at a single pressure at equilibrium.

From Figure 2.1 we know that the critical temperature is about $T_c = 304$ K. At the critical temperature, not only the first derivative but also the second derivative should be zero, i.e.,

$$\frac{\partial P}{\partial V_m} = 0 \tag{2.15}$$

$$\frac{\partial^2 P}{\partial V_m^2} = 0. \tag{2.16}$$

Hence, suppose V_c is the molar volume at the critical temperature. Rearranging and taking the derivatives of Eq. (2.14), we have

$$2aV_c^{-3} = \frac{RT_c}{(V_c - b)^2} \tag{2.17}$$

$$6aV_c^{-4} = \frac{2RT_c}{(V_c - b)^3}. \tag{2.18}$$

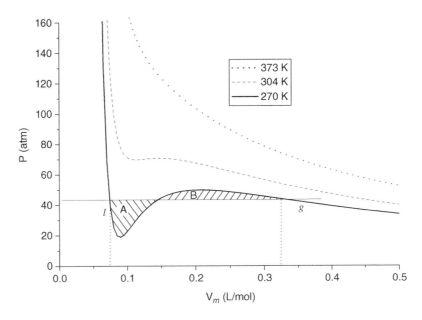

Figure 2.1 $P - V_m$ plot of van der Waals equations of state for CO_2. van der Waals constants $a = 3.592$ atm L^2 mol^{-2} and $b = 0.04267$ L mol^{-1} (Atkins 1978), and $R = 0.0821$ atm L mol^{-1} K^{-1}. l and g represent the coexistence liquid and gas states at $T = 270$ K.

Immediately, we obtain

$$a = \frac{9V_c RT_c}{8} \quad \text{and} \tag{2.19}$$

$$b = \frac{V_c}{3}. \tag{2.20}$$

Hence, a and b can also be presented as functions of P_c and T_c as follows:

$$a = \frac{27(RT_c)^2}{64P_c} \quad \text{and} \tag{2.21}$$

$$b = \frac{RT_c}{8P_c}, \tag{2.22}$$

where

$$P_c = \frac{RT_c}{V_c - b} - a\left(\frac{1}{V_c}\right)^2 = \frac{3RT_c}{8V_c}. \tag{2.23}$$

On the other hand, if the constants a and b are known, T_c, P_c, and V_c can be calculated from the above expressions.

In Figure 2.1, we see that the horizontal line at $p = 44.8$ atm is intersected three times by the EOS curve for $T = 270$ K. The first intersection at small V_m (high density), i.e., point l in the diagram, represents the liquid. The second intersection occurs in a "non-physical" portion of the EOS because the positive slope at that intersection implies that increasing the pressure would cause higher molar volume corresponding to expansion of the vapor (Sukop and Thorne 2006). The final intersection (point g) represents the vapor phase. Hence, at $T = 270$ K (corresponding to $p = 44.8$ atm), two phases (labeled l and g in the diagram) may coexist at equilibrium. For the liquid and gas, V_m is approximately 0.076 L/mol and 0.321 L/mol, respectively.

$P - V$ diagrams are very useful for finding the coexisting densities of the liquid and vapor phases. To find the densities analytically, the Maxwell construction should be used. The Maxwell construction can be stated as (Boltzmann 1964/1995)

$$\int_{V_{m,l}}^{V_{m,g}} P dV_m = p_0(V_{m,g} - V_{m,l}), \qquad (2.24)$$

where P is the pressure in the EOS and p_0 is a constant pressure. In Figure 2.1, the above equation means that the area bounded by the EOS curve from point l to point g, the two vertical dotted lines and the V_m axis (i.e., $\int_{V_{m,l}}^{V_{m,g}} P dV_m$) should be equal to to the rectangular area bounded by the horizontal line $p = 44.8$ atm, the two vertical dotted lines and the V_m axis (i.e., $p_0(V_{m,g} - V_{m,l})$). This also means the two areas labeled A and B in Fig. 2.1 are equal. Hence, the Maxwell construction is also known as an "equal area rule".

Using the EOS and the Maxwell construction to solve for p_0, $V_{m,l}$, and $V_{m,g}$ analytically may be challenging. Sukop and Thorne (2006) provide a simplified trial-and-error solution for the Maxwell construction. Here we suggest a method to solve the problem using Matlab. As an example, a concise Matlab code for solving the van der Waals EOS at $T = 270$ K consists of the following (written by Jianlin Zhao, China University of Petroleum, Eastern China):

```
clear;
clc;

syms Pr Tr P T Pc Tc R w a b alpha Vm Vml Vmv rho;

a=3.592
b=0.04267
R=0.0821
T=270

ezplot(R*T/(Vm-b)-a/(Vm*Vm),[0,1]);

title('vdw EOS');
xlabel('Vm');
ylabel('P');

 P=42
 deltaP=1
 S1=1.0;
 S2=0.0;
 while(abs((S1-S2)/S1)>0.00000001)
```

```
Vm=solve('0.0821*270/(Vm-0.04267)-3.592/(Vm*Vm)-P=0','Vm');
    %to get solutions
Vm1=subs(Vm(1));
Vm2=subs(Vm(2));
Vm3=subs(Vm(3));
Vm1=real(Vm1)
Vm2=real(Vm2)
Vm3=real(Vm3)

S1=R*T*log(Vm1-b)+a/(Vm1)-(R*T*log(Vm2-b)+a/(Vm2))   % S1 is
  the integration area
S2=P*(Vm1-Vm2)        %S2 is a rectangular area
if(S1>S2)    %compare the areas
    P=P+deltaP;
else
    P=P-deltaP;
end
deltaP=deltaP/1.2
end
  P
  Vm1
  Vm2
  Vm3
```

It is noted that in this scheme the integrated form of $\int P dV_m$ should be obtained in advance, i.e.,

$$\int P dV_m = RT \ln (V_m - b) + \frac{a}{V_m}. \tag{2.25}$$

When running this Matlab code, the range for V_m, P, and the increment δP should be chosen carefully to ensure that Matlab is able to obtain the correct result. For simple EOS like the van der Waals, the analytical integrated form can be obtained. One additional case for the Carnahan–Starling (C–S) EOS is shown in Eq. (2.31).

On the other hand, the analytical integrated form may be complicated. Alternatively, the numerical integration of the EOS can be implemented in Matlab directly as shown here, with the van der Waals EOS taken as an example:

```
fcs = @(Vm) 1*R*T./(Vm.-b)-a*(1./Vm).^2;
% Numerical integration results
S1=integral(fcs,Vm2,Vm1)
```

This means just replacing the analytical 'S1' calculation with the above numerical 'S1' calculation.

Next we introduce some popular EOS that are subsequently investigated in the context of the SC model.

The van der Waals (vdW) EOS is the simplest and most famous cubic EOS:

$$p = \frac{\rho RT}{1 - b\rho} - a\rho^2, \tag{2.26}$$

where $a = \frac{27(RT_c)^2}{64 p_c}$, $b = \frac{RT_c}{8 p_c}$, R is the gas constant, and T is the temperature.

The material constants a and b in the van der Waals EOS may differ for every non-ideal liquid–vapor pair. However, there exists an invariant form applicable to all fluids. Suppose the non-dimensional variables are $\tilde{T} = \frac{T}{T_c}, \tilde{p} = \frac{p}{p_c}$, and $\tilde{\rho} = \frac{\rho}{\rho_c}$. We then have

$$\tilde{p}p_c = \frac{\tilde{\rho}\rho_c R \tilde{T} T_c}{1 - b\tilde{\rho}\rho_c} - a(\tilde{\rho}\rho_c)^2. \tag{2.27}$$

Substituting $a = \frac{27(RT_c)^2}{64 p_c}, b = \frac{RT_c}{8p_c}$, and $\frac{p_c R T_c}{p_c} = \frac{8}{3}$ (see Eq. (2.23)) into the above equation, we have the invariant form:

$$\tilde{p} = \frac{8\tilde{\rho}\tilde{T}}{3 - \tilde{\rho}} - 3\tilde{\rho}^2. \tag{2.28}$$

The invariance is an example of the principle of corresponding states (Yuan and Schaefer 2006). Two fluids existing in corresponding states means that they have the same reduced pressure, reduced density, and reduced temperature.

The Redlich–Kwong (R–K) EOS was introduced in 1949 and provided a considerable improvement over other equations of the time. It takes the following form:

$$p = \frac{\rho R T}{1 - b\rho} - \frac{a\rho^2}{\sqrt{T}(1 + b\rho)}, \tag{2.29}$$

with $a = \frac{0.42748 R^2 T_c^{2.5}}{p_c}$ and $b = \frac{0.08662 R T_c}{p_c}$. While superior to the van der Waals EOS, it performs poorly with respect to the liquid phase. Hence it may be unable to calculate vapor–liquid equilibria accurately.

Carnahan and Starling (1969) modified the repulsive term of the van der Waals EOS and obtained more accurate expressions for hard sphere systems. The C–S EOS is given by

$$p = \rho R T \frac{1 + b\rho/4 + (b\rho/4)^2 - (b\rho/4)^3}{(1 - b\rho/4)^3} - a\rho^2, \tag{2.30}$$

with $a = 0.4963 R^2 T_c^2 / p_c$ and $b = 0.18727 R T_c / p_c$. Suppose $\theta = \frac{b\rho}{4}$ and $v' = \frac{1}{\theta} = \frac{4}{b\rho} = \frac{4v}{b}$, we have the integrated form of $\int p\,dv$

$$\int p\,dv = \int \rho R T \left[1 + \frac{-2\theta^2 + 4\theta}{(1 - \theta)^3} \right] - a\rho^2\,dv$$

$$= \frac{a}{v} + R T \ln\, v + R T \int \frac{-2 + 4v'}{(v' - 1)^3} \frac{4}{b}\,dv$$

$$= \frac{a}{v} + R T \ln\, v + R T \int \frac{-2 + 4v'}{(v' - 1)^3}\,dv'$$

$$= \frac{a}{v} + R T \ln\, v + R T \frac{-4v' + 3}{(v' - 1)^2}$$

$$= \frac{a}{v} + R T \ln\, v + R T \frac{-16bv + 3b^2}{(4v - b)^2}. \tag{2.31}$$

2.2.1 Parameters in EOS
Difference between the *T* in sound speed and the real temperature

The isothermal (athermal) sound speed c_s is fixed once the velocity set is chosen, e.g., $c_s = \frac{1}{\sqrt{3}}c$ for D2Q9 model (He and Luo 1997; Lallemand and Luo 2003). The sound speed does not depend on the temperature T or any adjustable parameters in the LBM model (Lallemand and Luo 2003). In the derivation of different velocity models (He and Luo 1997) $c_s = \sqrt{RT}$ but this T is not directly relative to temperature. While in the EOS, the temperature T can represent the real temperature by the reduced temperature $\tilde{T}$. Hence, there are no connections between the T in $c_s = \sqrt{RT}$ and the temperature in an EOS.

The parameters in the EOS in our example simulations are listed in Table 2.1. The values are all given in lattice units. Of course parameters different from those in Table 2.1 can be chosen. The effect of the parameters on surface tension is discussed in Section 2.8.4. Here the lattice unit for temperature is tu (temperature unit). That means the parameters a, b, and R have the units of $lu^5/(mu\ ts^2)$, lu^3/mu, and $lu^2/(ts^2\ tu)$, respectively, in the van der Waals EOS. After the parameters are chosen, the corresponding critical parameters for the vdW, C–S, and R–K EOSs are determined. These are listed in Table 2.2. They are all in lattice units. In the following study, if not specified usually the C–S EOS is used because it allows simulations to achieve the highest density ratios (Yuan and Schaefer 2006).

According to the choice of the main parameters in the EOS (see Table 2.1), we also draw three typical curves in Figure 2.2 for the C–S EOS. From the figure we can see that when T is equal to the critical temperature $T_c = 0.0943$ tu, there is only one phase. When the temperature is lower than T_c, there will be two

Table 2.1 Parameters in EOS (following Yuan and Schaefer (2006)).

EOS	a	b (lu^3/mu)	R ($lu^2/(ts^2$ tu))
vdW	$\frac{9}{49}$ lu^5/(mu ts^2)	$\frac{2}{21}$	1
R–K	$\frac{2}{49}$ lu^5 tu$^{0.5}$/(mu ts^2)	$\frac{2}{21}$	1
C–S	1 lu^5/(mu ts^2)	4	1

Table 2.2 Critical properties in our study for vdW, R–K, and C–S EOS.

EOS	T_c (tu)	ρ_c (mu/lu^3)	p_c (mu/(lu ts^2))
vdW	0.5714	3.5000	0.7500
R–K	0.1961	2.9887	0.1784
C–S	0.0943	0.1136	0.0044

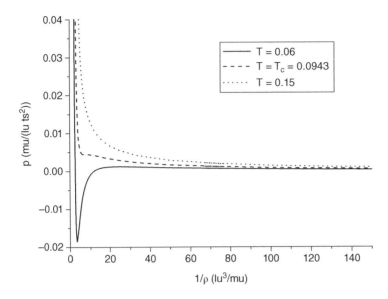

Figure 2.2 $P - \frac{1}{\rho}$ diagram of C–S EOS (Eq. (2.30)). The constants are $a = 1$ lu/(mu ts²), $b = 4$ lu³/mu, and $R = 1$ lu²/(ts² tu). Curves of $T = 0.06$ tu, $T = 0.0943$ tu, and $T = 0.15$ tu are plotted.

coexisting phases. In Figure 2.2, the curve of $T = 0.06$ tu, i.e., $\frac{T}{T_c} = 0.64$, is shown. In this curve, the two coexisting liquid and gas densities are about 0.381 mu/lu³ and 0.0022 mu/lu³, respectively. When the temperature is higher than T_c, the fluid system reaches supercritical state.

2.3   Thermodynamic consistency

In He and Doolen (2002), "thermodynamically consistent" means that "...a theory for non-equilibrium transport phenomena has to recover to thermodynamic theory for equilibrium state...". It implies that an LBM model correctly incorporates all thermodynamic quantities, including internal energy, free energy, chemical potential, and entropy (He and Doolen 2002).

First the free energy is introduced briefly. According to the theory of Rowlinson and Widom (1982) and Evans (1979), the pressure tensor can be derived from a free-energy functional:

$$\Psi = \int \left[\psi(\rho, T) + \frac{\kappa}{2} |\nabla \rho|^2 \right] d\mathbf{r}, \qquad (2.32)$$

where ψ is the bulk free-energy density at temperature T, $d\mathbf{r}$ represents a tiny volume fraction, and κ is a constant.

The pressure tensor should be related to the free energy in this way (Evans 1979; He and Doolen 2002):

$$p_{\alpha\beta} = p_0 \delta_{\alpha\beta} + \kappa \partial_\alpha \rho \partial_\beta \rho, \tag{2.33}$$

with

$$p_0 = p - \kappa\rho\nabla^2\rho - \frac{\kappa}{2}|\nabla\rho|^2, \tag{2.34}$$

where

$$p = \rho\psi'(\rho) - \psi(\rho) \tag{2.35}$$

is the EOS of the fluid and $\psi'(\rho) = \frac{d\psi}{d\rho}$.

From the above three equations, we have (Gross et al. 2011):

$$p_{\alpha\beta} = \left[p - \kappa\rho\nabla^2\rho - \frac{1}{2}\kappa(\nabla\rho)^2\right]\delta_{\alpha\beta} + \kappa\partial_\alpha\rho\partial_\beta\rho. \tag{2.36}$$

In this book, "thermodynamic consistency" specifically means that the N–S equation recovered from the Lattice Boltzmann equation correctly contains the above pressure tensor, which contains a non-ideal EOS that defines the analytical coexistence curve.

2.3.1   The SCMP LBM EOS

Comparing Eqs (2.36) and (2.11) with Eq. (2.10), we know that if

$$\kappa = -\frac{1}{2}Gc_s^4$$

$$\text{and} \quad \psi \propto \rho, \tag{2.37}$$

the pressure tensor derived from the SC LBM is consistent with this thermodynamic one (Eq. (2.36)). Hence, Eq. (2.37) is the thermodynamic consistency condition for the SC LBM.

However, in the original SC model (Shan and Chen 1993), $\psi(\rho) = \rho_0[1 - \exp(-\rho/\rho_0)]$, (Eq. (2.5)), which does not satisfy $\psi \propto \rho$. That means the original SC model is not thermodynamically consistent.

Shan and Chen (1994) proposed

$$\psi(\rho) = \psi_0 \exp(-\rho_0/\rho), \tag{2.38}$$

where ψ_0 and ρ_0 are arbitrary constants. They state that in this way, the coexistence curve of the LBM model becomes more consistent with the thermodynamic theory. For an isothermal process, the Maxwell construction is satisfied. According to the two conditions (Eq. (2.37)) we require for thermodynamic consistency, the SC model with this $\psi(\rho)$ is still not thermodynamically consistent because $\psi \propto \rho$ is not satisfied (He and Doolen 2002).

Sukop and Thorne (2006) discussed many examples using the SC EOS

$$p = c_s^2\rho + \frac{c_s^2 G}{2}\psi^2(\rho), \tag{2.39}$$

with $\psi(\rho) = \psi_0 \exp(-\rho_0/\rho)$ (Shan and Chen 1994). The critical G, i.e., G_c, which is analogous to the critical temperature, and the corresponding ρ_c are obtained through

$$\frac{\partial p}{\partial \rho} = c_s^2 + G_c c_s^2 \psi^2 \frac{\rho_0}{\rho_c^2} = 0 \tag{2.40}$$

$$\frac{\partial^2 p}{\partial \rho^2} = G_c c_s^2 \left(2\psi^2 \frac{\rho_0^2}{\rho_c^4} - 2\psi^2 \frac{\rho_0}{\rho_c^3} \right) = 0, \tag{2.41}$$

where the subscript c denotes the critical state. In the derivation, it is noted that

$$\frac{\partial \psi}{\partial \rho} = \psi_0 \exp(-\rho_0/\rho)\frac{\rho_0}{\rho^2} = \psi\frac{\rho_0}{\rho^2}. \tag{2.42}$$

Solving the above two equations we have

$$\rho_c = \rho_0 \tag{2.43}$$

$$G_c = \frac{-\rho_0}{[\psi(\rho_c)]^2} \tag{2.44}$$

The $P - \rho$ diagram of this SC EOS with $\psi_0 = 4$ and $\rho_0 = 200$ is shown in Figure 2.3. According to Eq. (2.44), the critical $G_c = \frac{-200}{(4e^{-1})^2} = -92.4$ mu lu^{-3}. The critical behavior at G_c is also shown in Figure 2.3.

There are many applications using this SC model to study multiphase fluids in Sukop and Thorne (2006). Here we demonstrate how to incorporate other EOS into this LBM. Using EOS other than the original SC EOS, the model is able to achieve higher density ratios (Yuan and Schaefer 2006).

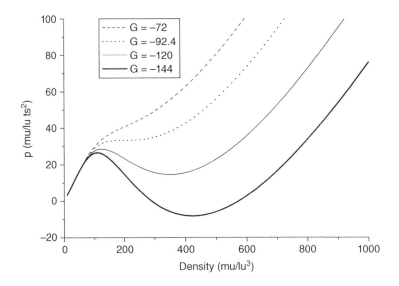

Figure 2.3 $P - \rho$ plot of the SC equations of state. $\psi(\rho) = \psi_0 \exp(-\rho_0/\rho)]$ with $\psi_0 = 4$ and $\rho_0 = 200$.

2.3.2   Incorporating other EOS into the SC model
Inter-particle Force Model A

For Inter-particle Force Model A, from Eq. (2.39), we can see that the thermo-dynamic p is a function of ψ. Alternatively, if the expression of p (EOS) is already known, we can use the following definition of ψ (He and Doolen 2002; Sankaranarayanan et al. 2002)

$$\psi = \sqrt{\frac{2(p - c_s^2 \rho)}{c_s^2 G}} \qquad (2.45)$$

to replace Eq. (2.5) or (2.38) in order to incorporate different EOS into the SC LBM (Yuan and Schaefer 2006). For example, if we want to incorporate the van der Waals EOS into the SC LBM, the p in Eq. (2.45) should be substituted by

$$p = \frac{\rho T}{1 - \rho b} - a\rho^2. \qquad (2.46)$$

Here again we can see that $\psi(\rho)$ is usually not proportional to ρ. Hence, incorporating other EOS into the SC model does not lead to a thermodynamically consistent model. However, as stated in the introduction, in many applications this thermodynamic inconsistency is not relevant or of primary importance.

Inter-particle Force Model B

For Inter-particle Force Model B, the corresponding pressure tensor can be derived similarly (Sbragaglia et al. 2007). Here we derive the formula briefly. First, the Inter-particle Force Model B is

$$\mathbf{F}_{int}(\mathbf{x}, t) = -\psi(\mathbf{x}, t) \sum_i w_i [G_1 \psi(\mathbf{x} + \mathbf{e}_i \Delta t, t) + G_2 \psi(\mathbf{x} + 2\mathbf{e}_i \Delta t, t)] \mathbf{e}_i. \qquad (2.47)$$

Applying the Taylor expansion on the above equation to $O(\Delta t)$, we have

$$\begin{aligned}
F_\alpha &= -G_1 \psi \sum_i w_i \psi(\mathbf{x} + e_{i\alpha} \Delta t) e_{i\alpha} - G_2 \psi \sum_i w_i \psi(\mathbf{x} + 2e_{i\alpha} \Delta t) e_{i\alpha} \\
&= -G_1 \psi \left[\sum_i w_i e_{i\alpha} \psi + \Delta t \sum_i w_i e_{i\alpha} e_{i\beta} \partial_\beta \psi \right] \\
&\quad -G_2 \psi \left[\sum_i w_i e_{i\alpha} \psi + 2\Delta t \sum_i w_i e_{i\alpha} e_{i\beta} \partial_\beta \psi \right] + \dots . \\
&\approx -\frac{G_1 + 2G_2}{2} \Delta t c_s^2 \partial_\alpha \psi^2.
\end{aligned} \qquad (2.48)$$

We can see the derivation is very similar to that in Appendix A in Benzi et al. (2006) or Eq. (2.8). Hence, the corresponding EOS for Inter-particle Force Model B is

$$p = c_s^2 \rho + c_s^2 \frac{G_1 + 2G_2}{2} \psi^2(\rho). \qquad (2.49)$$

We propose using the following formula to incorporate different EOS into Inter-particle Force Model B (Huang et al. 2007):

$$\psi = \sqrt{\frac{2(p - c_s^2 \rho)}{c_s^2(G_1 + 2G_2)}}.$$ (2.50)

We will use this formulation to compare different inter-particle force models in Section 2.9.

2.4   Analytical surface tension

2.4.1   Inter-particle Force Model A

The following derivation of the analytical surface tension in the SC LBM is identical to those described in Shan and Chen (1994) and Benzi et al. (2006). We repeat the essential steps for the reader's convenience. In the following 2D discussion we assume that a flat interface parallel to the x axis separates liquid and gas. The phase interface implicitly defined by $\rho = (\rho_l + \rho_g)/2$ is chosen as the origin of the y axis. For Inter-particle Force Model A, the normal component of the pressure tensor p_{yy} obtained from Eq. (2.10) is

$$p_{yy} = c_s^2 \rho + \frac{1}{2}Gc_s^2\psi^2 + \frac{1}{2}Gc_s^4\left(\psi\partial_{yy}\psi - \frac{1}{2}\partial_y\psi\partial_y\psi\right).$$ (2.51)

In both phases, far from the interface the pressure p_0 satisfies the following relation (Benzi et al. 2006; Shan and Chen 1994):

$$p_0 = c_s^2\rho_g + \frac{1}{2}Gc_s^2\psi^2(\rho_g) = c_s^2\rho_l + \frac{1}{2}Gc_s^2\psi^2(\rho_l).$$ (2.52)

One obtains the density profile $\frac{d\rho}{dy}$ by solving Eqs (2.51) and (2.52) assuming $\frac{d\rho}{dy} = 0$ at $y = \pm\infty$. A change of variables simplifies Eq. (2.51). In order to get a formal solution, let $(d\rho/dy)^2 = z$ and notice

$$\frac{d^2\rho}{dy^2} = \frac{1}{2}\frac{dz}{d\rho}.$$ (2.53)

Then Eq. (2.51) can be transformed as follows (Benzi et al. 2006; Shan and Chen 1994):

$$p_{yy} = c_s^2\rho + \frac{1}{2}Gc_s^2\psi^2 + \frac{1}{4}Gc_s^4\frac{\psi^2}{\psi'}\frac{d}{d\rho}\left(z\frac{\psi'^2}{\psi}\right),$$ (2.54)

where $\psi' = \partial\psi/\partial\rho$. By direct integration, and using the definition of Eq. (2.11), one can obtain the following solution for $z(\rho)$ (Benzi et al. 2006; Shan and Chen 1994):

$$z(\rho) = \frac{4\psi}{Gc_s^4(\psi'^2)}\int_{\rho_g}^{\rho}(p_{yy} - c_s^2\rho - \frac{1}{2}Gc_s^2\psi^2)\frac{\psi'}{\psi^2}d\rho.$$ (2.55)

Suppose

$$p(\rho) = c_s^2\rho + \frac{1}{2}Gc_s^2\psi^2$$ (2.56)

is the EOS formula. p_{yy} should be a constant along the y direction after the interface reaches an equilibrium state, which is approximately equal to p_0 (refer to Eq. (2.52)). Hence we have

$$z(\rho) = \frac{4\psi}{Gc_s^4(\psi'^2)} \int_{\rho_g}^{\rho} (p_0 - p(\rho)) \frac{\psi'}{\psi^2} d\rho$$
$$= \frac{4\psi}{Gc_s^4(\psi'^2)} \left[\left(-\frac{p_0}{\psi} + \frac{p(\rho)}{\psi} \right) |_{\rho_g}^{\rho} - \int_{\rho_g}^{\rho} \frac{dp(\rho)}{d\rho} \frac{1}{\psi} d\rho \right]. \tag{2.57}$$

To satisfy the boundary condition (i.e., Eq. (2.52)), we require

$$\int_{\rho_g}^{\rho_l} (p_0 - p(\rho)) \frac{\psi'}{\psi^2} d\rho = 0. \tag{2.58}$$

The quantities p_0, ρ_l, and ρ_g can be simultaneously obtained numerically from Eq. (2.58). Then by solving Eq. (2.55), the density profile in the vicinity of the interface can be obtained. Finally the surface tension can be computed from (Benzi et al. 2006; Shan and Chen 1994):

$$\sigma = -\frac{1}{2} Gc_s^4 \int_{-\infty}^{+\infty} (\partial_y \psi)^2 dy = -\frac{1}{2} Gc_s^4 \int_{\rho_l}^{\rho_g} \psi'^2 [z(\rho)]^{\frac{1}{2}} d\rho. \tag{2.59}$$

We give an example to illustrate how to calculate the above integration. Suppose the EOS is the C–S EOS. Then Eq. (2.56) should be replaced by Eq. (2.30) when calculating Eq. (2.57). The parameters a, b, and R in the EOS are listed in Table 2.1 and in this example $T = 0.75 T_c$. Through solving Eq. (2.58), one can obtain the coexisting densities of liquid and gas, $\rho_l = 0.332$ and $\rho_g = 0.00861$. The corresponding $v_l = \frac{1}{\rho_l} = 3.011$ and $v_g = \frac{1}{\rho_g} = 116.093$.

After v_l and v_g are known, the integration Eq. (2.59) can be calculated numerically. Suppose the range $v \in [v_l, v_g]$ is discretized into 1000 points, then $z(\rho)$ and ψ' at each point can be calculated. Figure 2.4 shows $z(\rho), \psi'$ and the integration $\int_{\rho_l}^{\rho} \psi'^2 [z(\rho)]^{\frac{1}{2}} d\rho$ as functions of $v = \frac{1}{\rho}$.

From Figure 2.4 we can see that the integration $\int_{\rho_l}^{\rho_g} \psi'^2 [z(\rho)]^{\frac{1}{2}} d\rho$ is about -0.375. Hence, finally, the surface tension $\sigma = -\frac{1}{2} Gc_s^4 \int_{\rho_l}^{\rho_g} \psi'^2 [z(\rho)]^{\frac{1}{2}} d\rho = 0.00694$ (refer to Eq. (2.59)).

2.4.2 Inter-particle Force Model B

For Inter-particle Force Model B, one can obtain the pressure tensor by an analogous procedure. The result is similar to Eq. (2.10) (Sbragaglia et al. 2007):

$$p_{\alpha\beta}^* = \left[c_s^2 \rho + \frac{1}{2} A_1 c_s^2 \psi^2 + \frac{1}{2} A_2 c_s^4 (\psi \Delta \psi + \frac{1}{2} |\nabla \psi|^2) \right] \delta_{\alpha\beta} - \frac{1}{2} A_2 c_s^4 \partial_\alpha \psi \partial_\beta \psi, \tag{2.60}$$

where A_1 and A_2 are constants related to G_1 and G_2 in Eq. (2.7):

$$A_1 = G_1 + 2G_2 \quad \text{and} \quad A_2 = G_1 + 8G_2. \tag{2.61}$$

The corresponding surface tension is

$$\sigma^* = -\frac{1}{2} A_2 c_s^4 \int_{-\infty}^{+\infty} (\partial_y \psi)^2 dy. \tag{2.62}$$

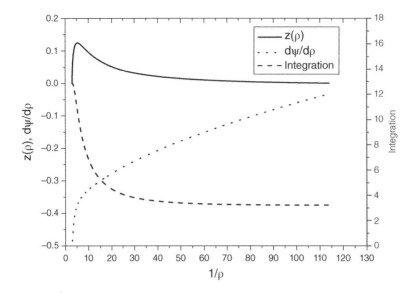

Figure 2.4 $z(\rho)$, ψ', and the integration $\int_{\rho_l}^{\rho} \psi'^{2}[z(\rho)]^{\frac{1}{2}} d\rho$ as functions of $v = \frac{1}{\rho}$. The C–S EOS (Eq. (2.30)) with $T = 0.75T_c$ is used. At the equilibrium state, $v_l = \frac{1}{\rho_l} = 3.011$ and $v_g = \frac{1}{\rho_g} = 116.093$ coexist.

2.5 Contact angle

The wetting and spreading of fluids on solid surfaces is critical in industrial and natural processes ranging from petroleum extraction from reservoirs to liquid/air interaction in human lungs. When a droplet comes into contact with a solid, there is a contact line that demarcates the point of contact between wetting and non-wetting fluids and the solid surface. The contact angle between the fluids and the surface can be calculated through Young's equation (Eq. (2.63)) provided the interfacial tensions between the fluid components σ_{12} and between each component and the solid surface σ_{s1}, σ_{s2} are known (Finn 2006, Young 1805):

$$\cos \theta = \frac{\sigma_{s2} - \sigma_{s1}}{\sigma_{12}}. \tag{2.63}$$

Figure 2.5 shows a schematic diagram of the contact angle. In the case of SCMP fluid, the interfacial tension is the surface tension and the interfacial tension between the liquid and vapor phases of the substance and the solid phase are relevant.

There have been some studies on wetting and spreading phenomena using the SCMP SC model. Raiskinmäki et al. (2000) simulated spreading of small droplets on smooth and rough solid surfaces using the 3D SCMP LBM. Sukop and

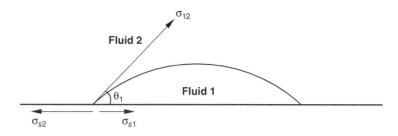

Figure 2.5 Contact angle.

Thorne (2006) used force balance arguments to successfully predict the parameters needed to simulate $0°, 90°$, and $180°$ contact angles.

Benzi et al. (2006) presented a mesoscopic model for the interaction between a solid wall and an SCMP fluid. They derived an analytical expression for the contact angle that covers the range of contact angles from $0°$ to $180°$. There are also many works on wetting and spreading phenomena using the multi-component multiphase SC model (Huang et al. 2007; Kang et al. 2002, 2005; Martys and Chen 1996; Pan et al. 2004; Schaap et al. 2007), which will be addressed in the next chapter.

The desired contact angle can also be obtained conveniently through changing a parameter ρ_w, which represents a fluid density on the wall that has the sole purpose of setting the contact angle (Benzi et al. 2006). The adhesion force between the gas/liquid phase and solid walls is calculated by Eq. (2.64). Here we assume the solid phase has density ρ_w, i.e., $\psi(\rho(x_w)) = \psi(\rho_w)$,

$$\mathbf{F}_{ads}(\mathbf{x}, t) = -G\psi(\rho(\mathbf{x}, t)) \sum_a w_a \psi(\rho_w) s(\mathbf{x} + \mathbf{e}_a \Delta t, t)\mathbf{e}_a. \qquad (2.64)$$

An indicator function $s(\mathbf{x} + \mathbf{e}_a \Delta t, t)$ that is equal to 1 or 0 for a solid or a fluid domain node, respectively, acts as a switch that turns the adhesion force on or off. "Wall density" ρ_w is not really relevant to the "true" density of solid phase, but rather is a free parameter used here to tune different wall properties (Benzi et al. 2006).

In our simulations, any lattice node in the computational domain represents either a solid node or a fluid node. For solid nodes, the bounce-back algorithm instead of the collision step is implemented before the streaming step to mimic a no-slip wall boundary condition.

Figure 2.6 demonstrates that different contact angles can be obtained by adjusting ρ_w. In these simulations, the computational domain is 200×100, the upper and lower boundaries are solid walls, and the east and west boundaries are periodic. Here the R–K EOS (Eq. (2.29)) is used with $T = 0.85T_c$. The liquid phase density is $\rho_l = 6.06$ and the gas phase density is $\rho_g = 0.5$. When the parameter ρ_w varies between ρ_l and ρ_g, the contact angle varies between

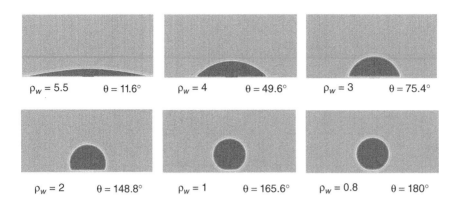

| $\rho_w = 5.5$ | $\theta = 11.6°$ | $\rho_w = 4$ | $\theta = 49.6°$ | $\rho_w = 3$ | $\theta = 75.4°$ |
| $\rho_w = 2$ | $\theta = 148.8°$ | $\rho_w = 1$ | $\theta = 165.6°$ | $\rho_w = 0.8$ | $\theta = 180°$ |

Figure 2.6 Different contact angles obtained through adjusting the parameter ρ_w. The R–K EOS (Eq. (2.29)) with $T = 0.85T_c$, $\rho_l = 6.06$, and $\rho_g = 0.5$ was used in the simulations. Source: Huang and Lu (2009). Reproduced with permission of AIP Publishing LLC.

$0°$ and $180°$. The surface tension σ in the above case is about 0.16 mu/ts^2 in the numerical simulation and its value is adjustable through changes in the parameter b in the EOS.

The following 3D simulations use the C–S EOS with $\frac{T}{T_0} = 0.875$ mentioned in Section 2.2. The computational domain is $50 \times 50 \times 40$. The upper and lower boundaries are walls. Periodic boundary conditions are applied to the other boundaries. In the simulations, a spherical droplet is initially in contact with the bottom wall. The coexisting densities at this temperature are approximately $\rho_l = 0.265$ mu/lu^3 and $\rho_g = 0.038$ mu/lu^3. In Figure 2.7 we can see that when ρ_w are specified as 0.08, 0.10, 0.12, and 0.18 mu/lu^3, the equilibrium contact angles we obtain are approximately $150°, 90°, 80°$, and $30°$, respectively. However, no explicit formula has been proposed to predict the contact angle as a function of ρ_w in advance for the SCMP model.

2.6 Capillary rise

The SCMP SC LBM can effectively simulate capillary rise in simple capillaries and capillary phenomena in more complex porous media. The dimensionless Bond number Bo relates capillary and gravitational forces, and allows the simulation of direct analogues of real systems in the same way as that the Reynolds number allows the simulation of equivalent flow regimes (Sukop and Thorne 2006).

Assuming a zero contact angle, we can use the 2D Young–Laplace equation to determine the pressure difference across a curved (2D) interface:

$$\Delta p = \frac{\sigma}{r}, \tag{2.65}$$

where σ is the surface tension and r is the half-width of the 2D capillary tube.

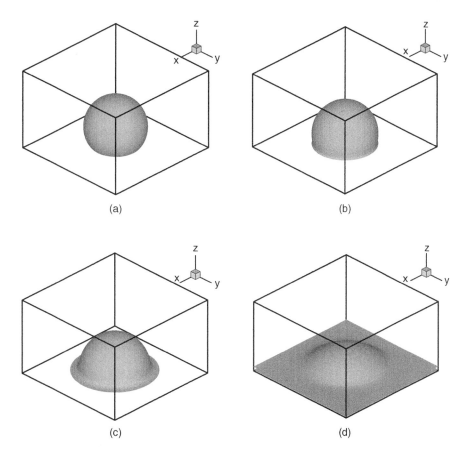

Figure 2.7 Different equilibrium contact angles obtained by adjusting the parameter ρ_w. C–S EOS with $\frac{T}{T_0} = 0.875$ was used in the simulations. (a) $\rho_w = 0.08$, (b) $\rho_w = 0.1$, (c) $\rho_w = 0.12$, (d) $\rho_w = 0.18$. The contour of $\rho = 0.09$ is shown in all frames. The unit of the density is mu/lu^3. The coexisting densities are approximately $\rho_l = 0.265$ and $\rho_g = 0.038$.

The hydrostatic pressure difference between the top and bottom of a column of incompressible liquid of height h in a gravitational field g is

$$\Delta p = (\rho_l - \rho_g)gh. \tag{2.66}$$

Equating the right-hand sides of Eqs (2.65) and (2.66) gives

$$h = \frac{\sigma}{(\rho_l - \rho_g)gr}. \tag{2.67}$$

The dimensionless Bond number Bo reflects the balance between gravitational and capillary forces and is

$$Bo = \frac{r^2(\rho_l - \rho_g)g}{\sigma}. \tag{2.68}$$

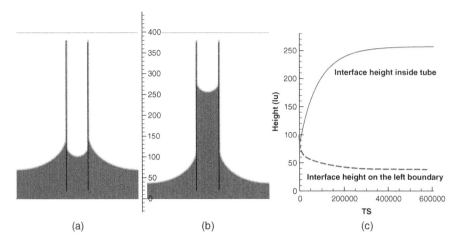

(a)   (b)   (c)

Figure 2.8 Capillary rise. (a) Capillary rise at $t = 10,000$ ts, (b) equilibrium rise height at $t = 600,000$ ts, and (c) the height of liquid inside and outside (on the left boundary) the capillary tube as a function of time. The R–K EOS with $\frac{T}{T_0} = 0.9$ was used in the simulations. The coexisting densities are approximately $\rho_l = 5.44$, $\rho_g = 0.81$, and $\rho_w = 5.40$ mu/lu^3.

The Bond number can be used to define an LBM capillary rise simulation equivalent to a real capillary system. Imagine a real capillary slit 0.002 m in width ($r = 0.001$ m) in contact with a pool of water with density 1000 kg/m^3 overlain by air with a density $\rho_{air} = 1.23$ kg/m^3 and gravity $g = 9.8$ m/s^2. The surface tension of water at 23°C is 72.13×10^{-3} N/m. According to Eq. (2.67), the expected capillary rise $h = 7.36 \times 10^{-3}$ m.

The Bond number for the real system is $Bo = r/h = 1/7.36$. Defining an analogous LBM system can begin with capillary tube size. Suppose the maximum capillary rise we want in our simulation is 200 lu. Then, in our specific case, the radius $r = 200$ lu/7.36 = 27 lu.

As shown in Figure 2.8(a), we chose a domain size of 300 × 400 with a wall on the bottom. To avoid an open boundary on the top, another wall is placed on the upper boundary. It will not affect the rise height. Periodic boundary conditions are applied on the left and right boundaries. The length of the tube is about 360 lu, and it is placed 20 lu height above the bottom wall.

In our LBM simulation, the R–K EOS (Eq. (2.29)) is used. Here $T = 0.9T_0$, the corresponding coexisting densities are $\rho_l = 5.44$ and $\rho_g = 0.81$, and the numerical surface tension is $\sigma = 0.096$ when $\tau = 1$. Hence, according to

$$g = \frac{\sigma}{(\rho_l - \rho_g)hr}, \tag{2.69}$$

the gravitational acceleration in our simulation should be $g = 3.84 \times 10^{-6}$ lu/ts^2 for our model problem. Since the liquid is completely wetting, the density of the wall nodes is set to be $\rho_w = 5.40$.

We begin with the liquid filling the region below $y = 100$ lu both inside and outside of the capillary tube. The results are shown in Figure 2.8. Figure 2.8(a) shows the capillary rise at $t = 10,000$ ts when the liquid begins to rise. Figure 2.8(c) shows the interface heights inside and outside (on the left boundary) of the capillary as a function of time. We can see that the liquid column inside the tube rises quickly at the beginning but later slows down and finally reaches an equilibrium state. The equilibrium height difference between the inside (256.7 lu) and outside of the tube (38.3 lu) is 218.4 lu, which is about 9% higher than the target value of 200 lu computed analytically above. In part this is due to the presence of compressibility in the SC model, i.e., the gas phase is more compressible than the liquid phase.

2.7 Parallel flow and relative permeabilities

For immiscible two-phase flows in porous media, a typical situation is that the wetting fluid attaches and moves along the solid surface, while the non-wetting phase flows in the center of the pores. In this case, the liquid phase is wetting and the gas phase is non-wetting. The velocity of the non-wetting phase depends in part on the dynamic viscosity ratio of the non-wetting and wetting fluids, i.e., $M = \frac{\eta_n}{\eta_w}$.

Here we simulate immiscible two-phase co-current flow between two parallel plates (see Figure 2.9). In the simulation, the periodic boundary condition is applied on the inlet/outlet boundaries. No-slip (bounce-back) boundary conditions are applied on the upper and lower plates. The kinematic viscosities for non-wetting and wetting fluids are identical, i.e., $v_n = v_w = c_s^2(\tau - 0.5)$, hence $M = \frac{\rho_n}{\rho_w}$.

In the simulation, as illustrated in Figure 2.9, the wetting phase flows in the region $a < |y| < b$ and the non-wetting phase flows in the central region $0 < |y| < a$.

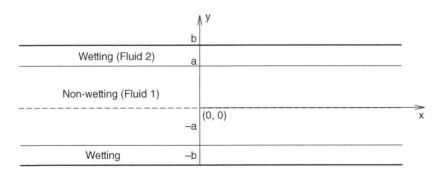

Figure 2.9 Co-current immiscible two-phase flow in a 2D channel. The wetting phase (Fluid 2) flows along upper and lower plate while non-wetting phase (Fluid 1) flows in the center region.

Obviously, the saturation of wetting fluid in this simulation is $S_w = 1 - \frac{a}{b}$ and $S_n = \frac{a}{b}$. Assuming a Poiseuille-type flow in the channel, the analytical solution for the velocity profile between the parallel plates is given in Appendix A.

From the analytical solution, the four relative permeabilities as a function of the non-wetting saturation can be obtained as

$$
\begin{aligned}
k_{11} &= \frac{Q_1(\mathcal{G}_2=0)}{Q_{10}} = S_n^3 + 3MS_n^2 - 3MS_n^3, \\
k_{21} &= \frac{Q_2(\mathcal{G}_2=0)}{Q_{10}} = \frac{3}{2}MS_n(1 - S_n)^2, \\
k_{12} &= \frac{Q_1(\mathcal{G}_1=0)}{Q_{20}} = \frac{3}{2}S_n(1 - S_n)^2, \\
k_{22} &= \frac{Q_2(\mathcal{G}_1=0)}{Q_{20}} = (1 - S_n)^3.
\end{aligned}
\tag{2.70}
$$

Q is volumetric flow rate. From Eq. (2.70) we can see that $k_{i,2} \in [0, 1]$ when saturation $S_w \in [0, 1]$, while $k_{i,1}$ may be higher than 1 when saturation $S_w \in [0, 1]$ because $k_{i,1}$ is not only a function of S_w but also M.

Figure 2.10 shows the velocity profiles for $M = 18$ and $M = \frac{1}{18}$ with $S_w = 0.5$. In the figure, velocity profiles in (a) and (c), and (b) and (d) are obtained through applying body force $\mathcal{G} = 3 \times 10^{-7}$ mu lu/ts^2 only on Fluid 1 and Fluid 2, respectively. The mesh used in the simulation is 2×100. The velocity profile calculated from the LBM agrees well with the analytical one. The errors of the velocity profile obtained from the LBM and the analytical one in (a), (b), (c), and (d) are 7.5%, 3.7%, 12.4%, and 3.2%. The velocity profiles in (a) and (c) can be used to calculate the relative permeability tensor components k_{11} and k_{21}, while those in (b) and (d) are used to calculate the k_{12} and k_{22} components.

The SC LBM simulation using $\tau = 0.6$ is found to give more accurate velocity profiles than $\tau = 1$ (Huang and Lu 2009).

Figure 2.11 illustrates the $k_{i,j}$ as a function of the S_w when $M < 1$. Again, the LBM results agree well with the analytical curves. As expected from Eq. (2.70), the relative permeabilities of both phases are less than 1.

2.8   Forcing term in the SC model

Incorporation of the forcing term in the LBM generally means an extra term S_i added to the LBE to mimic the body force F_α in the following incompressible N–S equations,

$$
\begin{aligned}
&\partial_t \rho + \partial_\alpha \rho u_\alpha = D_t \rho + \rho \partial_\alpha u_\alpha = 0, \\
&\partial_t \rho u_\alpha + \rho u_\beta \partial_\beta u_\alpha = -\partial_\alpha p + \rho v \partial_\beta (\partial_\beta u_\alpha + \partial_\alpha u_\beta) + F_\alpha,
\end{aligned}
\tag{2.71}
$$

where D_t is the material derivative, which means $D_t \rho = \partial_t \rho + u_\alpha \partial_\alpha \rho$.

For both single-phase and multiphase flows, correct treatment of the forcing term in LBM is an important issue. Guo et al. (2002) and Buick and Greated (2000) discussed various schemes in the literature before 2002. However, in those studies the connections between the forcing schemes of the SC-type

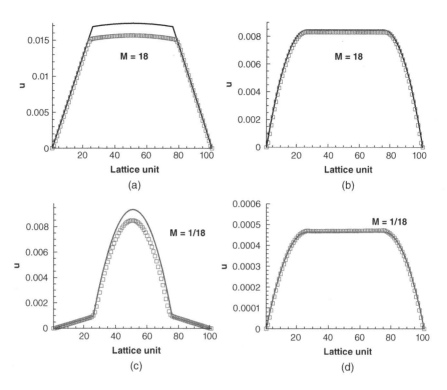

Figure 2.10 Velocity profiles u_x in the 2D channel with $\tau = 0.6$. The solid lines are analytical solutions and the squares are the LBM results. In (a) and (b) the wetting phase is less viscous ($M = 18$), while in (c) and (d) the wetting phase is more viscous ($M = \frac{1}{18}$). $\mathcal{G} = 3 \times 10^{-7}$ lu/ts^2 was applied only on (a) and (c), the central non-wetting Fluid 1 in Figure 2.9, and (b) and (d), the external wetting Fluid 2 in Figure 2.9, respectively. Velocity u has units lu/ts.

multiphase models (Shan and Chen 1993) and Luo (1998) were not investigated. Based on a straightforward theoretical analysis, we demonstrate that the Shan and Chen (1993) and Luo (1998) methods are identical when neglecting terms of order higher than $O(\frac{\mathbf{F}^2 \Delta t \tau}{\rho})$ (Huang et al. 2011a). More recently, Kupershtokh et al. (2009) proposed a so-called "exact difference method". After a simple analysis, we demonstrate that this forcing scheme is identical to the one used in Shan and Chen (1993) and Luo (1998) up to terms of order $O(\frac{(\mathbf{F}\Delta t)^2}{\rho})$ (Huang et al. 2011a).

For simulations of two-phase flows, the forcing-term strategy is of critical importance. Some popular LBM multiphase models such as the color-gradient based LBM (Gunstensen et al. 1991; Rothman and Keller 1988; Tölke et al. 2006) and the free-energy-based LBM (Swift et al. 1995) usually do not involve forcing terms explicitly. However, the popular models proposed by He et al. (1999), Lee and Lin (2005), and Shan and Chen (1993) depend on accurate forcing strategies in their models. Different discretization schemes for the forcing term in the model

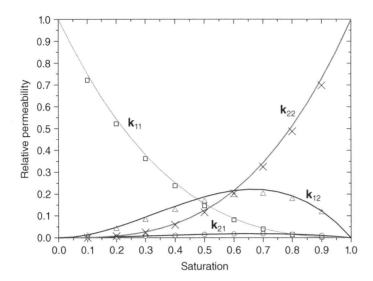

Figure 2.11   Relative permeabilities (k_{ij}) as function of wetting saturation for two-phase flow in a 2D channel, $M = \frac{1}{12}$. Lines show analytical solution and matched points from simulations. Source: Huang and Lu (2009). Reproduced with permission of AIP Publishing LLC.

of He et al. (1999) have been discussed by Wagner (2006) and Kikkinides et al. (2008) in detail. Kikkinides et al. (2008) discovered that in order to make the He et al. (1999) model thermodynamically consistent, the surface tension is limited to a very narrow range. In addition, they showed that only through a special discretization of the forcing term can thermodynamic consistency be achieved for that model (Kikkinides et al. 2008).

In Shan and Chen (1994) the surface tension obtained from the SC model with $\tau = 0.6\Delta t$ is consistent with analytical solutions. Yet, the effect of varying τ on the results is not addressed. By carrying out simulations with different τ values for two typical EOS, we will see that both the resulting density ratio and the surface tension strongly depend on τ when using the original SC model. The numerical tests will be performed both for the original SC model as well as variants with different forcing strategies.

In the following sections, we briefly review and analyze five forcing schemes. Then these schemes are incorporated into the SC model. The original SC model and the models with the forcing schemes of He et al. (1998) and Kupershtokh et al. (2009) are compared in detail. In the comparison, the surface tension and density ratios resulting from a typical EOS will be discussed in detail.

2.8.1   Schemes to incorporate the body force
Numerous schemes have been proposed to include the forcing term in the LBM. Five of those, from Shan and Chen (1993), Luo (1998), He et al. (1998), Ladd

and Verberg (2001), and Kupershtokh et al. (2009), are labeled I to V respectively and discussed below.

Scheme I
This scheme denotes the original strategy in the SC model and is presented in Section 2.1.1.

Scheme II
Luo (1998) described another scheme to incorporate the body force into the LBM. The source term to be added into the LBE is (Luo 1998)

$$S_i = w_i \left[\frac{1}{c_s^2}(e_{i\gamma} - u_\gamma) + \frac{1}{c_s^4}e_{i\alpha}u_\alpha e_{i\gamma} \right] F_\gamma \tag{2.72}$$

and the velocity is defined according to Eq. (2.1).

We also note the scaled forcing term introduced by Junk et al. (2005). In the scaled strategy, $S_i = \lambda S_i(\mathbf{x}, t) + (1 - \lambda)S_i(\mathbf{x} + e_i\Delta t, t + \Delta t)$, where λ should satisfy $0 \le \lambda \le 1$. However, how to choose λ for a specific flow problem is not illustrated explicitly in Junk et al. (2005). In their practical tests, usually $\lambda = 1$ is adopted (Junk et al. 2005). If $\lambda = 1$, it is identical to Scheme II. Here we do not intend to discuss the effect of parameter λ and this scaled forcing strategy is not discussed in our study because the choice of λ seems arbitrary.

Scheme III
The idea of the scheme proposed by He et al. (1998) is simple. On the left-hand side of Eq. (1.1) there is a force term $\mathbf{F} \cdot \nabla_\xi f$. Suppose f^{eq} is the leading part of f and the gradient of f^{eq} has a major contribution to the gradient of f. Then one finds (He et al. 1998) $\mathbf{F} \cdot \nabla_\xi f \approx \mathbf{F} \cdot \nabla_\xi f^{eq} = -\mathbf{F} \cdot \frac{\xi-\mathbf{u}}{c_s^2}f^{eq}$.

Considering discrete lattice effects, the corresponding formula of the forcing term according to He et al. (1998) is

$$S_i = \left(1 - \frac{1}{2\tau}\right) \frac{1}{\rho c_s^2} F_\gamma(e_{i\gamma} - u_\gamma)f_i^{eq} \tag{2.73}$$

and the velocity should be calculated as:

$$u_\alpha = \Sigma_i f_i e_{i\alpha} + \frac{F_\alpha \Delta t}{2\rho}. \tag{2.74}$$

Scheme IV
Ladd and Verberg (2001) proposed that the forcing term in the LBM should be expanded in a power series in the particle velocity, i.e.,

$$S_i = w_i \left[A + B_\gamma \frac{1}{c_s^2}(e_{i\gamma}) + C_{\alpha\gamma} \frac{1}{2c_s^4}(e_{i\alpha}e_{i\gamma} - c_s^2\delta_{\alpha\gamma}) \right], \tag{2.75}$$

and the velocity should be defined by Eq. (2.74), where A, B_γ, and $C_{\alpha\gamma}$ are determined by a Chapman–Enskog expansion.

Table 2.3 Overview of the forcing schemes.

Scheme	Equilibrium velocity u_α^{eq}	Physical velocity u_α^*	Additional forcing term?	Group
I	$\sum_i f_i e_{i\alpha} + \frac{\tau F_a}{\rho}$	$\sum_i f_i e_{i\alpha} + \frac{F_a \Delta t}{2\rho}$	No	A
II	$\sum_i f_i e_{i\alpha}$	$\sum_i f_i e_{i\alpha}$	Yes	A
III	$\sum_i f_i e_{i\alpha} + \frac{F_a \Delta t}{2\rho}$	$\sum_i f_i e_{i\alpha} + \frac{F_a \Delta t}{2\rho}$	Yes	B
IV	$\sum_i f_i e_{i\alpha} + \frac{F_a \Delta t}{2\rho}$	$\sum_i f_i e_{i\alpha} + \frac{F_a \Delta t}{2\rho}$	Yes	B
V	$\sum_i f_i e_{i\alpha}$	$\sum_i f_i e_{i\alpha} + \frac{F_a \Delta t}{2\rho}$	Yes	A

Based on a careful derivation from the LBE to the N–S equations, Guo et al. (2002) suggested

$$S_i = \left(1 - \frac{1}{2\tau}\right) w_i \left[\frac{1}{c_s^2}(e_{i\gamma} - u_\gamma) + \frac{1}{c_s^4} e_{i\alpha} u_\alpha e_{i\gamma}\right] F_\gamma. \tag{2.76}$$

Scheme V

Recently, Kupershtokh et al. (2009) proposed an "exact difference method" (EDM), which is claimed to be derived directly from the Boltzmann equation. In this scheme, the source term in the LBE should be

$$S_i = f_i^{eq}(\rho, \mathbf{u}^{eq} + \Delta\mathbf{u}) - f_i^{eq}(\rho, \mathbf{u}^{eq}), \tag{2.77}$$

where

$$\Delta\mathbf{u} = \frac{\mathbf{F}\Delta t}{\rho}. \tag{2.78}$$

In this scheme, the true fluid velocity $u_\alpha^* = \sum_i f_i e_{i\alpha} + \frac{F_a \Delta t}{2\rho}$ and the "equilibrium" velocity $u_\alpha^{eq} = \sum_i f_i e_{i\alpha}$ are not identical (Kupershtokh et al. 2009).

2.8.2   Scheme overview

From the introduction above we can see that three things may be modified by the presence of a body force: (i) the equilibrium velocity, (ii) the physical velocity, and/or (iii) the additional forcing term in the LBE. An overview of the five forcing schemes is shown in Table 2.3. In Schemes II, III, and IV, the true fluid velocity and the "equilibrium" velocity are identical, but for Schemes I and V the velocities are slightly different. Scheme I is the only scheme that does not need an additional forcing term in the LBE.

We show below that Schemes I, II, and V are identical and Schemes III and IV are identical with very minor differences. The former three and the latter two schemes are thus classified into two groups, A and B, respectively.

2.8.3   Theoretical analysis

We will start by comparing the scheme in the original SC model (Scheme I) to that of Luo (1998) (Scheme II).

In the original SC model, the forcing term is incorporated by defining $\mathbf{u}^{eq} = \mathbf{u} + \frac{1}{\rho}\tau\mathbf{F} = \frac{1}{\rho}(\Sigma_i f_i \mathbf{e}_i + \tau\mathbf{F})$ and in the collision step $\mathbf{u}^{eq}$ is substituted into f_i^{eq}. We note that the LBE can be rewritten as

$$
\begin{aligned}
f_i(\mathbf{x} + \mathbf{e}_i\Delta t, t + \Delta t) &= f_i(\mathbf{x}, t) - \frac{\Delta t}{\tau}(f_i(\mathbf{x}, t) - f_i^{eq}(\mathbf{u}^{eq})) \\
&= f_i(\mathbf{x}, t) - \frac{\Delta t}{\tau}(f_i(\mathbf{x}, t) - f_i^{eq}(\mathbf{u})) + [\frac{\Delta t}{\tau}(f_i^{eq}(\mathbf{u}^{eq}) - f_i^{eq}(\mathbf{u}))].
\end{aligned}
\tag{2.79}
$$

In the above equation the terms in the square brackets can be regarded as the source term in Eq. (1.2). Hence, in the SC model the explicit source term is

$$
\begin{aligned}
S_i &= \frac{\Delta t}{\tau}(f_i^{eq}(\mathbf{u}^{eq}) - f_i^{eq}(\mathbf{u})) \\
&= \frac{\Delta t}{\tau}\left\{ w_i\rho\left[1 + \frac{1}{c_s^2}e_{i\alpha}\left(u_\alpha + \frac{F_\alpha\tau}{\rho}\right) + \frac{1}{2c_s^4}e_{i\alpha}\left(u_\alpha + \frac{F_\alpha\tau}{\rho}\right)e_{i\beta}\left(u_\beta + \frac{F_\beta\tau}{\rho}\right) - \frac{1}{2c_s^2}\left(u_\alpha + \frac{F_\alpha\tau}{\rho}\right)^2 \right] \right. \\
&\left. \qquad - w_i\rho\left[1 + \frac{1}{c_s^2}e_{i\alpha}u_\alpha + \frac{1}{2c_s^4}e_{i\alpha}u_\alpha e_{i\beta}u_\beta - \frac{1}{2c_s^2}u_\alpha u_\alpha \right] \right\} \\
&= w_i\left[\frac{1}{c_s^2}(e_{i\alpha} - u_\alpha) + \frac{1}{c_s^4}e_{i\beta}u_\beta e_{i\alpha} \right] F_\alpha\Delta t + w_i\rho\frac{\Delta t}{\tau}\left\{ \frac{1}{2c_s^4}e_{i\alpha}e_{i\beta}\frac{F_\alpha F_\beta}{\rho^2}\tau^2 - \frac{1}{2c_s^2}\left(\frac{F_\alpha\tau}{\rho}\right)^2 \right\}.
\end{aligned}
\tag{2.80}
$$

If terms of $O(\frac{\Delta t}{\tau}(\frac{F_\alpha F_\beta\tau^2}{\rho}))$ and higher are neglected, Scheme I is identical to Scheme II (Luo 1998). We note that through expanding the distribution function in the Boltzmann equation on the basis of the Hermite orthogonal polynomials in velocity space, Shan et al. (2006) derived a correction of $f_i^{eq}(\mathbf{u}^{eq})$. According to their study (Shan et al. 2006), in order to obtain second-order accuracy, the *corrected* $f_i^{eq}(\mathbf{u}^{eq})'$ should be $f_i^{eq}(\mathbf{u}^{eq}) - w_i\rho\frac{\Delta t}{\tau}\{\frac{1}{2c_s^4}e_{i\alpha}e_{i\beta}\frac{F_\alpha F_\beta}{\rho^2}\tau^2 - \frac{1}{2c_s^2}(\frac{F_\alpha\tau}{\rho})^2\}$. We observe that this correction is consistent with our present simple analysis because with this correction the terms $O(\frac{\Delta t}{\tau}(\frac{F_\alpha F_\beta\tau^2}{\rho}))$ in Eq. (2.80) can be canceled.

In our numerical simulations presented below we found that the terms $O(\frac{\Delta t}{\tau}(\frac{F_\alpha F_\beta\tau^2}{\rho}))$ and higher have a very minor effect on a single-phase flow. However, in the following SC multiphase flow simulations, the terms $O(\frac{\Delta t}{\tau}(\frac{F_\alpha F_\beta\tau^2}{\rho}))$ are found to be significant. Starting from $\sum_i S_i e_{i\alpha}e_{i\beta} = u_\alpha F_\beta + u_\beta F_\alpha + \frac{\Delta t}{\tau}\{F_\alpha F_\beta\frac{\tau^2}{\rho}\}$, through a Chapman–Enskog expansion one can readily see the resulting N–S equations with an extra body force of order $O(\nabla \cdot \frac{\Delta t}{\tau}\{F_\alpha F_\beta\frac{\tau^2}{\rho}\})$ on the right-hand side of the equations. In other words, what the original SC multiphase model really mimics is Eq. (2.71) with a non-physical extra non-linear force term on the right-hand side.

We continue with comparison of Schemes V (Kupershtokh et al. 2009) and II (Luo 1998). In Scheme V, the source term is given by

$$
\begin{aligned}
S_i &= f_i^{eq}(\rho, \mathbf{u} + \Delta\mathbf{u}) - f_i^{eq}(\rho, \mathbf{u}) \\
&= w_i\rho\left[1 + \frac{1}{c_s^2}e_{i\alpha}\left(u_\alpha + \frac{F_\alpha\Delta t}{\rho}\right) + \frac{1}{2c_s^4}e_{i\alpha}\left(u_\alpha + \frac{F_\alpha\Delta t}{\rho}\right)e_{i\beta}\left(u_\beta + \frac{F_\beta\Delta t}{\rho}\right) - \frac{1}{2c_s^2}\left(u_\alpha + \frac{F_\alpha\Delta t}{\rho}\right)^2 \right] \\
&\qquad - w_i\rho\left[1 + \frac{1}{c_s^2}e_{i\alpha}u_\alpha + \frac{1}{2c_s^4}e_{i\alpha}u_\alpha e_{i\beta}u_\beta - \frac{1}{2c_s^2}u_\alpha u_\alpha \right] \\
&= w_i\left[\frac{1}{c_s^2}(e_{i\alpha} - u_\alpha) + \frac{1}{c_s^4}e_{i\beta}u_\beta e_{i\alpha} \right] F_\alpha\Delta t + w_i\rho\left\{ \frac{1}{2c_s^4}e_{i\alpha}e_{i\beta}\frac{F_\alpha F_\beta}{\rho^2}\Delta t^2 - \frac{1}{2c_s^2}\left(\frac{F_\alpha\Delta t}{\rho}\right)^2 \right\}.
\end{aligned}
\tag{2.81}
$$

Similar to the case of Eq. (2.80), if omitting terms of $O(\frac{F_\alpha F_\beta \Delta t^2}{\rho})$, we observe that this formula is identical to the one proposed in the study of Luo (1998). Thus, for this scheme we encounter the same problem as the SC model analyzed above, i.e., it involves an undesirable extra non-linear force in the N–S equations. Hence we proved that Schemes I, II, and V are identical if the terms $O(\frac{F_\alpha F_\beta \Delta t^2}{\rho})$ are omitted in Schemes I and V.

Next, we will compare Schemes III (He et al. 1998) and IV (Guo et al. 2002; Ladd and Verberg 2001) in Group B. In Scheme III (He et al. 1998), the source term is given as

$$
\begin{aligned}
S_i &= \left(1 - \frac{1}{2\tau}\right) \frac{1}{\rho c_s^2} F_\gamma (e_{i\gamma} - u_\gamma) f_i^{eq} \\
&= \left(1 - \frac{1}{2\tau}\right) \frac{1}{\rho c_s^2} F_\gamma (e_{i\gamma} - u_\gamma) w_i \rho \left[1 + \frac{1}{c_s^2} e_{i\alpha} u_\alpha + \frac{1}{2c_s^4} e_{i\alpha} u_\alpha e_{i\beta} u_\beta - \frac{1}{2c_s^2} u_\alpha u_\alpha \right] \\
&= \left(1 - \frac{1}{2\tau}\right) w_i \left[\frac{1}{c_s^2}(e_{i\gamma} - u_\gamma) + \frac{1}{c_s^4} e_{i\alpha} u_\alpha e_{i\gamma} \right] F_\gamma \\
&\quad + \left(1 - \frac{1}{2\tau}\right) w_i \left[-\frac{1}{c_s^4} e_{i\alpha} u_\alpha F_\gamma u_\gamma + \frac{1}{2c_s^6} e_{i\alpha} u_\alpha e_{i\beta} u_\beta F_\gamma (e_{i\gamma} - u_\gamma) - \frac{1}{2c_s^4} u_\alpha u_\alpha F_\gamma (e_{i\gamma} - u_\gamma) \right].
\end{aligned}
$$
(2.82)

In comparison to the schemes of Ladd and Verberg (2001) and Guo et al. (2002), i.e., Eq. (2.76), in Eq. (2.82) there are extra terms S_i'. Omitting terms of $O(u^3)$, we find the extra terms are

$$
S_i' = \left(1 - \frac{1}{2\tau}\right) w_i \left[-\frac{1}{c_s^4} e_{i\alpha} u_\alpha F_\gamma u_\gamma + \frac{1}{2c_s^6} e_{i\alpha} u_\alpha e_{i\beta} u_\beta F_\gamma e_{i\gamma} - \frac{1}{2c_s^4} u_\alpha u_\alpha F_\gamma e_{i\gamma} \right]. \qquad (2.83)
$$

By means of simple algebra, we found that $\Sigma_i S_i' = 0, \Sigma_i S_i' e_{i\kappa} = 0$ and $\Sigma_i S_i' e_{i\kappa} e_{i\delta} = 0$. For the derivation of the N–S equations from the LBE, only the zero- to second-order momenta of S_i are used. The S_i in Eq. (2.82) satisfy $\Sigma_i S_i = 0$, $\Sigma_i S_i e_{i\kappa} = (1 - \frac{1}{2\tau})F_\kappa$ and $\Sigma_i S_i e_{i\kappa} e_{i\delta} = (1 - \frac{1}{2\tau})(u_\kappa F_\delta + u_\delta F_\kappa)$. These formulae can be used to derive the N–S equations correctly (Guo et al. 2002). Hence, the extra terms S_i' in Scheme III do not affect the derivation. In other words, neglecting terms of order $O(u^3)$, Schemes III and IV are identical.

Thus, we have shown that the schemes proposed by Luo (1998), Kupershtokh et al. (2009), and Shan and Chen (1993) are identical when omitting terms of $O(\frac{F_\alpha F_\beta \Delta t^2}{\rho})$ and the schemes proposed by He et al. (1998), Ladd and Verberg (2001), and Guo et al. (2002) are identical when omitting terms of $O(u^3)$.

Through numerical tests of the single-phase unsteady Taylor–Green vortex flow, we have confirmed that for single-phase flow, the five schemes result in almost identical accuracy (Huang et al. 2011a). We consider multiphase numerical tests in the next section.

2.8.4 Numerical results and discussion

In this section we review the properties of the original SC model and the SC model in combination with the other four forcing strategies. All simulations in this section use Inter-particle Force Model A except those in the last subsection.

Cases of a cylindrical droplet and cases of a flat interface were simulated. For cases of a flat interface, the computational domain is $N_x \times N_y = 10 \times 200$, with periodic boundary condition in both directions. The central region ($50 \leq y \leq 150$) and the other region were filled with liquid and gas, respectively. The density field was initialized as $\rho(y) = \rho_{gas} + \frac{\rho_{liquid} - \rho_{gas}}{2}[\tanh(\frac{2(y-50)}{W}) - \tanh(\frac{2(y-150)}{W})]$, where W is the initial interface thickness. W is always chosen as 5 lu in the initial conditions of our simulations. The density profile is smooth when the above function is used to initialize the density field. This type of initialization is expected to be more stable in simulations (allowing lower τ) compared to initialization using step functions. The units of the density and the surface tension are mu/lu^3 and mu/ts^2, respectively.

For the cylindrical droplet cases, if not specified, the computational domain consists of 200×200 grid nodes and a central circular area with liquid was initialized in the domain and the other part was initialized with lower density gas. Periodic boundary conditions were applied on all boundaries. The analytical densities from the EOS at a specified temperature and zero velocity were used as initial conditions. The density field was initialized as

$$\rho(x,y) = \frac{\rho_{liquid} + \rho_{gas}}{2} - \frac{\rho_{liquid} - \rho_{gas}}{2}\left[tanh\left(\frac{2(\sqrt{(x-x_1)^2 + (y-y_1)^2} - R_0)}{W}\right)\right],$$
(2.84)

where (x_1, y_1) is the center point of the domain and R_0 is the initial radius of the droplet. The convergence criterion is

$$\frac{\Sigma|(u(t) - u(t - 2000\Delta t))| + |(v(t) - v(t - 2000\Delta t)|}{\Sigma|u(t)| + |v(t)|} < 10^{-7}$$
(2.85)

and the summation is taken over the whole computational domain. We checked that this criterion is sufficient to allow the flat interface and the droplet to reach the equilibrium state.

The effective surface tension of a numerical scheme is obtained from the simulation results of a droplet immersed in the gas, which can be described by the Laplace law. After the drop radius R_0 and the pressure difference between the inside (p_{in}) and the outside (p_{out}) of the drop are measured, the surface tension σ can be determined from $\frac{\sigma}{R_0} = p_{in} - p_{out}$, where R_0 is the distance between the center of the circle and a point where $\rho = (\rho_{in} + \rho_{out})/2$.

The single-phase LBM is usually second-order accurate in space. The spatial accuracy of the SC model using the five schemes was tested for the cases of a cylindrical droplet (Huang et al. 2011a). It was found that Schemes I, III, IV, and V are approximately second-order accurate in space. This conclusion is independent of the τ value (Huang et al. 2011a).

Droplet-size effects have also been tested through the simulation of cases with different initial drop sizes (Huang et al. 2011a). In terms of the surface tension, the difference in the two test cases with different initial radii is less than a few percent. Hence, drop size effects are negligible for many types of problems (Huang et al. 2011a).

Results of many simulations with different τ and $\frac{T}{T_c}$ show that in terms of spurious current, densities, and surface tension, Schemes III and IV are identical, which is consistent with our theoretical analysis. Hence, in the following, from Group B, only Scheme III will be discussed.

Comparison between schemes in Group A

First Schemes I and V are compared.

For Scheme V, it is interesting that Kupershtokh et al. (2009) claimed that "For a flat interface in the stationary case, we can obtain the maximum density ratio larger than 10^7 for the vdW and mKM EOS, and even larger than 10^9 for the C–S EOS." Here applying Scheme V coupled with the Inter-particle Force Model A and the C–S EOS for the cases of a cylindrical droplet, we indeed achieve a high-density ratio. The result is shown in Figure 2.12. From the figure, we can see that when $\frac{T}{T_c} = 0.5$, the density ratio is about $\frac{\rho_l/\rho_c}{\rho_g/\rho_c} = \frac{4.00}{7.36\times10^{-4}} \approx 5000$. However, there are large discrepancies between the LBM result and the analytical solution, particularly for lower $\frac{T}{T_c}$. That may be due to the undesirable extra terms in Scheme V (refer to Eq. (2.81)). Kupershtokh et al. (2009) suggested that using the Inter-particle Force Model B and finding an optimal parameter can make the LBM simulation result match the analytical solution. For more details, refer to Section 2.9.

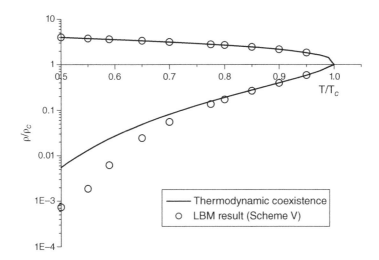

Figure 2.12 Coexistence curve for the C–S EOS. The solid line is the analytical solution obtained from the Maxwell construction (corresponding to thermodynamic consistency). The numerical results were obtained from Scheme V coupled with Inter-particle Force Model A and the C–S EOS ($\tau = \Delta t$).

Table 2.4 Density ratios and surface tension obtained from the SC model with Schemes I and V for $\tau/\Delta t = 0.6$ (C–S EOS is used).

$\frac{T}{T_c}$	Scheme I		Scheme V		Analytical	
	ρ_l/ρ_g	$\sigma(\times 10^{-3})$	ρ_l/ρ_g	$\sigma(\times 10^{-3})$	ρ_l/ρ_g	$\sigma(\times 10^{-3})$
0.95	0.210/0.0650	0.4892	0.211/0.0669	0.8622	0.210/0.0655	0.5690
0.90	0.246/0.0401	1.421	0.249/0.0447	2.550	0.247/0.0430	1.638
0.85	0.277/0.0228	2.766	0.280/0.0300	4.713	0.278/0.0278	3.061
0.80	N/A	—	0.307/0.0191	7.267	0.306/0.0167	4.834
0.75	N/A	—	0.334/0.0116	10.251	0.332/0.00863	6.932
0.70	N/A	—	N/A	—	0.357/0.00312	9.386

It is worth mentioning that for $\tau = \Delta t$, the results of Schemes I and V are found to be identical (down to an accuracy of 10^{-15}) for the same parameter sets, while for $\tau \neq \Delta t$ the results of the two schemes show significant deviations from each other. This τ-effect will be discussed in the following subsection.

The study of Huang et al. (2011a) showed that the results obtained by simulations of the cylindrical droplet and the flat interface are consistent. Here only the density ratios and surface tensions obtained from the cylindrical droplet are discussed. The results of Schemes I and V are shown in Table 2.4. Analytical solutions for the density ratios and the surface tension are also given. If not specified, the analytical solutions are obtained through the mechanical stability condition (Benzi et al. 2006), which means the densities ρ_l and ρ_g satisfy Eq. (2.58). On the other hand, in the case of thermodynamic consistency, ρ_l and ρ_g can be obtained from the Maxwell equal-area construction based on a $p - v$ diagram of the EOS (which requires $\int_{v_l}^{v_g} pdv = p_0(v_g - v_l)$, where $v = \frac{1}{\rho}$ and p_0 is a constant for a specific temperature). From Eq. (2.58), we can see that only if $\psi(\rho) \propto \rho$ is the SC model thermodynamically consistent (Benzi et al. 2006). However, most EOS do not satisfy this constraint. Hence the coexistence curves obtained from the mechanical stability condition and the thermodynamic equilibrium show small discrepancies.

For the densities of liquid and gas, Table 2.4 shows that the numerical results are consistent with the analytical ones at higher $\frac{T}{T_c}$. For decreasing temperatures, the difference between numerical and analytical ρ_g increases. For Scheme I, the maximum difference between the numerical (cases of a cylindrical droplet) and analytical ρ_g is approximately 28% at $\frac{T}{T_c} = 0.85$. For Scheme V, the maximum difference is approximately 34% at $\frac{T}{T_c} = 0.75$.

This discrepancy between numerical and analytical ρ_g may be caused by the terms of $O(\frac{(\mathbf{F}\Delta t)^2}{\rho})$ in Eqs (2.80) and (2.81). We also note that the coefficient multiplying the terms of $O(\frac{(\mathbf{F}\Delta t)^2}{\rho})$ is different in Schemes I and V. Hence the stability behavior and densities of the liquid and gas of the two schemes are different.

This also means that the terms of $O(\frac{(\mathbf{F}\Delta t)^2}{\rho})$ significantly affect the performance of Schemes I and V.

As shown in Table 2.4, for surface tension the numerical results of Scheme I are generally consistent with the analytical ones (maximum difference at $\frac{T}{T_c} = 0.85$ is approximately 10%). The surface tensions obtained with Scheme V at $\tau = 0.6\Delta t$ are significantly different from the analytical ones for any $\frac{T}{T_c}$ (maximum difference is approximately 50% compared with the analytical one).

The density ratios obtained with Scheme II (not shown) are found to be approximately consistent with the analytical results. However, the numerical surface tension is negative, which is unphysical. The stability of this scheme is also found to be the worst among the five schemes. For example, the lowest temperature for $\tau = \Delta t$ is about $T = 0.925T_c$. That may be due to the lack of terms of $O(\frac{(\mathbf{F}\Delta t)^2}{\rho})$ in Scheme II compared to Schemes I and V. Hence, although Scheme II works well for single-phase flows, it seems not to be so successful for multiphase flows. In the following, we will not discuss Scheme II further.

τ effect

In this section, the dependence of the surface tension and the density ratio on the relaxation time τ is discussed in detail. Comparisons are carried out between the original SC model (Scheme I) and the SC model in combination with Schemes III and V. The comparisons focus on the resulting surface tension, which is a very important property in multiphase flow simulations.

To investigate forcing Scheme III, we simulated cases for different temperatures and τ. Figure 2.13 illustrates the surface tension, the density of liquid and gas, and the density ratio as a function of $\frac{T}{T_c}$. Figure 2.13(a), (b), and (c) demonstrates that ρ_{liquid}, ρ_{gas}, and σ do not change with τ for a given temperature. The numerical ρ_{gas} agrees better with the analytical solution obtained from the mechanical stability condition than with that from thermodynamic coexistence. Because the numerical ρ_{gas} is smaller than the analytical ρ_{gas}, the density ratio of simulated liquid and gas is significantly larger than the analytical one (Figure 2.13(d)). The small discrepancy between the analytical and numerical ρ_{gas} may be caused by compressibility in the SC model or by the effect of spurious currents and requires further investigation. In Figure 2.13(c) the surface tensions obtained for $\tau = 0.6\Delta t, 0.8\Delta t$, and $1.0\Delta t$ are independent of τ and all agree well with the theoretical σ, which is obtained from the mechanical stability condition.

For Scheme I, Figure 2.14 illustrates ρ_{liquid}, ρ_{gas}, σ, and density ratio as a function of $\frac{T}{T_c}$. In Figure 2.14(c) the surface tensions obtained from $\tau = 0.7\Delta t$ and $0.52\Delta t$ are consistent with the analytical values while those values obtained from cases of $\tau = 1.0\Delta t, 1.5\Delta t$, and $2\Delta t$ are much larger than the analytical ones. Hence, the surface tension calculated from the original SC model depends on the value of τ, which is unphysical. This numerical result is consistent with our theoretical analysis in Section 2.8.3 because the undesirable terms are small when τ is small.

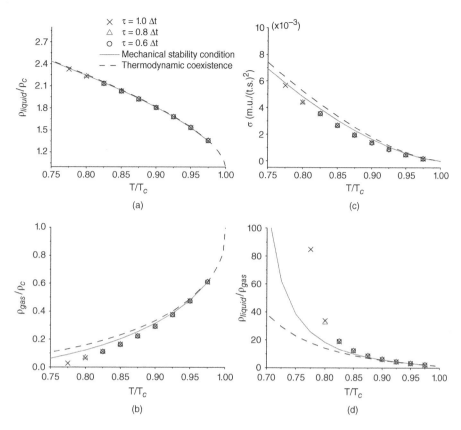

Figure 2.13 Results obtained from the SC model with C-S-EOS and forcing Scheme III for different values of τ. (a) Reduced density of liquid, (b) reduced density of gas, (c) surface tension, and (d) liquid-to-gas density ratio as a function of T/T_c. Reprinted from Huang et al. (2011a), copyright (2011), with permission from APS.

In Figure 2.14(b) the ρ_{gas} obtained from $0.52\Delta t < \tau < \Delta t$ seems more consistent with the analytical solution than $\tau = 1.5\Delta t$ and $2.0\Delta t$. When $\tau = 1.5\Delta t$ and $2.0\Delta t$, the numerical ρ_{gas} is significantly different from the analytical ρ_{gas} and the density ratios of liquid to gas severely deviate from the analytical ones (Figure 2.14(d)). From Figure 2.14(d) it is found when $\tau \approx 1$ (cross symbol in the subfigure), the simulation remains very stable even at $\frac{T}{T_c} = 0.5$. The maximum density ratio that this scheme can achieve is about 5000 when $\tau = \Delta t$.

Finally, we discuss Scheme V. Figure 2.15 illustrates the reduced density of liquid and gas, surface tension, and the density ratio as a function of T/T_c. At the different temperatures, ρ_{gas} values obtained with different τ seem almost identical. However, from Figure 2.15(d) we can see that at lower temperatures, for example $\frac{T}{T_c} < 0.65$, the density ratio is still affected significantly by τ. From Figure 2.15(c) we can see that the surface tensions obtained from different τs are highly consistent. However, all surface tensions deviate from the analytical ones.

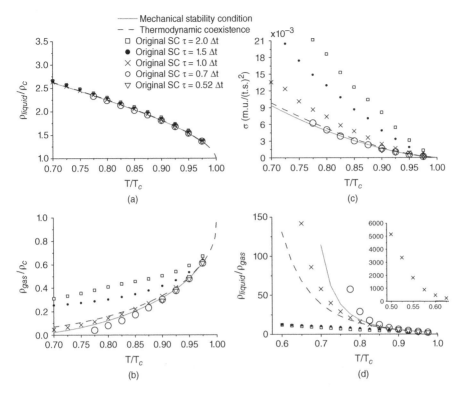

Figure 2.14 Results obtained from the original SC model (Scheme I) for different values of τ. (a) Reduced density of liquid, (b) reduced density of gas, (c) surface tension, and (d) liquid-to-gas density ratio as a function of T/T_c for C–S EOS. Reprinted from Huang et al. (2011a), copyright (2011), with permission from APS.

In summary, Scheme III is independent of τ. The numerical results (density of gas, surface tension) of Scheme I strongly depend on τ. For $\tau = 0.7$ the observed surface tension is consistent with its analytical counterpart and liquid/vapor density ratios up to 5000 can be simulated with $\tau = 1$. For Scheme V, the surface tension is almost not affected by τ but the values deviate from the analytical ones. The density ratio obtained from Scheme V is slightly affected by τ. It is noted that the conclusions about the τ effect are not only applicable to the C–S EOS but also to the other EOS (Huang et al. 2011a).

Maximum density ratio and numerical stability

In this section we discuss numerical stability for multiphase flow simulations. The numerical stability depends on forcing schemes. For example, when using the original SC model (Scheme I) with $\tau = \Delta t$, the simulation remains stable even at $\frac{T}{T_c} = 0.50$ (corresponding to an equilibrium density ratio of 5000, refer to Figure 2.14(d)). The maximum density ratio of the SC model with Scheme III is limited

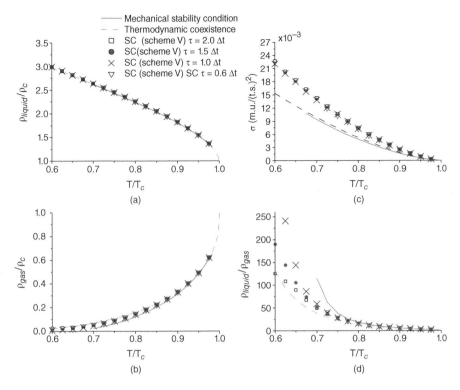

Figure 2.15 Results obtained by the SC model in combination with Scheme V for different values of τ. (a) Reduced density of liquid, (b) reduced density of gas, (c) surface tension, and (d) density ratio as a function of T/T_c for C–S EOS. Reprinted from Huang et al. (2011a), copyright (2011), with permission from APS.

to $0.3175/0.003748 = 84.78$ at $\frac{T}{T_c} = 0.775$ (refer to Figure 2.13(d)). For Scheme V, the maximum density ratio is limited to $0.4195/0.0004677 = 896.94$ at $\frac{T}{T_c} = 0.575$. This means that the numerical stability of the original SC model with $\tau = \Delta t$ is substantially larger in comparison to the stability with forcing Schemes III and V. Thus in flow simulations where the maximum density ratio is of primary concern, the original SC model may be a good choice.

For different τ, the minimum $\frac{T}{T_c}$s that Schemes I, III, and V can simulate are shown in Figure 2.16. The region to the upper right of each line corresponds to the stable region of each scheme. Generally speaking, the numerical stability of Scheme III is not as good as that of Schemes I and V. The numerical stability depends on the scheme partially because the schemes recover different macroscopic equations. Although Schemes I and V have better numerical stability for two-phase flows, they do not recover the N–S equations exactly. The extra terms in the momentum equation (compared to the N–S equations) may help to stabilize Schemes I and V.

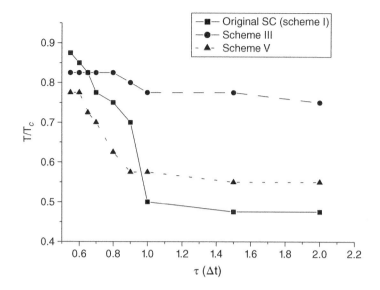

Figure 2.16   Stability of forcing Schemes I, III, and V as a function of τ. The region to the upper right of each line is the corresponding stable region of each scheme. The C–S EOS is used with $a = 1, R = 1$, and $b = 4$. Reprinted from Huang et al. (2011a), copyright (2011), with permission from APS.

Interface thickness

When numerical instabilities appear, the interface usually only occupies 2 or 3 lu. Since the SC multiphase model is a diffuse-interface method, the interface thickness is an important factor in simulations. Usually the thicker the simulated interface, the more stable the simulation will be. However, a very thick interface is not consistent with physical reality.

In the SC multiphase models the interfaces form spontaneously from the model force balances and we are not able to adjust the interface thickness explicitly and independently of other model parameters. The interface thickness depends on not only the temperature and τ, but also the formula of the particular EOS. For example, parameters a, R, and b in the C–S EOS may affect the interface thickness and numerical stability. The present choice of $a = 1, R = 1$, and $b = 4$ is appropriate because the interface thickness is approximately 5 lu for most T and τ; this represents a workable compromise between the desire to maintain a sharp interface and the need for numerical stability.

To study the effect of variations in these parameters on interface thickness, first $R = 1$ and $b = 4$ are fixed and the parameter a is allowed to change. We found that the interface widths are approximately 2, 3, 5, 7, and 11 lu for $a = 4, 2, 1, 0.5$, and 0.2, respectively, for a case of $\frac{T}{T_c} = 0.85$ and $\tau = 0.6\Delta t$ using Scheme III. Thus a large a leads to a thin interface, as expected from the EOS.

To study the effect of b, $a = 1$ and $R = 1$ are fixed in simulations. The coexisting equilibrium densities of the liquid and gas are affected by the parameter b but the liquid/gas density changes are not independent and the density ratio of liquid to gas does not change. We simulated a case of $\frac{T}{T_c} = 0.85$ and $\tau = 0.6\Delta t$ using Scheme III. When $b = 0.5$ the simulation is not possible. When $b = 1.0$ the interface is very sharp and the width of the interface is about 3 lu. When $b = 20$ the interface occupies approximately 10 lu, which is often very thick relative to the radii of drops and bubbles one might wish to simulate.

If we only change the gas constant R and fix the values of $a = 1$ and $b = 4$, the density ratio and surface tension and thickness do not change with R. That can be explained by the following. We note that the C–S EOS in Eq. (2.30) can be rewritten as

$$p = \frac{1.5092a}{b^2} \left\{ \theta \left(\frac{T}{T_c} \right) \left[1 + \frac{-2\theta^2 + 4\theta}{(1 - \theta)^3} \right] - 10.601\theta^2 \right\}, \qquad (2.86)$$

where $\theta = \frac{b}{4}\rho$. In the above derivation the relationship between the critical temperature and the parameters a, b, and R, i.e., $T_c = \frac{0.3773a}{bR}$ for the C–S EOS, is used. Obviously the parameter R enters this EOS only through $\frac{T}{T_c}$ and consequently will not affect the phase-separation property at a specific $\frac{T}{T_c}$.

Overall, after the parameters in an EOS are properly chosen, the selection of forcing schemes will significantly affect the numerical stability when the SC model is used.

2.9 Multirange pseudopotential (Inter-particle Force Model B)

We also carried out simulations using the Inter-particle Force Model B (Sbragaglia et al. 2007). These results were compared with the analytical result for surface tension obtained from Eq. (2.62). Similar behaviors to those in Figures 2.13 and 2.14 were observed, i.e., the SC model in combination with Scheme III and the original SC model (for $\tau < 0.7\Delta t$) are consistent with the analytical solutions. Figure 2.17 shows the surface tension obtained from the SC model in combination with Scheme III. The numerical results agree well with the analytical ones for different combinations of the parameters A_1 and A_2.

In Kupershtokh et al. (2009) an approximation for the inter-particle force is proposed as

$$\mathbf{F}_{int}(\mathbf{x}, t) = -G \left[(1 - 2A)\psi(\mathbf{x}, t) \sum_i w_i \psi(\mathbf{x} + \mathbf{e}_i \Delta t, t)\mathbf{e}_i + A \sum_i w_i \psi^2(\mathbf{x} + \mathbf{e}_i \Delta t, t)\mathbf{e}_i \right],$$
$$(2.87)$$

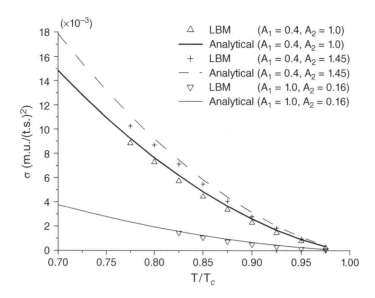

Figure 2.17 Surface tension comparison between the SC model with forcing Scheme III and the analytical solution for different A_1 and A_2. In the simulations, the inter-particle force including next-nearest neighbors for the C–S EOS is used and $\tau = \Delta t$. Reprinted from Huang et al. (2011a), copyright (2011), with permission from APS.

which differs slightly from the one proposed by Sbragaglia et al. (2007) (i.e., Eq. (2.7)) but the idea is similar. A Taylor expansion of Eq. (2.87) reveals

$$F_\alpha(\mathbf{x}, t) \approx -\frac{G}{2}\Delta t c_s^2 \partial_\alpha \psi^2 - \frac{G}{2}\Delta t^3 c_s^4 \psi \cdot \partial_\alpha \nabla^2 \psi - \{ GA\Delta t^3 c_s^4 (\partial_\alpha \psi \nabla^2 \psi + 2\partial_\beta \psi \partial_\alpha \partial_\beta \psi) \}. \tag{2.88}$$

Thus the difference between the above equation and Eq. (2.8) lies in the terms inside the braces in Eq. (2.88). The surface tension can be changed by modifying parameter A.

In Sbragaglia et al. (2007) the authors argue that the surface tension can be changed without affecting the density ratio. However, in our work we found the density ratio changes when modifying the ratio of $\frac{A_2}{A_1}$. In the following, from Group B, only Scheme III will be discussed. An example for the SC model with forcing Scheme III with $\tau = \Delta t$ and the C–S EOS is illustrated in Table 2.5. From the table we can see that for a relatively large ratio, $\frac{T}{T_c} = 0.85$, the density ratio changes are small, yet for $\frac{T}{T_c} = 0.775$ they are more pronounced.

If the deviation of the surface tension from the analytical solution and the density ratio dependence on τ are ignored, one can adjust the surface tension to "force" the equilibrium densities to satisfy the thermodynamic coexistence condition, which essentially corresponds to mimicking thermodynamic consistency with the SC model. For the original SC model, this adjustment is

Table 2.5 Density ratio for different surface tensions (SC model with forcing Scheme III, $\tau = \Delta t$, C–S EOS).

$\frac{T}{T_c}$	$\sqrt{\frac{A_2}{A_1}}$	ρ_l/ρ_g	$\sigma(\times 10^{-3})$
0.85	1.0	0.2768/0.02250	2.683
0.85	1.458	0.2785/0.02587	4.199
0.85	1.904	0.2795/0.02775	5.461
0.775	1.0	0.3175/0.003745	5.661
0.775	1.458	0.3191/0.008159	7.939
0.775	1.904	0.3201/0.01052	10.57

Reprinted from Huang et al. (2011a), copyright (2011), with permission from APS.

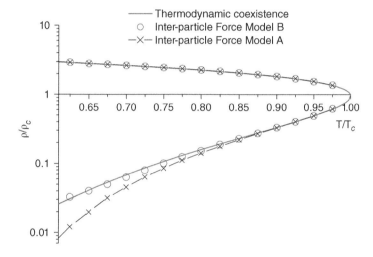

Figure 2.18 Coexistence curve for the C–S EOS. The solid line is the analytical solution obtained from the Maxwell construction (corresponding to thermodynamic consistency). The numerical results were obtained from the original SC model with $\tau = \Delta t$. Reprinted from Huang et al. (2011a), copyright (2011), with permission from APS.

relatively easy: Figure 2.18 shows the coexistence curve for the C–S EOS. The numerical results are obtained for the original SC model with $\tau = \Delta t$. The result for Inter-particle Force Model A shows a large discrepancy from the thermodynamic coexistence condition. However, using Inter-particle Force Model B we can adjust the densities to match the thermodynamic coexistence well by modifying the parameters A_1 and A_2. In this simulation

$$A_1 = 1.0, \quad A_2 = 0.16, \quad G_1 = 1.28, \quad \text{and } G_2 = -0.14. \tag{2.89}$$

The relationship that these parameters should satisfy is illustrated in Eq. (2.61). These optimal parameters were chosen through trial and error.

2.10   Conclusions

In this chapter the SCMP SC model is introduced and the incorporation of different EOS is described. The thermodynamical inconsistency is discussed. Cases of contact angle, capillary rise, and parallel flow in channels are simulated. The results show that the SCMP SC model is a good tool to simulate two-phase flow with a density ratio less than several hundred.

Our theoretical analysis identified the relations between forcing schemes proposed by Shan and Chen (1993) (Scheme I), Luo (1998) (Scheme II), He et al. (1998) (Scheme III), Ladd and Verberg (2001) and Guo et al. (2002) (Scheme IV), and Kupershtokhet al. (2009) (Scheme V). We demonstrated that Schemes I, II, and V are identical when terms of $O(\frac{F_\alpha F_\beta \Delta t^2}{\rho})$ in Schemes I and V are omitted. Schemes III and IV are identical if terms of $O(u^3)$ in Scheme III are neglected.

Comparing numerical results with the analytical solutions for a typical EOS, the surface tension obtained from the SC model using the forcing Scheme III (Group B) is consistent with the Maxwell construction and essentially independent of the value of τ. However, Scheme II (Group A) is unsuccessful in simulating two-phase fluid interactions. For the original SC model (Scheme I in Group A), the density ratio and surface tension depend on the specific value of τ and large discrepancies between the numerical and analytical surface tension were observed for $\tau > 0.7\Delta t$. For Scheme V (Group A), the density ratio and the surface tension are almost τ independent but the values deviate from the analytical ones.

Although the SC model has been proven to be thermodynamically inconsistent, it is possible to restore thermodynamic consistency by adjusting the strength of surface tension in order to match the coexistence curve obtained by the Maxwell construction using the multirange pseudopotential formula (Sbragaglia et al. 2007).

For numerical stability in two-phase flow simulations, Scheme III is less stable than Schemes I and V.

2.11   Appendix A: Analytical solution for layered multiphase flow in a channel

Here the analytical solution for the velocity profile of layered two-phase flow in a channel is given. The flow is illustrated in Figure 2.9. The whole flow in the y direction is from $y = -b$ to $y = b$. In the figure the wetting (Phase 2) phase flows along upper and lower plate while non-wetting phase (Phase 1) flows in the center. When the constant body forces G_1 and G_2 are applied to Phase 1 and Phase 2, respectively, the fluid flow of each phase is governed by the following equations:

$$\rho_1 \nu_1 \nabla^2 u_1 = G_1 \quad \text{and} \quad \rho_2 \nu_2 \nabla^2 u_2 = G_2,$$

where $u_1(y)$ and $u_2(y)$ are the velocities of Phase 1 and Phase 2, respectively. The boundary conditions are

$$\begin{cases} \left(\partial_y u_1\right)\big|_{y=0} = 0, \quad \rho_1 \nu_1 (\partial_y u_1)\big|_{y=a} = \rho_2 \nu_2 (\partial_y u_2)\big|_{y=a} \\ u_1\big|_{y=a} = u_2\big|_{y=a} \quad u_2\big|_{y=b} = 0 \end{cases}.$$

Solving the above equations with the boundary conditions we obtain the analytical solutions:

$$u_1 = A_1 y^2 + C_1 \quad \text{and} \quad u_2 = A_2 y^2 + B_2 y + C_2$$

where

$$A_1 = -\frac{G_1}{2\rho_1 \nu_1}, \quad A_2 = -\frac{G_2}{2\rho_2 \nu_2},$$

$$B_2 = -2A_2 a + 2MA_1 a,$$

$$C_1 = (A_2 - A_1)a^2 - B_2(b - a) - A_2 b^2, \quad C_2 = -A_2 b^2 - B_2 b,$$

and $M = \frac{\eta_1}{\eta_2}$ is the viscosity ratio.

With above solutions, it is easy to get the relative permeabilities, as illustrated in Eq. (2.70). The following is an example of how to get k_{11} in Eq. (2.70). As we know, the volumetric flow rate for steady single-phase Darcy flow is

$$Q_{10} = \int_0^b (A_1 y^2 - A_1 b^2) dy = -\frac{2}{3} b^3 A_1$$

and the Phase 1 volumetric flow rate in layered two-phase flow is

$$Q_1(G_2 = 0) = \int_0^a u_1 \, dy$$

$$= \int_0^a (A_1 y^2 - A_1 a^2 - 2MA_1 a(b - a)) dy$$

$$= -\frac{2}{3} A_1 a^3 - 2MA_1 a^2 b + 2MA_1 a^3, \tag{2.90}$$

where only G_1 is applied to Phase 1 and $G_2 = 0$. Hence, k_{11} is

$$k_{11} = \frac{Q_1(G_2 = 0)}{Q_{10}} = S_n^3 + 3MS_n^2 - 3MS_n^3.$$

The other relative permeabilities in Eq. (2.70) can also be obtained.

2.12   Appendix B: FORTRAN code to simulate single component multiphase droplet contacting a wall, as shown in Figure 2.7(c)

```
NOTES:
! 1. The whole project is composed of head.inc, main.for,
!  streamcollision.for, force.for, and output.for.
! 2. head.inc is a header file, where some common blocks are defined.
! 3. In the file: 'params.in' is a file you can put the critical
!  parameters.
!  Do not include "c=======..." in the file. The file begins with a
!  parameter.
! 4. Before running this project, please create a new sub-folder
!  'out'
!  under the working directory.
! 5. Here C-S EOS is given as an example. The other EOS can be
!  included similarly.
! 6. Please create a new folder 'out' in the working directory
!  before running the code


head.inc:
C==========================================================
! Domain size is lx \times ly \times \lz
      integer lx,ly,lz
      PARAMETER(lx=51,ly=51,lz=41)
      common/AA/ G,tau,Gads(2),ex(0:18),ey(0:18),ez(0:18),opp(18)
      real*8 G,tau,Gads
      integer ex,ey,ez, opp

      common/b/ error,vel,xc(0:18),yc(0:18),zc(0:18),t_k(0:18)
      real*8 error,vel,xc,yc,zc,t_k

      common/vel/ c_squ,cc,TT0,rho_w,Ca,RR, Nwri
      real*8 c_squ,cc,TT0, rho_w, Ca,RR
      integer Nwri

      common/app/ t_0,t_1,t_2, rho_h, rho_l
      real*8 t_0,t_1,t_2, rho_h, rho_l

main.for
C==========================================================
      program D3Q19LBM
      implicit none
      include "head.inc"
      integer t_max,time,k
! This array defines which lattice positions are occupied by fluid
!  nodes (obst=0)
! or solid nodes (obst=1)
      integer obst(lx,ly,lz)
! Velocity components
      real*8  u_x(lx,ly,lz),u_y(lx,ly,lz), u_z(lx,ly,lz)
! Pressure and density
      real*8  p(lx,ly,lz),rho(lx,ly,lz)
```

```
! The real fluid density
! which may differ from the velocity components in the above; refer
!    to SC model
       real*8  upx(lx,ly,lz),upy(lx,ly,lz),upz(lx,ly,lz)

! The force components: Fx, Fy, Fz for the interaction between
!    fluid nodes.
! Sx, Sy, Sz are the interaction (components) between the fluid nodes
!    and solid nodes
! ff is the distribution function
       real*8  ff(0:18,lx,ly,lz),Fx(lx,ly,lz),Fy(lx,ly,lz),
     &    Fz(lx,ly,lz), Sx(lx,ly,lz),Sy(lx,ly,lz), Sz(lx,ly,lz)

! TT0W is the value of T/T0;   RHW and RLW are the coexisting
!    densities
! in the specified T/T0.
! For initialization, \rho_l (lower density)
! and \rho_h (higher density) are supposed to be known.
       real*8   TT0W(12), RHW(12), RLW(12)
!-------------------------------------------
! Author: Haibo Huang, Huanghb@ustc.edu.cn
!-------------------------------------------
! The below data define the D3Q19 velocity model, xc(ex), yc(ey),
!    zc(ez)
! are the components of e_{ix}, e_{iy}, and e_{iz}, respectively.
       data xc/0.d0, 1.d0, -1.d0, 0.d0, 0.d0, 0.d0, 0.d0, 1.d0, 1.d0,
     &   -1.d0, -1.d0, 1.d0, -1.d0, 1.d0, -1.d0, 0.d0, 0.d0,
     &   0.d0, 0.d0  /,
     &    yc/0.d0, 0.d0, 0.d0, 1.d0, -1.0d0, 0.d0, 0.d0, 1.d0, -1.d0,
     &   1.d0, -1.d0, 0.d0, 0.d0, 0.d0, 0.d0, 1.d0, 1.d0,
     &   -1.d0, -1.d0/,
     &    zc/0.d0, 0.d0, 0.d0, 0.d0, 0.d0, 1.d0, -1.d0, 0.d0, 0.d0,
     &   0.d0, 0.d0, 1.d0, 1.d0, -1.d0, -1.d0, 1.d0, -1.d0,
     &   1.d0, -1.d0/
        data ex/0, 1, -1, 0, 0, 0, 0, 1, 1, -1, -1, 1, -1, 1, -1, 0, 0,
     &   0, 0 /,
     &    ey/0, 0, 0, 1, -1, 0, 0, 1, -1, 1, -1, 0, 0, 0, 0, 1, 1,
     &   -1, -1/,
     &    ez/0, 0, 0, 0, 0, 1, -1, 0, 0, 0, 0, 1, 1, -1, -1, 1, -1,
     &   1, -1/

! This array gives the opposite direction for e_1, e_2, e_3,
!    .....e_18
! It implements the simple bounce-back rule we use in the collision
!    step
! for solid nodes (obst=1)
       data opp/2,1,4,3,6,5,10,9,8,7,14,13,12,11,18,17,16,15/

! C-S EOS
! RHW and RLW are the coexisting densities in the corresponding
!    specified T/T0.
       data TT0W/0.975d0, 0.95d0, 0.925d0, 0.9d0, 0.875d0, 0.85d0,
     &     0.825d0,  0.8d0, 0.775d0, 0.75d0, 0.7d0, 0.65d0 /,
     &      RHW/ 0.16d0, 0.21d0,  0.23d0, 0.247d0, 0.265d0, 0.279d0,
     &     0.29d0, 0.314d0, 0.30d0,  0.33d0, 0.36d0, 0.38d0 /,
     &      RLW/0.08d0, 0.067d0, 0.05d0, 0.0405d0, 0.038d0, 0.032d0,
     &     0.025d0, 0.0245d0, 0.02d0, 0.015d0, 0.009d0, 0.006d0/
```

```
! Speeds and weighting factors
      cc = 1.d0
      c_squ = cc *cc / 3.d0
      t_0 =  1.d0 / 3.d0
      t_1 =  1.d0 / 18.d0
      t_2 =  1.d0 / 36.d0
! Weighting coefficient in the equilibrium distribution function
      t_k(0) = t_0
      do 1 k =1,6
       t_k(k) = t_1
    1 continue
      do 2 k =7,18
       t_k(k) = t_2
    2 continue

! Please specify which temperature
! and corresponding \rho_h, \rho_l in above 'data' are chosen.
! Initial T/T0, rho_h, and rho_l for the C-S EOS are listed in above
!  'data' section
      k = 5    ! important
      TT0 = TT0W(k)
      rho_h = RHW(k)
      rho_l = RLW(k)


C======================================================================
c      Initialisation
C======================================================================
      write (6,*) '@@@⊔⊔3D LBM for single component multiphase @@@'
      write (6,*) '@@@ lattice size lx = ',lx
      write (6,*) '@@@                 ly = ',ly
      write (6,*) '@@@                 lz = ',lz
      call read_parameters(t_max)
      call read_obstacles(obst)

      call init_density(obst,u_x ,u_y ,rho ,ff )
      open(40,file='./out/residue.dat')
C======================================================================
! Begin iterations
      do 100 time = 1, t_max
        if ( mod(time, Nwri) .eq. 0 .or. time. eq. 1) then
        write(*,*) time
        call write_results2(obst,rho,p,upx,upy,upz,time)
        end if

        call stream(obst,ff )    ! streaming (propagation) step

! Obtain the macro variables
      call getuv(obst,u_x ,u_y, u_z, rho, ff )

! Calculate the actual velocity
      call calcu_upr(obst,u_x,u_y,u_z,Fx,Fy,Fz,
     &    Sx,Sy,Sz,rho, upx,upy,upz)

! Calculate the interaction force between fluid nodes,
! and the interaction force between solid and fluid nodes.
      call calcu_Fxy(obst,rho,Fx,Fy,Fz,Sx,Sy,Sz,p)
```

```
!      BGK model (a single relaxation parameter) is used
       call collision(tau,obst,u_x,u_y,u_z,rho ,ff ,Fx ,Fy ,Fz,
     &    Sx,Sy, Sz )    ! collision step ,

   100 continue
c===== End of the main loop
       close(40)
       write (6,*) '@@@@** end **@@@@'
       end
c-------------------------------------
       subroutine read_parameters(t_max)
       implicit none
       include "head.inc"
       integer  t_max
       real*8 visc

       open(1,file='./params.in')
! Initial radius of the droplet.
       read(1,*) RR
! \rho_w in calculation of fluid-wall interaction
       read(1,*) rho_w
! Relaxation parameter, which is related to viscosity
       read(1,*) tau
! Maximum iteration specified
       read(1,*) t_max
! Output data frequency (can be viewed with TECPLOT)
       read(1,*) Nwri
       close(1)
        visc =c_squ*(tau-0.5)
        write (*,'("kinematic viscosity=",f12.5, "lu^2/ts",
     &    2X, "tau=", f12.7)') visc, tau

       end

!-------------------------------------------------------
! Initialize which nodes are wall node (obst=1) and
! which are fluid nodes (obst=0)
       subroutine read_obstacles(obst)
       implicit none
       include "head.inc"

       integer  x,y,z,obst(lx,ly,lz)
         do 11 z = 1, lz
          do 10 y = 1, ly
           do 40 x = 1, lx
             obst(x,y,z) =  0
             if(z .eq. 1)   obst(x, y,1) = 1
    40      continue
    10    continue
    11 continue

       end
!-------------------------------------------------
       subroutine init_density(obst,u_x,u_y,rho,ff)
       implicit none
       include "head.inc"
```

```
      integer i,j,x,y,z,k,n,obst(lx,ly,lz)
      real*8  u_squ,u_n(0:18),fequi(0:18),u_x(lx,ly,lz),u_y(lx,ly,lz),
     & rho(lx,ly,lz),ff(0:18,lx,ly,lz),u_z(lx,ly,lz)

      do 12 z = 1, lz
        do 11 y = 1, ly
          do 10 x = 1, lx
          u_x(x,y,z) = 0.d0
          u_y(x,y,z) = 0.d0
          u_z(x,y,z) = 0.d0
          rho(x,y,z) = rho_l
          if(real(x-lx/2)**2+real(y-ly/2)**2+real(z-5)**2< RR**2) then
             rho(x,y,z) = rho_h
          endif
10        continue
11     continue
12 continue

      do 82 z = 1, lz
        do 81 y = 1, ly
          do 80 x = 1, lx
            u_squ    =   u_x(x,y,z)*u_x(x,y,z) + u_y(x,y,z)*u_y(x,y,z)
     &             +   u_z(x,y,z) *u_z(x,y,z)
            do 60 k = 0,18
              u_n(k)    = xc(k)*u_x(x,y,z) + yc(k)*u_y(x,y,z)
     &                  + zc(k) *u_z(x,y,z)
              fequi(k) = t_k(k)* rho(x,y,z) * ( cc*u_n(k) / c_squ
     &                  + (u_n(k)*cc) *(u_n(k)*cc) / (2.d0 * c_squ *c_squ)
     &                  - u_squ / (2.d0 * c_squ)) + t_k(k)  * rho(x,y,z)
              ff(k,x,y,z)= fequi(k)
60          continue
80        continue
81      continue
82 continue
      end
!-----------------------------------------------------


streamcollision.for
C============================================================

      subroutine stream(obst,ff)
      implicit none
      include "head.inc"
      integer  k,obst(lx,ly,lz)
      real*8 ff(0:18,lx,ly,lz),f_hlp(0:18,lx,ly,lz)
      integer  x,y,z,x_e,x_w,y_n,y_s,z_n,z_s

      do 12 z = 1, lz
        do 11 y = 1, ly
          do 10 x = 1, lx
!
          z_n = mod(z,lz) + 1
          y_n = mod(y,ly) + 1
          x_e = mod(x,lx) + 1
```

```
            z_s = lz - mod(lz + 1 - z, lz)
            y_s = ly - mod(ly + 1 - y, ly)
            x_w = lx - mod(lx + 1 - x, lx)

c......... Propagation
            f_hlp(1 ,x_e,y  ,z  ) = ff(1,x,y,z)
            f_hlp(2 ,x_w,y  ,z  ) = ff(2,x,y,z)
            f_hlp(3 ,x  ,y_n,z  ) = ff(3,x,y,z)
            f_hlp(4 ,x  ,y_s,z  ) = ff(4,x,y,z)
            f_hlp(5 ,x  ,y  ,z_n) = ff(5,x,y,z)
            f_hlp(6 ,x  ,y  ,z_s) = ff(6,x,y,z)
            f_hlp(7 ,x_e,y_n,z  ) = ff(7,x,y,z)
            f_hlp(8 ,x_e,y_s,z  ) = ff(8,x,y,z)
            f_hlp(9 ,x_w,y_n,z  ) = ff(9,x,y,z)
            f_hlp(10,x_w,y_s,z  ) = ff(10,x,y,z)
            f_hlp(11,x_e,y  ,z_n) = ff(11,x,y,z)
            f_hlp(12,x_w,y  ,z_n) = ff(12,x,y,z)
            f_hlp(13,x_e,y  ,z_s) = ff(13,x,y,z)
            f_hlp(14,x_w,y  ,z_s) = ff(14,x,y,z)
            f_hlp(15,x  ,y_n,z_n) = ff(15,x,y,z)
            f_hlp(16,x  ,y_n,z_s) = ff(16,x,y,z)
            f_hlp(17,x  ,y_s,z_n) = ff(17,x,y,z)
            f_hlp(18,x  ,y_s,z_s) = ff(18,x,y,z)
   10    continue
   11    continue
   12 continue
c-------------------Update distribution function
        do 22 z = 1, lz
          do 21 y = 1, ly
            do 20 x = 1, lx
              do k =1, 18
                 ff(k,x,y,z) = f_hlp(k,x,y,z)
              enddo
   20       continue
   21    continue
   22 continue

        return
        end

c------------------------------------------
        subroutine getuv(obst,u_x,u_y,u_z,rho,ff)
        include "head.inc"
        integer x,y,obst(lx,ly,lz)
        real*8  u_x(lx,ly,lz),u_y(lx,ly,lz),rho(lx,ly,lz),
      & ff(0:18,lx,ly,lz),u_z(lx,ly,lz)

        do 12 z = 1, lz
          do 11 y = 1, ly
            do 10 x = 1, lx
              rho(x,y,z) = 0.d0

        if(obst(x,y,z) .eq. 0 ) then
              do 5 k = 0 ,18
                 rho(x,y,z) = rho(x,y,z) + ff(k,x,y,z)
    5       continue
c----------------------
```

```
      if(rho(x,y,z) .ne. 0.d0) then

      u_x(x,y,z)=(ff(1,x,y,z)+ ff(7,x,y,z)+ ff(8,x,y,z) +
     &            ff(11,x,y,z) + ff(13,x,y,z)
     &            -(ff(2,x,y,z) + ff(9,x,y,z) + ff(10,x,y,z)+
     &            ff(12,x,y,z) + ff(14,x,y,z) ))/rho(x,y,z)

      u_y(x,y,z) = (ff(3,x,y,z) + ff(7,x,y,z) + ff(9,x,y,z) +
     &            ff(15,x,y,z) + ff(16,x,y,z)
     &            -(ff(4,x,y,z) + ff(8,x,y,z) + ff(10,x,y,z) +
     &            ff(17,x,y,z) + ff(18,x,y,z) ) /rho(x,y,z)

      u_z(x,y,z)= (ff(5,x,y,z) + ff(11,x,y,z) + ff(12,x,y,z)+
     &            ff(15,x,y,z) + ff(17,x,y,z)
     &            -(ff(6,x,y,z) + ff(13,x,y,z) + ff(14,x,y,z) +
     &            ff(16,x,y,z) + ff(18,x,y,z) ) /rho(x,y,z)
      endif

      endif

 10      continue
 11   continue
 12 continue
      end
c----------------------------------------

      subroutine calcu_upr(obst,u_x,u_y,u_z,
     & Fx,Fy,Fz,Sx,Sy,Sz,rho,upx,upy,upz)
      implicit none
      include "head.inc"
      integer x,y,z ,obst(lx,ly,lz)
      real*8  u_x(lx,ly,lz),u_y(lx,ly,lz),rho(lx,ly,lz),
     % upx(lx,ly,lz), upy(lx,ly,lz), upz(lx,ly,lz),
     & u_z(lx,ly,lz), Fx(lx,ly,lz), Fy(lx,ly,lz), Fz(lx,ly,lz),
     & Sx(lx,ly,lz), Sy(lx,ly,lz), Sz(lx,ly,lz)

      do  9 z = 1, lz
       do 10 y = 1, ly
        do 11 x = 1, lx

      if(obst(x,y,z) .eq. 0) then

      upx(x,y,z) = u_x(x,y,z) + (Fx(x,y,z)+Sx(x,y,z))/2.d0/ rho(x,y,z)
      upy(x,y,z) = u_y(x,y,z) + (Fy(x,y,z)+Sy(x,y,z))/2.d0/ rho(x,y,z)
      upz(x,y,z) = u_z(x,y,z) + (Fz(x,y,z)+Sz(x,y,z))/2.d0/ rho(x,y,z)

      else
      upx(x,y,z) = u_x(x,y,z)
      upy(x,y,z) = u_y(x,y,z)
      upz(x,y,z) = u_z(x,y,z)
      endif

 11      continue
 10      continue
 9    continue
     end
c----------------------------------------
```

```fortran
      subroutine collision(tauc,obst,u_x,u_y,u_z,
     &  rho,ff,Fx,Fy,Fz, Sx, Sy, Sz)
!
      implicit none
      include "head.inc"
      integer  l,obst(lx,ly,lz)
      real*8 u_x(lx,ly,lz),u_y(lx,ly,lz),ff(0:18,lx,ly,lz),rho(lx,ly,lz)
      real*8 Fx(lx,ly,lz),Fy(lx,ly,lz), Sx(lx,ly,lz), Sy(lx,ly,lz)
      real*8 Fz(lx,ly,lz),u_z(lx,ly,lz),Sz(lx,ly,lz),temp(18)

      integer  x,y,z,k
      real*8   u_n(0:18),fequ(0:18),fequ2(0:18),u_squ,tauc,ux,uy,uz

      do 4 z = 1, lz
       do 5 y = 1, ly
        do 6 x = 1, lx
         if(obst(x,y,z) .eq. 1) then
          do k =1, 18
           temp(k) =ff(k,x,y,z)
          enddo
          do k =1, 18
           ff(opp(k),x,y,z) = temp(k)
          enddo
         endif

         if(obst(x,y,z) .eq. 0) then
          ux = u_x(x,y,z) +tauc * ( Fx(x,y,z)+Sx(x,y,z) ) / rho(x,y,z)
          uy = u_y(x,y,z) +tauc * ( Fy(x,y,z)+Sy(x,y,z) ) / rho(x,y,z)
          uz = u_z(x,y,z) +tauc * ( Fz(x,y,z)+Sz(x,y,z) ) / rho(x,y,z)

           u_squ = ux * ux + uy * uy +uz *uz

        do 10 k = 0,18
c..........Equillibrium distribution function
           u_n(k)  = xc(k)*ux + yc(k)*uy + zc(k)*uz
           fequ(k) = t_k(k)* rho(x,y,z) * ( cc*u_n(k) / c_squ
     &                + (u_n(k)*cc) *(u_n(k)*cc) / (2.d0 * c_squ *c_squ)
     &                - u_squ / (2.d0 * c_squ)) + t_k(k) * rho(x,y,z)

c..........Collision step
       ff(k,x,y,z) = fequ(k) + (1.d0-1.d0/tauc)*(ff(k,x,y,z) -fequ(k))
  10     continue

         endif
  6     continue
  5    continue
  4   continue
      end
c------------------------------------------------
      subroutine getf_equ(rh,u,v,w,f_equ)
      include 'head.inc'
      real*8 rh, u,v,w,u_squ, f_equ(0:18),u_n(0:18)

         u_squ =u*u +v*v +w*w

      do 10 i =0,18
```

```
      u_n(i) = u *xc(i) +v *yc(i)+ w *zc(i)
         f_equ(i) = t_k(i) * rh *( u_n(i)/c_squ
   &                + u_n(i) *u_n(i) / (2.d0 * c_squ *c_squ)
   &                - u_squ / (2.d0 * c_squ)) + t_k(i) * rh
  10  continue
      end



force.for
C============================================================
      subroutine calcu_Fxy(obst,rho,Fx,Fy,Fz,Sx,Sy,Sz,p)
      implicit none
      include "head.inc"
      integer x,y,z,obst(lx,ly,lz),yn,yp,xn,xp,zp,zn, i,j,k
      real*8 Fx(lx,ly,lz),Fy(lx,ly,lz),Fz(lx,ly,lz)
   & ,psx(lx,ly,lz), sum_x, sum_y, sum_z, psx_w
   & ,rho(lx,ly,lz), Sx(lx,ly,lz), Sy(lx,ly,lz), Sz(lx,ly,lz)
   % , Fztemp, R,a,b, Tc, TT, alfa, omega, G1,p(lx,ly,lz)

!  Parameters in YUAN C-S EOS
         R = 1.0d0
         b = 4.d0
         a = 1.d0
         Tc = 0.3773d0*a/(b*R)
         TT= TT0 *Tc

         do 4 k = 1,lz
           do 5 j = 1,ly
             do 6 i = 1,lx
         if (obst(i,j,k ) .eq. 0 .and. rho(i,j,k).ne. 0.d0)  then

         if( (R*TT*
   &       (1.d0+(4.d0* rho(i,j,k)-2.d0* rho(i,j,k)* rho(i,j,k)
   &       )/(1.d0- rho(i,j,k))**3     )
   %         -a* rho(i,j,k) -1.d0/3.d0) .gt. 0.) then
           G1= 1.d0/3.d0
         else
           G1= -1.d0/3.d0
         endif

C++++++++++++++++++++++++++++++++++++++++++++++++++++++++++++++
         psx(i,j,k) = sqrt( 6.d0* rho(i,j,k) * ( R*TT*
   &       (1.d0+ (4.d0* rho(i,j,k)-2.d0*rho(i,j,k)*rho(i,j,k) )
   &       /(1.d0-rho(i,j,k))**3 )
   &         -a* rho(i,j,k) -1.d0/3.d0)
   &         /G1 )
c  Yuan C-S EOS
         p(i,j,k) = rho(i,j,k)/3.d0 + G1/6.d0 * psx(i,j,k) *psx(i,j,k)
       endif
    6 continue
    5 continue
    4 continue

      psx_w = sqrt( 6.d0* rho_w * ( R*TT*
   &       (1.d0+ (4.d0* rho_w-2.d0*rho_w * rho_w )
   &         /(1.d0- rho_w)**3   )
```

```
      &       -a* rho_w -1.d0/3.d0)
      &       /G1 )

        do 30 z = 1,lz
         do 20 y = 1,ly
          do 10 x = 1,lx
c.........interaction between neighbouring with periodic boundaries
              Fx(x,y,z) =0.d0
              Fy(x,y,z) =0.d0
              Fz(x,y,z) =0.d0

        if (obst(x,y,z) .eq. 0)  then

          sum_x = 0.d0
          sum_y = 0.d0
          sum_z = 0.d0

        do 11  k =1, 18
         xp=x+ex(k)
         yp=y+ey(k)
         zp=z+ez(k)
         if(xp .lt. 1)  xp = lx
         if(xp .gt. lx) xp =1
         if(yp .lt. 1)  yp = ly
         if(yp .gt. ly) yp =1
         if(zp .lt. 1)  zp = lz
         if(zp .gt. lz) zp =1

          if (obst(xp,yp,zp) .eq. 1)  then
! Interact with solid nodes (obst=1)
              sum_x = sum_x + t_k(k)*xc(k)
              sum_y = sum_y + t_k(k)*yc(k)
              sum_z = sum_z + t_k(k)*zc(k)
          else
! Interact with fluid nodes (obst=0)
              Fx(x,y,z)=Fx(x,y,z) +t_k(k)*xc(k)* psx(xp,yp,zp)
              Fy(x,y,z)=Fy(x,y,z) +t_k(k)*yc(k)* psx(xp,yp,zp)
              Fz(x,y,z)=Fz(x,y,z) +t_k(k)*zc(k)* psx(xp,yp,zp)
          endif

   11 continue
! Final wall-fluid interaction
          Sx(x,y,z) = -G1*sum_x *psx(x,y,z) *psx_w
          Sy(x,y,z) = -G1*sum_y *psx(x,y,z) *psx_w
          Sz(x,y,z) = -G1*sum_z *psx(x,y,z) *psx_w
! Final fluid-fluid interaction
          Fx (x,y,z)= -G1 *psx (x,y,z)* Fx(x,y,z)
          Fy (x,y,z)= -G1 *psx (x,y,z)* Fy(x,y,z)
          Fz (x,y,z)= -G1 *psx (x,y,z)* Fz(x,y,z)
        endif

   10 continue
   20 continue
   30 continue
      end
```

output.for
```
C========================================================

      subroutine  write_results2(obst,rho,p,upx,upy,upz, n)
      implicit none
      include "head.inc"
      integer  x,y,z,i,n
      real*8   rho(lx,ly,lz),upx(lx,ly,lz),upy(lx,ly,lz)
      real*8   upz(lx,ly,lz), p(lx,ly,lz)
      integer  obst(lx,ly,lz)

      character filename*16, B*6

      write(B,'(i6.6)') n
      filename='out/3D'//B//'.plt'

      open(41,file=filename)

      write(41,*) 'variables = x, y, z, rho, upx, upy, upz, p, obst'
      write(41,*) 'zone i=', lx, ', j=', ly, ', k=', lz, ', f=point'
      do 10 z = 1, lz
        do 10 y = 1, ly
          do 10 x = 1, lx
            write(41,9) x, y, z, rho(x,y,z),
     &  upx(x,y,z), upy(x,y,z), upz(x,y,z),
     &   p(x,y,z), obst(x,y,z)
   10 continue

    9 format(3i4, 5f15.8, i4)

      close(41)
      end
```

params.in
```
C===============================================
15.           ! RR droplet radius
0.12          ! rho_w, density of wall
1.0           ! tau(1)
10000         ! maximum time step
500           ! Nwri: output every Nwri time steps!
```

CHAPTER 3

Shan and Chen-type multi-component multiphase models

Shan and Chen (1993) proposed the multi-component Lattice Boltzmann model with inter-particle interaction. In the MCMP model the interaction between particles is a repulsive force that leads to phase separation and interface maintenance. The model has been widely used to simulate oil–water-like two-component flow in porous media or microchannels (Dong et al. 2010; Fan et al. 2001; Hilpert 2007; Hyväluoma et al. 2006; Koido et al. 2008; Li et al. 2005; Pan et al. 2004; Park and Li 2008; Porter et al. 2009; Schaap et al. 2007; Sinha et al. 2007; Spaid and Phelan Jr 1998; Vogel et al. 2005; Yang et al. 2002; Yu et al. 2007; Zhu et al. 2005). It was also extended to study bubble rise (Ngachin et al. 2015; Yang et al. 2001), droplet deformation and breakup (Sehgal et al. 1999; Xi and Duncan 1999), electrohydrodynamic drop deformation (Zhang and Kwok 2005), spinodal decomposition (Chin and Coveney 2002; González-Segredo et al. 2003), modeling of flowing soft systems (Benzi et al. 2009a), phase separation (Martys and Douglas 2001), and ternary amphiphilic fluid flow (Chen et al. 2000; Harting et al. 2005; Nekovee et al. 2000). The diffusion property of the MCMP SC model is discussed in Shan and Doolen (1996) and Sofonea and Sekerka (2001). In this chapter, the model is introduced briefly. For the reader's convenience, in Section 3.2 the derivation of pressure from the repulsive force expression is given in detail. An approximate method for specifying a contact angle is then given. Usually, this model is used to simulate two-component flow with density ratios near unity. It is unable to simulate flows with high kinematic viscosity ratios, for example > 10. Two-component flow in porous media is discussed at the end of this chapter.

3.1 Multi-component multiphase SC LBM

Here we implement the MCMP SC LBM (Shan and Doolen 1995) in two and three dimensions. In the model, one distribution function is introduced for each

Multiphase Lattice Boltzmann Methods: Theory and Application, First Edition.
Haibo Huang, Michael C. Sukop and Xi-Yun Lu.
© 2015 John Wiley & Sons, Ltd. Published 2015 by John Wiley & Sons, Ltd.
Companion Website: www.wiley.com/go/huang/boltzmann

of two or more fluid components. Here we work with just two components, but many more components could be considered. Each distribution function represents a fluid component and satisfies the following Lattice Boltzmann equation:

$$f_i^\sigma(\mathbf{x} + \mathbf{e}_i \Delta t, t + \Delta t) = f_i^\sigma(\mathbf{x}, t) - \frac{1}{\tau_\sigma}(f_i^\sigma(\mathbf{x}, t) - f_i^{\sigma,eq}(\mathbf{x}, t)), \qquad (3.1)$$

where $f_i^\sigma(\mathbf{x}, t)$ is the σth component density distribution function in the ith velocity direction and τ_σ is a relaxation time in the BGK model, which is related to the kinematic viscosity as $\nu_\sigma = c_s^2(\tau_\sigma - 0.5\Delta t)$. The equilibrium distribution function $f_i^{\sigma,eq}(\mathbf{x}, t)$ can be calculated as

$$f_i^{\sigma,eq}(\mathbf{x}, t) = w_i \rho_\sigma \left[1 + \frac{\mathbf{e}_i \cdot \mathbf{u}_\sigma^{eq}}{c_s^2} + \frac{(\mathbf{e}_i \cdot \mathbf{u}_\sigma^{eq})^2}{2c_s^4} - \frac{\mathbf{u}_\sigma^{eq2}}{2c_s^2} \right]. \qquad (3.2)$$

In Eqs (3.1) and (3.2) the $\mathbf{e}_i$s are the discrete velocities; for the D2Q9 and D3Q19 models used here, they are given in Chapter 1. The sound speed $c_s = \frac{c}{\sqrt{3}}$, where $c = \frac{\Delta x}{\Delta t}$. In Eq. (3.2) ρ_σ is the density of the σth component, which can be obtained from

$$\rho_\sigma = \sum_i f_i^\sigma. \qquad (3.3)$$

The macroscopic velocity $\mathbf{u}_\sigma^{eq}$ is given by (Shan and Chen 1993)

$$\mathbf{u}_\sigma^{eq} = \mathbf{u}' + \frac{\tau_\sigma \mathbf{F}_\sigma}{\rho_\sigma}, \qquad (3.4)$$

where $\mathbf{u}'$ is a velocity common to the various components and defined as

$$\mathbf{u}' = \frac{\sum_\sigma \left(\sum_i \frac{f_i^\sigma \mathbf{e}_i}{\tau_\sigma} \right)}{\left(\sum_\sigma \frac{\rho_\sigma}{\tau_\sigma} \right)}. \qquad (3.5)$$

This velocity is regarded as the whole fluid's velocity. In Eq. (3.4),

$$\mathbf{F}_\sigma = \mathbf{F}_{c,\sigma} + \mathbf{F}_{ads,\sigma}. \qquad (3.6)$$

It is the force acting on the σth component, here including fluid–fluid cohesion $\mathbf{F}_{c,\sigma}$ and fluid–solid adhesion $\mathbf{F}_{ads,\sigma}$. Note that while we refer to cohesion for consistency with the SCMP model in Chapter 2, the sign of the cohesion parameter G_c defined below is positive and the resulting interaction force is repulsive.

Each node in the computational domain is occupied by every σ component although one is dominant under most conditions, as described below. The minor components can be thought of as dissolved within the dominant component. With the techniques used here, the overall density of fluid ρ in the domain is approximately uniform because the densities are complementary in the sense that

$$\rho = \sum_\sigma \rho_\sigma. \qquad (3.7)$$

3.1.1   Fluid–fluid cohesion and fluid–solid adhesion

The cohesive force acting on the σth component is defined as (Martys and Chen 1996)

$$\mathbf{F}_{c,\sigma}(\mathbf{x},t) = -G_c\rho_\sigma(\mathbf{x},t)\sum_i w_i\rho_{\bar\sigma}(\mathbf{x}+\mathbf{e}_i\Delta t,t)\mathbf{e}_i, \qquad (3.8)$$

where the σ and $\bar\sigma$ denote two different fluid components and G_c is a parameter that controls the strength of the cohesion force.

The surface force acting on the σth component can be computed as follows (Martys and Chen 1996):

$$\mathbf{F}_{ads,\sigma}(\mathbf{x},t) = -G_{ads,\sigma}\rho_\sigma(\mathbf{x},t)\sum_i w_i s(\mathbf{x}+\mathbf{e}_i\Delta t)\mathbf{e}_i. \qquad (3.9)$$

Here $s(\mathbf{x}+\mathbf{e}_i\Delta t)$ is an indicator function that is equal to 1 or 0 for a solid or a fluid domain node, respectively. The interaction strength between each fluid and a wall can be adjusted by the parameters $G_{ads,\sigma}$. Early literature suggested that $G_{ads,\sigma}$ should be positive for non-wetting fluid and negative for wetting fluid (Kang et al. 2002; Martys and Chen 1996; Pan et al. 2004), but later we show that it is the difference between the $G_{ads,\sigma}$ that determines the contact angle. This approach is slightly different from the "wall density" in Chapter 2, but it is an equivalent alternative.

3.2   Derivation of the pressure

In the MCMP SC models the pressure is no longer the ideal one, i.e., $p = c_s^2\rho_i$. The formula for the pressure is directly related to the definition of the cohesion force. It will be illustrated in detail in the following.

The cohesive force acting on the σth component is usually defined as (Kang et al. 2002; Pan et al. 2004; Shan and Doolen 1995)

$$\mathbf{F}_{c,\sigma}(\mathbf{x},t) = -G_c\psi_\sigma(\mathbf{x},t)\sum_i \psi_{\bar\sigma}(\mathbf{x}+\mathbf{e}_i\Delta t,t)\mathbf{e}_i, \qquad (3.10)$$

where $\psi_\sigma(\rho_\sigma)$ is the effective mass (Shan and Doolen 1995) and is a function of local density. Usually, $\psi_\sigma = \rho_\sigma$ is used for simplicity (Kang et al. 2002; Pan et al. 2004; Shan and Doolen 1995). Here we can see there is a slight difference between Eq. (3.10) and Eq. (3.8). According to the definition of Eq. (3.10), the G_c in Eq. (3.10) is equal to the $G_c w_i$ in Eq. (3.8).

For the D2Q9 velocity model, we have

$$\sum_i w_i e_{i\alpha} = 0,$$

$$\sum_i w_i e_{i\alpha}e_{i\beta} = c_s^2\delta_{\alpha\beta},$$

$$\sum_i w_i e_{i\alpha} e_{i\beta} e_\gamma = 0, \quad \text{and}$$

$$\sum_i w_i e_{i\alpha} e_{i\beta} e_{i\gamma} e_{i\delta} = c_s^4 (\delta_{\alpha\beta}\delta_{\gamma\delta} + \delta_{\alpha\gamma}\delta_{\beta\delta} + \delta_{\alpha\delta}\delta_{\beta\gamma}). \tag{3.11}$$

Let us start with the Taylor expansion for $\rho_{\bar\sigma}(\mathbf{x} + \mathbf{e}_i \Delta t)$:

$$\rho_{\bar\sigma}(\mathbf{x} + \mathbf{e}_i \Delta t) = \rho_{\bar\sigma}(\mathbf{x}) + \partial_\beta \rho_{\bar\sigma}(\mathbf{x}) e_{i\beta} \Delta t + \frac{\Delta t^2}{2}\partial_\beta \partial_\gamma \rho_{\bar\sigma}(\mathbf{x}) e_{i\beta} e_{i\gamma} + \ldots . \tag{3.12}$$

Substituting Eq. (3.12) into Eq. (3.8) and using Eq. (3.11), we have

$$\mathbf{F}_{c,\sigma}(\mathbf{x}, t) = -G_c \rho_\sigma(\mathbf{x}, t)[c_s^2 \Delta t \delta_{\alpha\beta}\partial_\beta \rho_{\bar\sigma}(\mathbf{x})]$$

$$= -G_c \rho_\sigma(\mathbf{x}, t)[c_s^2 \Delta t \partial_\alpha \rho_{\bar\sigma}(\mathbf{x})] \tag{3.13}$$

This is Eq. (6) in Shan and Doolen (1995) for the D2Q9 model.

The momentum equation of order $O(\Delta t)$ that is recovered from the LBE is

$$\partial_t (\rho u_\alpha) + \partial_\beta (\rho u_\alpha u_\beta) + \partial_\alpha c_s^2 \rho = \sum_\sigma (F_{c,\sigma})_\alpha. \tag{3.14}$$

Here, to be consistent with the momentum equation in the α direction component, we take the α component of the $\mathbf{F}$, i.e., $(F_{c,\sigma})_\alpha$. It is noticed that

$$\sum_\sigma (F_{c,\sigma})_\alpha = -G_c \rho_\sigma(\mathbf{x}, t)[c_s^2 \Delta t \partial_\alpha \rho_{\bar\sigma}(\mathbf{x})] - G_c \rho_{\bar\sigma}(\mathbf{x}, t)[c_s^2 \Delta t \partial_\alpha \rho_\sigma(\mathbf{x})]$$

$$= -c_s^2 G_c \Delta t \partial_\alpha [\rho_\sigma(\mathbf{x}) \rho_{\bar\sigma}(\mathbf{x})]. \tag{3.15}$$

Compared to the Euler equation

$$\partial_t (\rho u_\alpha) + \partial_\beta (\rho u_\alpha u_\beta) + \partial_\alpha p = 0, \tag{3.16}$$

we have

$$p = c_s^2 \rho + \frac{1}{2}c_s^2 G_c \sum_{\sigma\bar\sigma} \rho_\sigma \rho_{\bar\sigma} = c_s^2 \rho + c_s^2 G_c \rho_\sigma \rho_{\bar\sigma} \tag{3.17}$$

because $\frac{1}{2}\sum_{\sigma\bar\sigma} \rho_\sigma \rho_{\bar\sigma} = \rho_\sigma \rho_{\bar\sigma}$.

3.2.1 Pressure in popular papers (2D)

In Shan and Doolen (1995), the coefficient G_c is defined as

$$G_c = \begin{cases} G, & |\mathbf{e}_i| = c, \\ 0, & \text{otherwise} \end{cases} \tag{3.18}$$

Following the above derivation procedure, substituting the Taylor expansion of $\psi_{\bar\sigma}(\mathbf{x} + \mathbf{e}_i \Delta t)$ into the force formula (Eq. (3.10)) we have

$$\sum_\sigma (\mathbf{F}_{c,\sigma})_\alpha = -G_c \rho_\sigma \sum_i \partial_\beta (\rho_{\bar\sigma} e_{i\beta} e_{i\alpha}) - G_c \rho_{\bar\sigma} \sum_i \partial_\beta (\rho_\sigma e_{i\beta} e_{i\alpha})$$

$$= -G_c \rho_\sigma \partial_\beta \rho_{\bar\sigma} \sum_i e_{i\beta} e_{i\alpha} - G_c \rho_{\bar\sigma} \partial_\beta \rho_\sigma \sum_i e_{i\beta} e_{i\alpha}$$

$$= -3c^2 G_c \partial_\alpha (\rho_\sigma \rho_{\bar\sigma}), \tag{3.19}$$

It is noted that for simplicity, here $\psi_\sigma(\rho^\sigma) = \rho^\sigma$ (Pan et al. 2004). Because the D2Q7 model is used (Shan and Doolen 1995) in the above derivation, the two formulae are used:

$$\sum_i e_{i\alpha} e_{i\beta} = 3c^2 \delta_{\alpha\beta}, \text{ and } \sum_i e_{i\alpha} e_{i\beta} e_{i\gamma} = 0. \tag{3.20}$$

Hence, the derived pressure is:

$$p = c_s^2 \rho + \frac{3}{2} c^2 G \sum_{\sigma\bar\sigma} \rho_\sigma \rho_{\bar\sigma}. \tag{3.21}$$

In Kang et al. (2002), for the D2Q9 model, the cohesion parameter is defined as

$$G_c = \begin{cases} G, & |\mathbf{e}_i| = c \\ G/4, & |\mathbf{e}_i| = \sqrt{2}c \\ 0, & |\mathbf{e}_i| = 0, \end{cases} \tag{3.22}$$

because $\sum_i G_c e_{i\alpha} e_{i\beta} = 3c^2 \delta_{\alpha\beta} G$, hence

$$p = c_s^2 \rho + \frac{3}{2} c^2 \sum_{\sigma\bar\sigma} G \rho_\sigma \rho_{\bar\sigma}. \tag{3.23}$$

It is noted that here G is nine times larger than our G_c in Eq. (3.8).

3.2.2 Pressure in popular papers (3D)

In (Schaap et al. 2007), the D3Q19 velocity model is used and the cohesion parameter G_c is

$$G_c = \begin{cases} 2G, & |\mathbf{e}_i| = c \\ G, & |\mathbf{e}_i| = \sqrt{2}c \\ 0, & |\mathbf{e}_i| = 0. \end{cases} \tag{3.24}$$

Because for the D3Q19 model (see Figure 1.1)

$$\sum_{i=1}^{6} e_{i\alpha} e_{i\beta} = 2c^2 \delta_{\alpha\beta} \tag{3.25}$$

and

$$\sum_{i=7}^{18} e_{i\alpha} e_{i\beta} = 8c^2 \delta_{\alpha\beta}, \tag{3.26}$$

we have

$$\begin{aligned} \sum_i G_c e_{i\alpha} e_{i\beta} &= \sum_{i=1}^{6} G_c e_{i\alpha} e_{i\beta} + \sum_{i=7}^{18} G_c e_{i\alpha} e_{i\beta} \\ &= \sum_{i=1}^{6} 2G e_{i\alpha} e_{i\beta} + \sum_{i=7}^{18} G e_{i\alpha} e_{i\beta} \\ &= 2G \times 2c^2 \delta_{\alpha\beta} + G \times 8c^2 \delta_{\alpha\beta} \\ &= 12c^2 \delta_{\alpha\beta} G. \end{aligned} \tag{3.27}$$

Hence,

$$p = c_s^2 \sum_\sigma \rho_\sigma + \frac{12}{2} c^2 \sum_{\sigma\bar{\sigma}} G \rho_\sigma \rho_{\bar{\sigma}} = c_s^2 \sum_\sigma \rho_\sigma + 12 c^2 G \rho_1 \rho_2. \qquad (3.28)$$

Pan et al. (2004) used the D3Q15 model and the adhesion parameter G_c is

$$G_c = \begin{cases} G, & |\mathbf{e}_i| = c \\ G/\sqrt{3}, & |\mathbf{e}_i| = \sqrt{3}c \\ 0, & |\mathbf{e}_i| = 0. \end{cases} \qquad (3.29)$$

For the D3Q15 model, similarly we have

$$\sum_{i=0}^{14} G_c e_{i\alpha} e_{i\beta} = \left(2 + \frac{8}{\sqrt{3}}\right) c^2 \delta_{ij} G. \qquad (3.30)$$

Hence,

$$p = c_s^2 \sum_\sigma \rho_\sigma + \frac{c^2}{2}\left(2 + \frac{8}{\sqrt{3}}\right) \sum_{\sigma\bar{\sigma}} G \rho_\sigma \rho_{\bar{\sigma}} = c_s^2 \sum_\sigma \rho_\sigma + \left(2 + \frac{8}{\sqrt{3}}\right) c^2 G \rho_1 \rho_2. \quad (3.31)$$

In the above, G_c should have the same dimension as $\frac{1}{\rho}$; otherwise, the dimension of terms of $c^2 G \rho_\sigma \rho_{\bar{\sigma}}$ is not consistent with pressure (mu/(lu ts^2)).

3.3   Determining G_c and the surface tension

In the SC model the parameter G_c controls the fluid–fluid interfacial tension and there is a threshold value $G_{c, crit} = \frac{1}{(\rho_1 + \rho_2)}$ for G_c (Martys and Chen 1996) below which an initially uniform mixed system of two immiscible fluids will yield a stable solution. ρ_1 and ρ_2 are the main and dissolved densities of Fluid 1 and Fluid 2, respectively. The critical value $G_{c,crit} = \frac{1}{10}$ for $\rho_1 + \rho_2 = 10$. To verify this we carried out two series of 2D simulations where we placed a pure bubble of Fluid 1 (ρ_1) inside a 100×100 square of Fluid 2 (ρ_2) with periodic boundaries; both fluids had equal total masses (i.e., the number of pixels occupied by Fluid 1 was equal to the number occupied by Fluid 2). The two simulation series had initial densities of $\rho_i = 2$ and 8 (where $\rho_i = \rho_1 + \rho_2$) and the parameter G_c was varied from zero until numerical instabilities occurred. Figure 3.1 shows the density of Fluid 1 (ρ_1) and a smaller dissolved density of Fluid 2 (ρ_2) inside the bubble, scaled according to initial density. From the figure we can see that for $0 \le G_c \rho_i \le 1.0$ the scaled densities are 0.5 for each component because the two fluids diffuse until a homogeneous solution is present at equilibrium. When $G_c \rho_i > 1.0$, the bubble filled with Fluid 1 becomes increasingly "pure", with the scaled density of Fluid 1 eventually exceeding 1 for $G_c \rho_i > 1.8$ due to the compressibility of the fluid. For various $G_c \rho_i$ values, the equilibrium ρ_1 and associated dissolved ρ_2 can be determined from Figure 3.1. In Figure 3.1 discrete points are obtained from

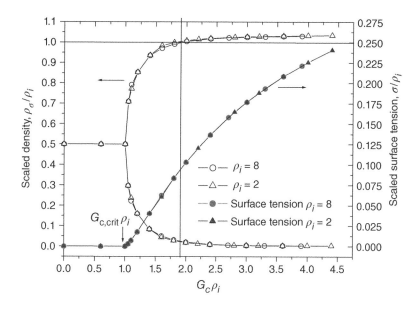

Figure 3.1 Scaled component densities (left y axis) and scaled lattice surface tension (right-hand y axis) as a function of the scaled cohesion parameter $G_c\rho_i$. The simulations were carried out for two sets of initial densities. The vertical line depicts the $G_c\rho_i$ value used in the rest of the study. $G_{c,\,crit}\rho_i$ depicts the critical $G_c\rho_i$ value above which stable phase separation is possible (see text). Reprinted from Huang et al. (2007), copyright (2007), with permission from APS.

numerical simulations and the lines connecting the points are drawn to guide the eye.

The pressure at position $\mathbf{x}$ can be determined from the densities as $p(\mathbf{x}) = \frac{\rho_1(\mathbf{x})+\rho_2(\mathbf{x})}{3} + \frac{G_c\rho_1(\mathbf{x})\rho_2(\mathbf{x})}{3}$ (refer to Eq. (3.17)). By measuring the component densities inside and outside a drop or bubble, the interfacial tension σ can be determined through the Laplace law $p(\mathbf{x}_{inside}) - p(\mathbf{x}_{outside}) = \frac{\sigma}{R}$, where R is the radius of the bubble. The scaled surface tension as a function of $G_c\rho_i$ is also illustrated in Figure 3.1. The relationship between σ and $G_c\rho_i$ is approximately linear when $1.0 < G_c\rho_i < 2.0$.

G_c should be chosen carefully. Larger G_c is preferable for multiphase simulations because it increases the interface sharpness and limits the "solubility" of the fluids within each other. On the other hand, smaller G_c is preferable because that alleviates some numerical difficulties and can reduce the compressibility of LBM fluids (Schaap et al. 2007). From Figure 3.1 we found that $1.6 < G_c\rho_i < 2.0$ is an appropriate compromise. Here we chose $G_c\rho_i = 1.8$.

Although the determination of equilibrium densities in Figure 3.1 is obtained from cases where both fluids have equal masses and the domains have periodic boundary conditions, Figure 3.1 is also valid for cases where the two fluids have different initial masses and for the presence of a solid boundary.

To demonstrate this, we prepared the following example. We initialized a small rectangular area (e.g., 4% of the whole 200×100 domain) near a wall with Fluid 1 surrounded by Fluid 2, and in each region the fluids have density 2.0 and dissolved density 0.6 (this dissolved density is arbitrary and an order of magnitude larger than the expected equilibrium dissolved density for the "main" density of ~2). In the simulation we set $G_c \rho_i = 1.8$, where $\rho_i = 2.0 + 0.6 = 2.6$, $G_c = 0.6923$, $G_{ads1} = -0.318$, and $G_{ads2} = 0.318$.

We observe that in all regions (except the interface area) the equilibrium main fluid density is 2.565 and dissolved density is 0.086, which agrees well with Figure 3.1 because for $G_c \rho_i = 1.8$, $\frac{\rho_{main}}{\rho_{dissolved}} = \frac{2.565}{0.086}$ is about $\frac{1}{0.03}$.

Finally, we considered cases with initial densities far from the expected equilibrium condition. Consider a subregion representing a proto-drop, for example. Suppose that subregion is initialized with density of Fluid 1 equal to 8.0 and dissolved density of Fluid 2 equal to 0.0, and the remainder of the domain is initialized with complementary densities of Fluid 1 equal to 0.0 and Fluid 2 equal to 8.0. Then the equilibrium area of Fluid 1 will be smaller than its initialized area and the density distributions will attain an equilibrium consistent with the coexistence curve. As an example, we initialized proto-drop regions with main density 8 and dissolved density 0.0, with $G_c = 0.225$. After 90,000 time steps, the equilibrium $\frac{\rho_{main}}{\rho_{dissolved}} = \frac{7.80}{0.24}$ agrees well with the expectations of Figure 3.1.

Hence, in the equilibrium state, the ratio of the main fluid density and dissolved density agrees well with the data in Figure 3.1. This conclusion is independent of initial masses, G_{ads1}, and G_{ads2} (Huang et al. 2007).

It is noted that in the above discussion the kinematic viscosities of the two components are identical (identical τ). For cases with different τ, please refer to Section 3.6 and Pan et al. (2004).

3.4   Contact angle

There have been many studies on wetting and spreading phenomena using the MCMP SC model. Martys and Chen (1996) studied multi-component fluids in complex 3D geometries. In their simulations the interaction force between a fluid and a wall was introduced, perhaps for the first time. They found that reasonably well-defined contact angles could be obtained by adjusting the interaction strength between each fluid and a surface such that one of the fluids wets the surface. Using the SC LBM Schaap et al. (2007) and Pan et al. (2004) studied the displacement of immiscible fluids in different porous media and estimated adhesion parameters through empirical calibration methods. In the above studies (Martys and Chen 1996; Pan et al. 2004; Schaap et al. 2007) no explicit relationship between contact angle and adhesion parameter has been proposed. Kang et al. (2002, 2005) studied displacement of immiscible droplets subject to gravitational forces in a 2D channel and a 3D duct, respectively. In their studies, the contact angles considered ranged from only 60 to 120° and the relationship between adhesion parameters and contact angles was estimated as linear. These early works on determining contact angles from adhesion parameters in the MCMP SC LBM were strictly empirical and limited to a relatively small range of contact angles. Sukop and Thorne (2006) use some simplistic arguments to successfully simulate 90° and 45° contact angles.

A number of automated procedures for measuring contact angles from images such as those that can be produced from LBM simulations are available as plug-ins for the US National Institutes of Health image processing software ImageJ.

3.4.1   Application of Young's equation to MCMP LBM

Young's equation for computing the contact angle contains interfacial tension values between the two fluids (σ_{12}) and between each fluid and the surface (σ_{S1} and σ_{S2}):

$$\cos\theta_1 = \frac{\sigma_{S2} - \sigma_{S1}}{\sigma_{12}} \tag{3.32}$$

This equation determines the contact angle θ_1 measured in Fluid 1 (Figure 2.5) (Adamson and Gast 1997).

In Huang et al. (2007) we proposed a straightforward application of Young's equation with substitution of the LBM cohesion parameter and a density factor $G_c\frac{\rho_1 - \rho_2}{2}$ for the fluid–fluid interfacial tension, and the adhesion parameters $G_{ads,1}$ and $G_{ads,2}$ from Eq. (3.9) for the corresponding fluid–solid interfacial tensions:

$$\cos\theta_1 = \frac{G_{ads,2} - G_{ads,1}}{G_c\dfrac{\rho_1 - \rho_2}{2}}. \tag{3.33}$$

Eq. (3.33) is simple to use and able to determine the contact angle using only the parameters G_c, the equilibrium main density ρ_1 and the associated dissolved density ρ_2, which can be determined as discussed in the above Section 3.3., and $G_{ads,1}$ and $G_{ads,2}$. Eq. (3.33) is similar to an equation proposed in Sukop and Thorne (2006) but is significantly improved since it gives more accurate predictions.

3.4.2   Contact angle measurement

Contact angles can be computed from measurements of the base and height of drops on a surface. If the base and height of a droplet are L and H, respectively, the radius of the droplet can be calculated from $R = \frac{4H^2 + L^2}{8H}$. Then, the contact angle can be estimated through the formula $\tan\theta_1 = \frac{L}{2(R-H)}$. The measurement of base and height involves some ambiguities that require resolution or at least a consistent approach.

The main difficulties in measuring drop base and height is how to define a precise location of the interface between the drop and the ambient fluid, since the simulated interface is actually several lattice units thick. We chose a cut-off density value as $\rho_{cut} = 0.5\rho_A$, where ρ_A is the density of the fluid that forms the drop. It is obvious that choosing ρ_{cut} too close to ρ_A or its associated dissolved density should be avoided since those values occur far from the center of the interface. Using a linear interpolation method, we determined the drop base width and the height of drop. Then we used the preceding formulae to determine the contact angle in a simulation.

3.4.3 Verification of proposed equation

To verify the proposed method for determining the contact angles, some 2D and 3D numerical simulations were carried out and 2D and 3D results from the literature (Kang et al. 2005; Schaap et al. 2007) were re-evaluated. First, multi-component fluids interacting with a surface were studied. A sampling of parameters $G_{ads,1}$ and $G_{ads,2}$ that we used to compute different contact angles with Eq. (3.33) is listed in Table 3.1. The results are illustrated in Figure 3.2. The density of the first fluid is shown in dark grey. Images represent 24,000 time steps from an initial condition of a 41×20 rectangle of the first fluid surrounded by the second fluid. In each region, the fluids have density 2 and dissolved density 0.06, which can be obtained from Figure 3.1.

For the 3D simulations, $G_c = 0.9$, $G_{ads,1} = -G_{ads,2}$ (values in Figure 3.3), and the initial densities are identical to those in the 2D simulations. The results are shown in Figure 3.3. The $G_{ads,2}$ value is labeled in each frame. The density of the first fluid is shown in grey. Frames represent 50,000 time steps from an initial condition of a $20 \times 20 \times 20$ cube of the first fluid surrounded by the second fluid.

Figure 3.4 shows the measured contact angles as a function of $G_{ads,2}$ when $G_{ads,1} = -G_{ads,2}$. The contact angles that are obtained from Eq. (3.33) are also shown for comparison. The contact angles calculated with Eq. (3.33) agree well with the measured ones over most of the range.

These results suggest that the prediction of Eq. (3.33) becomes somewhat less accurate at low contact angle. This prompted Huang et al. (2007) to investigate

Table 3.1 Adhesion parameters and contact angles (θ, degrees) for fluid 1 ($G_c = 0.9$ and $G_{ads,1} = -G_{ads,2}$).

Case	$G_{ads,2}$	θ (computed from Eq. (3.33))	θ (measured from Fig. 3.2)
a	−0.4	156.4	158.3
b	−0.3	133.4	135.1
c	−0.2	117.3	117.0
d	−0.1	103.2	103.2
e	0.1	76.8	75.3
f	0.2	62.7	59.5
g	0.3	46.6	40.6
h	0.4	23.6	18.9

Reprinted from Huang et al. (2007), copyright (2007), with permission from APS.

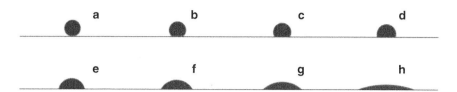

Figure 3.2 Simulations of different contact angles for multi-component fluids interacting with a surface (parameters G_c, $G_{ads,1}$, $G_{ads,2}$, and measured contact angles for each case are listed in Table 3.1). The density of the first fluid is shown in dark grey. Simulation domain is 200×100. Images represent 24,000 time steps from an initial condition of a 41×20 rectangle of the first fluid surrounded by the second fluid. In each region the fluids have density 2 and dissolved density 0.06. $\tau = 1$ for both fluids. Reprinted from Huang et al. 2007, copyright (2007), with permission from APS.

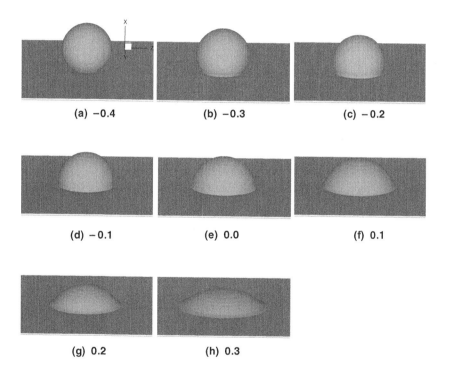

Figure 3.3 3D simulations of different contact angles for multi-component fluids interacting with a surface ($G_c = 0.9$, $G_{ads,1} = -G_{ads,2}$, $G_{ads,2}$ value is labeled in each frame). The density of the first fluid is shown in dark grey. Simulation domain is $96 \times 96 \times 48$. Images represent 50,000 time steps from an initial condition of a $20 \times 20 \times 20$ cube of the first fluid surrounded by the second fluid. In each region, the fluids have density 2 and dissolved density 0.06. $\tau = 1$ for both fluids.

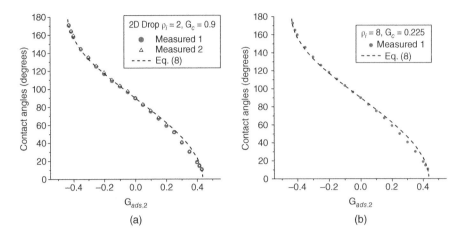

Figure 3.4 Contact angles for two fluids interacting with a surface, versus the value of the adhesion parameter, $G_{ads,2}$ ($G_{ads,1} = -G_{ads,2}$). Simulation domain is 200×100. Results represent 24,000 time steps from an initial condition of a 41×20 ("measured 1") and a 41×40 ("measured 2") rectangle of the first fluid surrounded by the second fluid. (a) in each region, the fluids have density 2 and dissolved density 0.06. (b) in each region, the fluids have density 8 and dissolved density 0.24. $\tau = 1$ for both fluids. Reprinted from Huang et al. (2007), copyright (2007), with permission from APS.

the range of applicability of Eq. (3.33). Figure 3.5 shows that there are large ranges of the parameters $G_{ads,1}$ and $G_{ads,2}$ where there is only a small difference between the contact angle computed using Eq. (3.33) and the contact angle measured using the algorithm described previously. In particular, the difference is less than 2° in the white area of Figure 3.5. The differences in predicted and observed contact angles become large where the difference in the magnitudes of G_{ads} is large, for example $G_{ads,1} = -0.7$ and $G_{ads,2} = 0$ or $G_{ads,1} = 0$ and $G_{ads,2} = -0.8$. Despite this, the best parameter sets that permit simulations to span the entire range of contact angles do not appear to fall along the line $G_{ads,2} = -G_{ads,1}$ but rather along $G_{ads,2} \approx -G_{ads,1} + 0.1$. The reasons for this apparent asymmetry are unclear and deserve further investigation, but the results provide good guidance for the selection of parameter sets. It is interesting to note that, in accordance with Eq. (3.33), it is the difference in the G_{ads} values rather than their signs that determines the contact angle. For example, for both parameter sets $G_{ads,1} = 0.1$ and $G_{ads,2} = -0.2$, and $G_{ads,1} = 0.4$ and $G_{ads,2} = 0.1$, the contact angle is 110.

In Huang et al. (2007) the numerical results from the literature (Kang et al. 2002; Schaap et al. 2007) are shown to be consistent with our simple formula (Eq. (3.33)). It should be noted that the definition of cohesion forces in the literature may be different from that in the present work. When G_c is defined on the basis of Eq. (3.8), the G_c of Schaap et al. (2007) would be 0.9 and both $G_{ads,2}$ and $G_{ads,1}$ would also 36 times larger than they are in Schaap et al. (2007). It should also

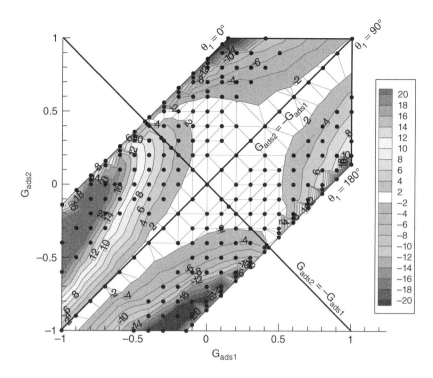

Figure 3.5 Difference (in degrees) between the contact angle computed using Eq. (3.33) and the contact angle measured using the algorithm described in the text as a function of $G_{ads,1}$ and $G_{ads,2}$ ($G_c = 0.9$). Black dots show the parameter values used in simulations. Reprinted from Huang et al. (2007), copyright (2007), with permission from APS.

be noted that G_c as used here equals nine times the G_c defined by Kang et al. (2002).

3.5   Flow through capillary tubes

A 3D drainage flow through a tube with a square-shaped pore geometry is simulated because the expected behavior of the fluid displacement can be calculated analytically according to the Laplace law.

The tube in our test has a square-shaped pore geometry with size 8×8 in the x, y plane. and the depth of the pore is 32 lu in the z direction (see Figure 3.7). An upper and lower four blank layers were added into the geometry to implement pressure inlet and outlet boundary conditions. Hence, the total domain size is $32 \times 32 \times 40$. The set up is almost identical to that in Pan et al. (2004) and our analysis follows a similar approach. The main component density is set to be 8.0 mu/lu^3 and the dissolved component has density of 0.16 mu/lu^3. The coefficient in the cohesion force $G_c = 0.225$. In our simulations, the hole is initially saturated by wetting fluid. Cases with different pressures on the upper inlet and lower outlet are simulated. The pressure boundary condition is set by the scheme proposed by Zou and He (1997).

In 3D simulations, according to the Laplace law, the capillary pressure is

$$p^c = 2\frac{\sigma}{R}, \tag{3.34}$$

where R is the radius of curvature of the interface. Because the width of this hole is 8 lu, here $R = 4$ lu. When $G_c = 0.225$ the surface tension $\sigma = 0.927$. When displacement occurs the minimum entry pressure is

$$p^c = 2\frac{0.927}{4} = 0.463. \tag{3.35}$$

Hence, the entry density difference between the inlet/outlet should be $\Delta\rho^c = \frac{p^c}{c_s^2} \approx$ 1.39 mu/lu³.

The non-dimensional pressure $\frac{p_c R}{\sigma}$ as a function of the wetting phase saturation S_w is shown in Figure 3.6. According to the Laplace law, the non-dimensional entry pressure should be $\frac{p_c R}{\sigma} = 2$ (see Eq. (3.34)). Figure 3.6 demonstrates that the non-wetting-phase entry pressure obtained from LBM simulation is slightly higher than the analytical one: $\frac{p_c R}{\sigma} = 2$, which is the minimum pressure needed to initiate the wetting-phase displacement. The LBM result is slightly higher than value of 2. This may be due to an insufficient computational domain (Pan et al. 2004) or the compressibility effect discussed in Section 3.8.

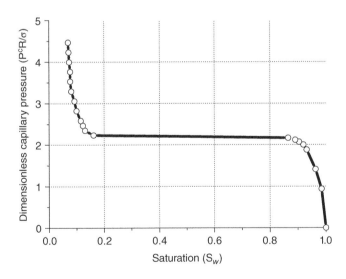

Figure 3.6 Primary drainage curve obtained by fluid displacement simulations in capillary tube with square-shaped cross-section. Reproduced by Jianlin Zhao, China University of Petroleum, Eastern China.

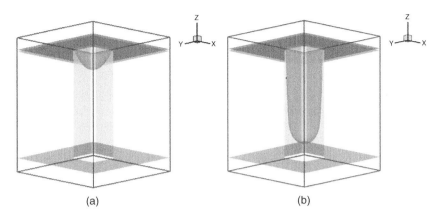

(a)   (b)

Figure 3.7 Primary drainage by fluid displacement simulations in a capillary tube with a square-shaped cross-section. (a) $\rho_{in} = 8.7$, $\rho_{out} = 7.3$, no displacement; (b) $\rho_{in} = 8.8$, $\rho_{out} = 7.2$, the hole is penetrated and the wetting fluid is displaced.

Snapshots of two cases are illustrated in Figure 3.7, which shows the equilibrium state when the inlet/outlet density difference $\Delta\rho = 8.7 - 7.3 = 1.4$ (corresponding pressure difference $\Delta p = c_s^2 \Delta\rho = 0.467$). Although this pressure difference reaches the minimum entry pressure, there is no displacement. It is consistent with what we see in Figure 3.6, i.e., minimum entry pressure is slightly higher than the analytical one. For the case where $\Delta\rho = \rho_{in} - \rho_{out} = 1.6$, the non-wetting fluid enters the hole and displaces the wetting fluid (Figure 3.7(b)). Figure 3.7(b) is a snapshot of the penetration process.

3.6   Layered two-phase flow in a 2D channel

Here we consider immiscible layered two-phase flow between two parallel plates, as described in Chapter 2. In the simulation, the non-wetting phase flows in the central region $0 < |y| < a$, while the wetting phase flows in the region $a < |y| < b$. The periodic boundary condition was applied on the left and right boundaries while the no-slip (bounce-back) boundary condition was applied on the upper and lower plates.

The kinematic viscosities for non-wetting and wetting fluids are $v_n = c_s^2(\tau_n - 0.5)$ and $v_w = c_s^2(\tau_w - 0.5)$, respectively. The saturations of wetting fluid in this study are $S_w = 1 - \frac{a}{b}$ and $S_n = \frac{a}{b}$. The computational domain is 10×100. Because the periodic boundary condition is used on the left and right boundaries, the computational lattice can be much smaller in the x direction.

Assuming a Poiseuille-type flow in the channel, the analytical solution for the velocity profile between the parallel plates can be obtained (refer to Appendix A in Chapter 2).

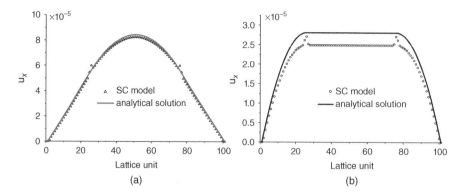

Figure 3.8 Velocity profiles comparison for a case of $M = 1$, $G = 1.5 \times 10^{-8}$, $S_w = 0.5$, $\tau_1 = 1.0$, and $\tau_2 = 1.0$ when the body force G is applied only to either Fluid 1 (a) or Fluid 2 (b). The numerical solution is obtained with the SC model. Source: Huang et al. 2011b. Reproduced with permission of Elsevier.

Figure 3.8 shows the velocity profiles at $x = 5$ for $M = 1$ and $S_w = 0.5$ that are obtained from the SC model. The numerical results are good but there are small jumps across the interface that are also noted in Kang et al. (2004).

To further evaluate the accuracy, the error between numerical and analytical solutions is defined as $Err(t) = \sum_i |u_0(y_i) - u(y_i, t)|$, where the summation is over the lattice nodes y_i in the slice $x = 5$, and u_0 is the analytical solution. The convergence criterion is $|\frac{Err(t) - Err(t - 10^4 \Delta t)}{Err(t - 10^4 \Delta t)}| < 10^{-4}$.

For the MCMP SC model it is difficult to achieve a specified viscosity ratio. This will be illustrated in detail in the following. To achieve a dynamic viscosity ratio less than 10, adjusting the relaxation time τ is one strategy in the MCMP SC model. That approach was followed in Pan et al. (2004).

In the SC model, it is not easy to obtain an exact dynamic viscosity ratio other than unity because the kinematic viscosity ratio and the density ratio cannot change independently in the model. As mentioned in Section 3.3, for a specific G_c and an equal τ for two components, the ratio of the main fluid density to the dissolved density in the whole computational domain (except the interface region) is a constant.

However, if the kinematic viscosities of the two components are different, the situation is different. We carried out a series of 2D simulations where we placed a pure bubble of Fluid 1 ($\rho_1 = 1.0$, $\tau_1 = 1.8$) inside a 100×100 square of Fluid 2 ($\rho_1 = 0$, $\tau_2 = 0.825$) with periodic boundaries; both fluids had equal total masses (i.e., the number of pixels occupied by Fluid 1 was equal to the number occupied by Fluid 2). The simulation series had initial densities of $\rho = \rho_1 + \rho_2 = 1.0$ in the whole domain and the parameter $G_c = 1.8$. It is found that the equilibrium main fluid (component 1) density and the dissolved (component 2) density inside the bubble are 1.259 and 0.009, respectively.

Table 3.2 Performance of the SC model for simulation of a layered two-phase flow with different viscosity ratios.

Case		SC
$M = 1$ $(G_1 = 0, G_2 \neq 0)$	Error	11.4%
	ts	80,000
$M = 1$ $(G_1 \neq 0, G_2 = 0)$	Error	3.07%
	ts	90,000
$M = 5$ $(G_1 = 0, G_2 \neq 0)$	Error	33.6%
	ts	80,000
$M = 5$ $(G_1 \neq 0, G_2 = 0)$	Error	33.4%
	ts	80,000

Source: Huang et al. 2011b. Reproduced with permission of Elsevier.

Outside the bubble, the main fluid (component 2) density and the dissolved (component 1) density are 0.910 and 0.153, respectively. Hence, the actual viscosity ratio is $M = \frac{\rho_1 c_s^2 (\tau_1 - 0.5)}{\rho_2 c_s^2 (\tau_2 - 0.5)} = \frac{1.259 \times (1.8 - 0.5)/3}{0.910 \times (0.825 - 0.5)/3} = 5.6$, if the more viscous component is non-wetting. Thus for the $M = 5$ cases the viscosity ratio in the SC model is actually $M = 5.6$ because to get the exact $M = 5$ is too difficult.

The errors between the analytical and LBM solutions at the final steady state are listed in Table 3.2. The number of time steps required for the velocities to converge for cases of $M = 1$ and $M = 5$ are also listed in Table 3.2. It is seen that the numerical error increases with the dynamic viscosity ratio M. The number of time steps required to reach the equilibrium state does not increase with M.

3.7 Pressure or velocity boundary conditions

3.7.1 Boundary conditions for 2D simulations

Here we use the pressure or velocity boundary conditions proposed by Zou and He (1997). Sukop and Thorne (2006) provide additional details.

For simplicity, we introduce how to handle boundary conditions for 2D cases first. To simulate 2D immiscible displacement, a velocity boundary condition and constant pressure (or density) boundary condition can be set for the upper and lower boundaries, respectively. In our example simulations, only non-wetting fluid displacing wetting fluid is considered and we suppose that the non-wetting fluid displaces the wetting fluid from top to bottom.

In the upper boundary the distribution functions f_4, f_7, and f_8 are unknown after the streaming step for the non-wetting fluid (majority component). Through the non-equilibrium bounce-back assumption (Zou and He 1997), one gets the density of the majority component as

$$\rho_n = \frac{f_0^n + f_1^n + f_3^n + 2(f_2^n + f_5^n + f_6^n)}{1 + u_i}, \tag{3.36}$$

where u_i is a specified inlet velocity of the non-wetting fluid. The unknown distribution functions can be obtained through (Zou and He 1997):

$$f_4^n = f_2^n - \frac{2}{3}\rho_n u_i$$

$$f_7^n = f_5^n + \frac{1}{2}(f_1^n - f_3^n) - \frac{1}{6}\rho_n u_i \qquad (3.37)$$

$$f_8^n = f_6^n + \frac{1}{2}(f_3^n - f_1^n) - \frac{1}{6}\rho_n u_i.$$

The pressure boundary condition for the lower boundary can be handled similarly (Zou and He 1997). Suppose ρ_w is specified as the density of the wetting component (majority component) on the lower outlet boundary node. One can get the outlet velocity of the wetting fluid as

$$u_o = 1 - \frac{1}{\rho_w}[f_0^w + f_1^w + f_3^w - 2(f_4^w + f_7^w + f_8^w)] \qquad (3.38)$$

and the unknown distribution functions f_2, f_5, and f_6 are

$$f_2^w = f_4^w + \frac{2}{3}\rho_w u_o$$

$$f_5^w = f_7^w + \frac{1}{2}(f_3^w - f_1^w) + \frac{1}{6}\rho_w u_o \qquad (3.39)$$

$$f_6^w = f_8^w + \frac{1}{2}(f_1^w - f_3^w) + \frac{1}{6}\rho_w u_o.$$

Alternatively, we may like to set the pressure inlet boundary condition for 2D displacement simulations. After the inlet density (majority component, i.e., non-wetting fluid) is specified as ρ_n, the inlet velocity is calculated as

$$u_i = -1 + \frac{1}{\rho_n}[f_0^n + f_1^n + f_3^n + 2(f_2^n + f_5^n + f_6^n)] \qquad (3.40)$$

and the unknown distribution functions f_4^n, f_7^n, and f_8^n (for non-wetting fluid) can be obtained as

$$f_4^n = f_2^n - \frac{2}{3}\rho_n u_i$$

$$f_7^n = f_5^n - \frac{1}{2}(f_3^n - f_1^n) - \frac{1}{6}\rho_n u_i \qquad (3.41)$$

$$f_8^n = f_6^n - \frac{1}{2}(f_1^n - f_3^n) - \frac{1}{6}\rho_n u_i.$$

We note that maintaining the velocity or density (pressure) of the minority component, which is usually set to be a very small value, say 10^{-8} mu/lu^3, on both the upper and lower boundaries is also important and requires a second set of boundary conditions.

3.7.2 Boundary conditions for 3D simulations

Here we suppose the most popular velocity model D3Q19 is used. To understand the following formulae more clearly, refer to Figure 1.1 for the velocity directions in the D3Q19 model. Suppose the non-wetting fluid displaces wetting fluid from top to bottom and here we set an upper pressure inlet boundary condition (ρ_n is specified). Similar to the D2Q9 cases, after the streaming step, the distribution functions f_6, f_{13}, f_{14}, f_{17}, and f_{18} on the upper boundary nodes are unknown. In the following we will derive the formulae to get these unknowns. First, the density of the non-wetting fluid can be obtained from the summation of the distribution function:

$$\rho_n = f_0 + f_1 + f_2 + f_3 + f_4 + f_5 + f_6 + f_7 + f_8 + f_9$$
$$+ f_{10} + f_{11} + f_{12} + f_{13} + f_{14} + f_{15} + f_{16} + f_{17} + f_{18}. \tag{3.42}$$

Second, the momenta of the fluid in the z, y, and x directions are

$$\rho_n u_i = (f_5 + f_{11} + f_{12} + f_{15} + f_{16}) - (f_6 + f_{13} + f_{14} + f_{17} + f_{18}), \tag{3.43}$$
$$0 = (f_3 + f_7 + f_8 + f_{15} + f_{17}) - (f_4 + f_9 + f_{10} + f_{16} + f_{18}), \tag{3.44}$$
$$0 = (f_1 + f_7 + f_9 + f_{11} + f_{13}) - (f_2 + f_8 + f_{10} + f_{12} + f_{14}), \tag{3.45}$$

where u_i is the z-direction inlet velocity, which is a negative value in our study. Note that the inlet velocities in the x and y directions are zero (see Eqs (3.44) and (3.45)). In the above Eqs (3.42) to (3.45) there are six unknowns (u_i, f_6, f_{13}, f_{14}, f_{17}, f_{18}). We have to add more constraints to solve the equation system. Following the idea of "non-equilibrium bounce-back" in the normal direction (the z direction), we have

$$f_5 - f_5^{eq} = f_6 - f_6^{eq}. \tag{3.46}$$

It is noted that f_5^{eq} and f_6^{eq} are

$$f_5^{eq} = \frac{1}{18}\rho_n \left[1 + 3u_i + \frac{9}{2}u_i^2 - \frac{3}{2}u_i^2 \right] \tag{3.47}$$

$$f_6^{eq} = \frac{1}{18}\rho_n \left[1 - 3u_i + \frac{9}{2}(-u_i)^2 - \frac{3}{2}(-u_i)^2 \right]. \tag{3.48}$$

Eq. (3.46) becomes

$$f_5 - f_6 = \frac{1}{3}\rho_n u_i. \tag{3.49}$$

Here we still need another equation to close the equation system. Substituting Eq. (3.49) into Eq. (3.43) yields

$$\frac{2}{3}\rho_n u_i = (f_{11} + f_{12} + f_{15} + f_{16}) - (f_{13} + f_{14} + f_{17} + f_{18}). \tag{3.50}$$

Extending the work of Zou and He (1997) for the D3Q15 model, here we propose the sixth equation, i.e., the final one to close the equation system:

$$\frac{1}{3}\rho_n u_i = (f_{11} + f_{12}) - (f_{13} + f_{14}). \tag{3.51}$$

From Eq. (3.50) we immediately have

$$\frac{1}{3}\rho_n u_i = (f_{15} + f_{16}) - (f_{17} + f_{18}). \tag{3.52}$$

Our proposal means that the difference between mass moving in the upward direction $(f_{11}, f_{12}, f_{15}, f_{16})$ and mass moving downward $(f_{13}, f_{14}, f_{17}, f_{18}, \text{see Figure} 1.1)$ is evenly distributed in the y, z and x, z planes. It appears to be a reasonable assumption. The above Eqs (3.42) to (3.45) and Eq. (3.49) and Eq. (3.51) comprise a closed equation system. The solutions are described in the following.
From Eqs (3.42) and (3.43) we have

$$\rho_n(1 + u_i) = f_0 + f_1 + f_2 + f_3 + f_4 + f_7 + f_8 + f_9 + f_{10}$$
$$+ 2(f_5 + f_{11} + f_{12} + f_{15} + f_{16}). \tag{3.53}$$

In our implementation first we calculate the inlet velocity from Eq. (3.53):

$$u_i = -1 + \frac{1}{\rho_n}(f_0^n + f_1^n + f_2^n + f_3^n + f_4^n + f_7^n + f_8^n + f_9^n + f_{10}^n)$$
$$+ \frac{2}{\rho_n}(f_5^n + f_{11}^n + f_{12}^n + f_{15}^n + f_{16}^n). \tag{3.54}$$

Then, the unknown distribution functions $f_6, f_{13}, f_{14}, f_{17},$ and f_{18} on the upper boundary nodes can be obtained as:

$$f_6^n = f_5^n - \frac{1}{3}\rho_n u_i,$$
$$f_{14}^n = f_{11}^n - \frac{1}{6}\rho_n u_i + \frac{1}{2}(-f_2^n - f_8^n - f_{10}^n + f_1^n + f_7^n + f_9^n),$$
$$f_{13}^n = f_{12}^n - \frac{1}{6}\rho_n u_i - \frac{1}{2}(-f_2^n - f_8^n - f_{10}^n + f_1^n + f_7^n + f_9^n), \tag{3.55}$$
$$f_{18}^n = f_{15}^n - \frac{1}{6}\rho_n u_i + \frac{1}{2}(f_3^n + f_7^n + f_8^n - f_4^n - f_9^n - f_{10}^n),$$
$$f_{17}^n = f_{16}^n - \frac{1}{6}\rho_n u_i - \frac{1}{2}(f_3^n + f_7^n + f_8^n - f_4^n - f_9^n - f_{10}^n).$$

If we set the lower pressure outlet boundary condition for 3D displacement simulations, then the outlet velocity is

$$u_o = 1 - \frac{1}{\rho_w}(f_0^w + f_1^w + f_2^w + f_3^w + f_4^w + f_7^w + f_8^w + f_9^w + f_{10}^w)$$
$$- \frac{2}{\rho_w}(f_6^w + f_{13}^w + f_{14}^w + f_{17}^w + f_{18}^w). \tag{3.56}$$

The unknown distribution functions f_5, f_{11}, f_{12}, f_{15}, and f_{16} on the lower boundary nodes can be obtained as:

$$f_5^w = f_6^w + \frac{1}{3}\rho_w u_o,$$

$$f_{11}^w = f_{14}^w + \frac{1}{6}\rho_w u_o - \frac{1}{2}(-f_2^w - f_8^w - f_{10}^w + f_1^w + f_7^w + f_9^w),$$

$$f_{12}^w = f_{13}^w + \frac{1}{6}\rho_w u_o + \frac{1}{2}(-f_2^w - f_8^w - f_{10}^w + f_1^w + f_7^w + f_9^w), \qquad (3.57)$$

$$f_{15}^w = f_{18}^w + \frac{1}{6}\rho_w u_o - \frac{1}{2}(f_3^w + f_7^w + f_8^w - f_4^w - f_9^w - f_{10}^w),$$

$$f_{16}^w = f_{17}^w + \frac{1}{6}\rho_w u_o + \frac{1}{2}(f_3^w + f_7^w + f_8^w - f_4^w - f_9^w - f_{10}^w).$$

These 2D and 3D boundary conditions are not limited to the MCMP SC model; they are also applicable to other popular single- and multi-component multiphase Lattice Boltzmann models, e.g., the RK model that is discussed in Chapter 4.

3.8   Displacement in a 3D porous medium

In this section the non-wetting fluid displacement of wetting fluid is simulated. The data for the porous medium came from Dr C.X. Pan, and is a 1/64 subdomain of a larger porous medium called GB1b (Pan et al. 2004). GB1b was generated using a random sphere pack containing 9,532 spheres. Additional details of the GB1b porous medium can be found in Pan et al. (2004). The 1/64 subdomain contained approximately 150 spheres and has 96^3 voxels. The mean pore throat radius is approximately 2.7 lattice units (Pan et al. 2004).

Figure 3.9 shows the dynamic distributions of non-wetting fluid during the displacement. The time interval between each frame is 3000 ts. Here, an upper and lower four blank layers were added into the geometry to implement pressure inlet (upper BC) and outlet (lower BC) boundary conditions. Hence, the total domain size was $96 \times 96 \times 104$. Except for the upper and lower boundaries, periodic boundary conditions are applied to the other four faces. The main component density is set to be 8.0 mu/lu^3 and the dissolved component has a density of 0.16 mu/lu^3. The coefficient in the cohesion force $G_c = 0.225$. To specify the wetting condition, we set $G_{ads,2} = -0.9$ and $G_{ads,1} = 0.9$, which means that Fluid 1 is totally non-wetting. In our simulations, the pore space is initially saturated by wetting fluid (Fluid 2). The inlet and outlet densities of the fluids are $\rho_{in} = 9.8$ and $\rho_{out} = 6.2$. The two fluids have relaxation times $\tau_1 = \tau_2 = 1$. From Figure 3.9

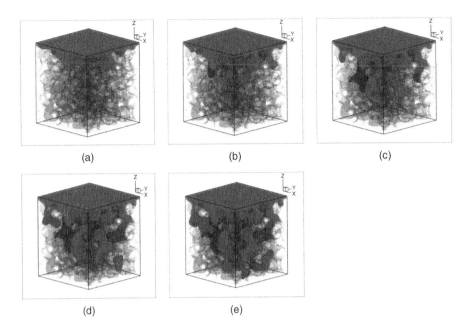

(a)   (b)   (c)

(d)   (e)

Figure 3.9 Dynamic distributions of the non-wetting phase during primary drainage for the GB1b porous medium system (Pan et al. 2004). The time interval between each frame is 3000 time steps. The transparent dark grey isosurface indicates the interface between the pore space and the solid phase. Dark grey and the pore space that is not dark grey represent the non-wetting and wetting fluid, respectively (produced by Jianlin Zhao, China University of Petroleum, Eastern China).

we can see the non-wetting fluid intrude into the wetting-fluid-saturated porous medium. With increasing time, more and more wetting fluid is displaced by the non-wetting fluid.

Although the above simulation results using the MCMP SC model are encouraging, there are still many challenges to face. The applications of this Lattice Boltzmann model to multiphase flow are " ... limited by the admissible range of fluid properties due to restricted numerical stability" (Pan et al. 2004). For example, when the viscosity ratio between phases becomes larger than 10, numerical instabilities may occur (Chin 2002).

Li et al. (2005) found that the MRT MCMP SC Lattice Boltzmann model can relieve the numerical stability issues and more accurate results can be obtained. However, the stability issue of the Lattice Boltzman methods has not been fully understood yet (Junk and Yong 2009, Junk et al. 2005).

Another issue we would like to discuss is the compressibility effect in these simulations using the MCMP SC model. In the above case, the inlet/outlet densities of the fluid ($\rho_{in} = 9.8$ and $\rho_{out} = 6.2$) are significantly different from the average density ($\rho = 8.0$). Obviously, the compressibility effect is prominent because $\Delta\rho = \rho_{in} - \rho_{out} = 3.6$ is comparable to the average density $\rho = 8.0$.

Of course, we are able to diminish the compressibility effect provided that the interfacial tension can be adjusted arbitrarily. In the above case, the interfacial tension $\sigma = 0.927$. If σ can be adjusted to be 0.0927, a case with $\Delta\rho = 0.36$ is an analogy for the above case because the non-dimensional parameter $\frac{p_c R}{\sigma} = \frac{c_s^2 \Delta\rho R}{\sigma}$ is identical, where R is a characteristic pore size. Hence, provided the interfacial tension is adjustable, the compressibility effect in the LBM simulation is expected to be alleviated. However, at the present stage we are unable to change the interfacial tension in the MCMP SC model arbitrarily.

CHAPTER 4

Rothman–Keller multiphase Lattice Boltzmann model

4.1 Introduction

Rothman and Keller (1988) proposed the first multi-component lattice gas model. The model was further developed by Gunstensen et al. (1991) through adding an extra binary fluid collision (perturbation operator) to the Lattice Boltzmann equation. Later, Grunau et al. (1993) introduced two free parameters in the rest equilibrium distribution function, and claimed the improved RK model was able to simulate flows with different densities. Latva-Kokko and Rothman (2005a) improved the recoloring step in the RK model, which is able to reduce the lattice pinning effect and decrease the spurious currents (Leclaire et al. 2012; Liu et al. 2012). The recoloring step is now widely used in applications of the RK model (Huang et al. 2011b; Latva-Kokko and Rothman 2007; Liu et al. 2012). The surface tension and diffusion properties of the model are discussed in Ginzbourg and Adler (1995) and Latva-Kokko and Rothman (2005b), respectively. Recently, Reis and Phillips (2007) developed a 2D nine-velocity RK model. In this model a revised binary fluid collision is proposed and is shown to be able to recover the term which accounts for interfacial tension in the N–S equations (Reis and Phillips 2007). In some studies (Lishchuk et al. 2003; Liu and Zhang 2011), the forcing term strategy is used to introduce the interfacial tension instead of the perturbation operator. Here we still adopt the perturbation operator because it is regarded as one of the main characteristics of the RK model.

In the literature, it is more common that cases with identical densities were simulated using the RK model, such as droplet deformation and breakup in simple shear flow (Liu et al. 2012), two-phase flow in porous media (Huang et al. 2014b; Langaas and Papatzacos 2001), droplet formation in a cross-junction microchannel (Wu et al. 2008), and high viscosity ratio two-phase parallel flow in a channel (Leclaire et al. 2012; Tölke 2001). Note in the latter two flows, only two fluids with identical densities but different kinematic viscosities are simulated.

Multiphase Lattice Boltzmann Methods: Theory and Application, First Edition.
Haibo Huang, Michael C. Sukop and Xi-Yun Lu.
© 2015 John Wiley & Sons, Ltd. Published 2015 by John Wiley & Sons, Ltd.
Companion Website: www.wiley.com/go/huang/boltzmann

The RK model is not able to handle general high-density ratio two-phase flows (Huang et al. 2013). In almost all validations of the model involving high-density ratios (Grunau et al. 1993; Leclaire et al. 2011; Tölke et al. 2002) only cases of stationary bubbles or droplets are simulated. In these studies, the interfacial tension was calculated by the Laplace law and then compared with the analytical value. Again, for cases of stationary bubbles or droplets, Leclaire et al. (2011) even claimed to have simulated a case of a density ratio between the fluids as high as 10,000. In addition to the cases of stationary droplet and bubble simulations, rising bubble cases (Liu et al. 2012; Tölke 2001; Tölke et al. 2002) are also used to perform validation of the RK method to some extent. However, usually only density ratios less than 4 are simulated (Liu et al. 2012; Reis and Phillips 2007; Tölke et al. 2001; Tölke 2002).

Some numerical studies (Huang et al. 2013; Rannou 2008) have reported that the RK model gave poor results for two-phase parallel flows with different densities in a channel. Through analysis of the recovered macroscopic equations Huang et al. (2013) found some extra terms in the recovered momentum equation. These extra terms may affect the numerical results significantly. For example, because of these terms the tangential shear stress condition in the interface vicinity is not satisfied properly. For cases of stationary bubbles or droplets, the extra terms may be negligible and not affect the result much in terms of interfacial tension.

This chapter is arranged in the following way. The RK model is introduced briefly. Through detailed theoretical analysis we show that usually the RK model introduces some extra unwanted terms in the recovered macroscopic momentum equation. For two-phase flows with different densities, the terms are usually unable to be eliminated. Through numerical study of two-phase parallel flows in a channel we confirm that the RK model for different density ratios is usually incorrect except for the stationary droplet and bubble cases. A scheme to eliminate the unwanted terms in Huang et al. (2013) is introduced.

We also discuss the issues of interfacial tension, isotropy, and spurious current of the RK model. Two dynamic multiphase flows, drainage and capillary filling, are simulated to demonstrate the validation of the model for high-kinematic-viscosity cases.

The pressure and velocity boundary conditions for single-phase flow are extended to simulate multiphase flows. The MRT RK model is applied to study two-phase flow in porous media. Two advantages of the MRT RK model are demonstrated. One is the significantly reduced spurious current and the other is better numerical stability. With these advantages two-component flows in porous media with both Ca and M ranging from 10^{-3} to 10^3 are simulated. All three typical flow patterns are reproduced and three domains in the M–Ca phase diagram (Lenormand et al. 1988) are obtained. Halliday et al. (2007) also developed the RK model by reducing the spurious current and facilitating simulational access to regimes of flow with a low capillary number and Reynolds number.

4.2 RK color-gradient model

In the RK model the particle distribution function (PDF) for fluid k is denoted by f_i^k. For two-phase flows, two distribution functions are defined, i.e., f_i^1, and f_i^2, where 1 and 2 refer to Fluids 1 and 2, respectively. Originally, Rothman and Keller (1988) used color—'red' or 'blue'—to identify the different fluids. Hence, this is commonly known as the color model. The total PDF at $(\mathbf{x}, t)$ is

$$f_i(\mathbf{x}, t) = \sum_k f_i^k(\mathbf{x}, t). \tag{4.1}$$

In the RK model there are three steps for each component: streaming, collision, and recoloring. The recoloring step is a characteristic of the RK model. Suppose an iteration begins from the streaming step. How the three steps construct a loop is illustrated in the following. The streaming step is similar to normal LBM streaming:

$$f_i^k(\mathbf{x} + \mathbf{e}_i \Delta t, t + \Delta t) = f_i^{k+}(\mathbf{x}, t), \tag{4.2}$$

where f_i^{k+} is the PDF after the recoloring step. In the above equation, $\mathbf{e}_i, i = 0, 1, \ldots, b$ are the discrete velocities of the velocity models. Here the D2Q9 velocity model ($b = 8$) is used.

The collision step can be written as (Latva-Kokko and Rothman 2005a)

$$f_i^{k*}(\mathbf{x}, t) = f_i^k(\mathbf{x}, t) + (\Omega_i^k)^1 + (\Omega_i^k)^2, \tag{4.3}$$

where $f_i^{k*}(\mathbf{x}, t)$ is the post-collision state. There are two collision terms in the equation, i.e., $(\Omega_i^k)^1$ and $(\Omega_i^k)^2$. Here the lattice BGK scheme is adopted and the first collision term is

$$(\Omega_i^k)^1 = -\frac{1}{\tau}(f_i^k(\mathbf{x}, t) - f_i^{k,eq}(\mathbf{x}, t)). \tag{4.4}$$

The equilibrium distribution function $f_i^{k,eq}(\mathbf{x}, t)$ is almost identical to the formula for common LBM except for the rest part:

$$f_i^{k,eq}(\mathbf{x}, t) = \rho_k \left(C_i + w_i \left[\frac{\mathbf{e}_i \cdot \mathbf{u}}{c_s^2} + \frac{(\mathbf{e}_i \cdot \mathbf{u})^2}{2c_s^4} - \frac{(\mathbf{u})^2}{2c_s^2} \right] \right), \tag{4.5}$$

where the density of the kth component is

$$\rho_k = \sum_i f_i^k \tag{4.6}$$

and the total density is

$$\rho = \sum_k \rho_k. \tag{4.7}$$

The momentum is

$$\rho \mathbf{u} = \sum_k \sum_i f_i^k \mathbf{e}_i. \tag{4.8}$$

In the above formula, the coefficients are (Grunau et al. 1993)

$$
C_i = \begin{cases} \alpha_k, & i = 0, \\ \dfrac{1 - \alpha_k}{5}, & i = 1, 2, 3, 4, \\ \dfrac{1 - \alpha_k}{20}, & i = 5, 6, 7, 8, \end{cases} \tag{4.9}
$$

where α_k is a parameter that is assumed to be able to adjust the density of fluids (Grunau et al. 1993, Reis and Phillips 2007). The density ratio is (Grunau et al. 1993, Reis and Phillips 2007)

$$
\kappa = \frac{\rho_1}{\rho_2} = \frac{1 - \alpha_2}{1 - \alpha_1}. \tag{4.10}
$$

The other parameters are $w_0 = \frac{4}{9}$, $w_i = \frac{1}{9}$, for $i = 1, 2, 3, 4$, and $w_i = \frac{1}{36}$, for $i = 5, 6, 7, 8$, where the is are the D2Q9 directions.

When the relaxation time parameters for the two fluids are very different, for example $\tau_1 = 0.501$ and $\tau_2 = 1.0$, $\tau(\mathbf{x})$ at the interface can be determined as follows. First let us define

$$
\psi(\mathbf{x}) = \frac{\rho_1(\mathbf{x}) - \rho_2(\mathbf{x})}{\rho_1(\mathbf{x}) + \rho_2(\mathbf{x})}. \tag{4.11}
$$

To make the relaxation parameter ($\tau(\mathbf{x})$) change smoothly at the interfaces between two fluids, here we adopt the interpolation scheme constructed by Grunau et al. (1993)

$$
\tau(\mathbf{x}) = \begin{cases} \tau_1 & \psi > \delta, \\ g_1(\psi) & \delta \ge \psi > 0, \\ g_2(\psi) & 0 \ge \psi \ge -\delta, \\ \tau_2 & \psi < -\delta, \end{cases} \tag{4.12}
$$

where $g_1(\psi) = s_1 + s_2\psi + s_3\psi^2$ and $g_2(\psi) = t_1 + t_2\psi + t_3\psi^2$, $s_1 = t_1 = 2\frac{\tau_1\tau_2}{\tau_1+\tau_2}$, $s_2 = 2\frac{\tau_1-s_1}{\delta}$, and $s_3 = -\frac{s_2}{2\delta}$, and $t_2 = 2\frac{t_1-\tau_2}{\delta}$ and $t_3 = \frac{t_2}{2\delta}$. Here $\delta \le 1$ is a free positive parameter that affects interface thickness and is usually set as $\delta = 0.98$. The viscosity of each component is

$$
v_k = c_s^2(\tau_k - 0.5)\Delta t, \tag{4.13}
$$

where $c_s^2 = \frac{1}{3}c^2$.

The second collision term is more complex and there are some different forms found in the literature (Gunstensen et al. 1991, Latva-Kokko and Rothman 2005a, Reis and Phillips 2007). An example is (Gunstensen et al. 1991, Latva-Kokko and Rothman 2005a):

$$
(\Omega_i^k)^2 = \frac{A_k}{2}|\mathbf{f}|(2 \cdot \cos^2(\lambda_i) - 1), \tag{4.14}
$$

where A_k is a parameter that affects the interfacial tension and λ_i is the angle between the color gradient $\mathbf{f}(\mathbf{x}, t)$ and the direction $\mathbf{e}_i$, and we have $\cos(\lambda_i) = \frac{\mathbf{e}_i \cdot \mathbf{f}}{|\mathbf{e}_i| \cdot |\mathbf{f}|}$ (Latva-Kokko and Rothman 2005a).

The color-gradient $\mathbf{f}(\mathbf{x}, t)$ is calculated as (Latva-Kokko and Rothman 2005a)

$$\mathbf{f}(\mathbf{x}, t) = \sum_i \mathbf{e}_i \sum_j [f_j^1(\mathbf{x} + \mathbf{e}_i \Delta t, t) - f_j^2(\mathbf{x} + \mathbf{e}_i \Delta t, t)]. \tag{4.15}$$

However, according to the study of Reis and Phillips (2007), the correct collision operator should be

$$(\Omega_i^k)^2 = \frac{A_k}{2} |\mathbf{f}| \left[w_i \frac{(\mathbf{e}_i \cdot \mathbf{f})^2}{|\mathbf{f}|^2} - B_i \right], \tag{4.16}$$

where $B_0 = -\frac{4}{27}$, $B_i = \frac{2}{27}$, for $i = 1, 2, 3, 4$, and $B_i = \frac{5}{108}$, for $i = 5, 6, 7, 8$. Using these parameters, the correct term due to interfacial tension in the N–S equations can be recovered (Reis and Phillips 2007).

The recoloring step is then implemented to achieve separation of the two fluids (Latva-Kokko and Rothman 2005a):

$$f_i^{1,+} = \frac{\rho_1}{\rho} f_i^* + \beta \frac{\rho_1 \rho_2}{\rho^2} f_i^{(eq)}(\rho, \mathbf{u} = 0) \cos(\lambda_i), \tag{4.17}$$

$$f_i^{2,+} = \frac{\rho_2}{\rho} f_i^* - \beta \frac{\rho_1 \rho_2}{\rho^2} f_i^{(eq)}(\rho, \mathbf{u} = 0) \cos(\lambda_i), \tag{4.18}$$

where $f_i^* = \sum_k f_i^{k*}$ and β usually takes any value between 0 and 1 (Latva-Kokko and Rothman 2005a).

After $f_i^1(\mathbf{x}, t)$, and $f_i^2(\mathbf{x}, t)$ are updated, the streaming steps (i.e. Eq. (4.2)) should be implemented for each component. Through iteration of the procedure illustrated above, two-phase flows can be simulated.

In the model A_k and β are the most important parameters that adjust interfacial properties. The interfacial thickness can be adjusted by β but the interfacial tension is independent of β and only determined by A_k and τ_1, τ_2 (Latva-Kokko and Rothman 2005a). The pressure in the flow field can be obtained from the density via the common equation of state

$$p = c_s^2 \rho. \tag{4.19}$$

Note that when components with identical densities are considered, the corresponding equilibrium distribution function is Eq. (4.5) with $C_i = w_i$. This is the common equilibrium distribution function usually used in the LBM. Hence, when two components have identical densities, the equilibrium distribution function has the same formula and it is not necessary to calculate both collision steps Eqs (4.4) and (4.16) separately for each component. The two collision steps become

$$(\Omega_i)^1 = -\frac{\Delta t}{\tau}(f_i(\mathbf{x}, t) - f_i^{eq}(\mathbf{x}, t)) \tag{4.20}$$

and

$$(\Omega_i)^2 = A|\mathbf{f}| \left[w_i \frac{(\mathbf{e}_i \cdot \mathbf{f})^2}{|\mathbf{f}|^2} - B_i \right],$$

(4.21)

where $A = \sum_k A_k/2$ and $f_i = \sum_k f_i^k$.

4.3 Theoretical analysis (Chapman–Enskog expansion)

Huang et al. (2013) showed that incorporation of the freedom of the rest particle equilibrium distribution in a revised RK model, i.e., C_i in Eq. (4.5) (Grunau et al. 1993), generally fails to recover the correct N–S equations. Here derivatives of the N–S equations in the flow region far from the interface are introduced in detail.

Reis and Phillips (2007) demonstrated that the term incorporating the interfacial tension in the recovered macroscopic momentum equation is only introduced by $(\Omega_i^k)^2$. The collision step $(\Omega_i^k)^2$ and the recoloring step do not affect the other terms in the derived momentum equation (Eq. (4.45)) (Reis and Phillips 2007). Hence, in the following $(\Omega_i^k)^2$ and the recoloring step are not considered and the interfacial tension does not appear in the N–S equations.

Applying the Taylor expansion, we have

$$f_i^k(\mathbf{x} + \mathbf{e}_i \Delta t, t + \Delta t) = \sum_{n=0}^{\infty} \frac{1}{n!} \Delta t^n D_t^n f_i^k(\mathbf{x}, t).$$

(4.22)

where $D_t^n \equiv (\partial_t + e_{i\gamma}\partial_\gamma)^n$.

Retaining terms to $O((\Delta t)^2)$, the common Lattice Boltzmann equation (Eq. (4.2) without the second collision term $(\Omega_i^k)^2$ and recoloring step) becomes

$$\Delta t \left(\frac{\partial}{\partial t} + e_{i\alpha} \frac{\partial}{\partial x_\alpha} \right) f_i^k + \frac{\Delta t^2}{2} \left(\frac{\partial}{\partial t} + e_{i\alpha} \frac{\partial}{\partial x_\alpha} \right)^2 f_i^k + O((\Delta t)^2)$$

$$= -\frac{1}{\tau}(f_i^k - f_i^{k,eq}).$$

(4.23)

In the Chapman–Enskog expansion, variables f_i^k and ∂_t are supposed to be decomposed into several components within different time scales because the variables are thought of as multiscale variables, that is

$$f_i^k = f_i^{k,(0)} + \Delta t f_i^{k,(1)} + \Delta t^2 f_i^{k,(2)} + \cdots$$

(4.24)

and

$$\partial_t = \partial_{t_1} + \Delta t \partial_{t_2} + \cdots.$$

(4.25)

Substituting the above two equations into Eq. (4.23) and rearranging, we have

$$O(\epsilon) : (f_i^{k,(0)} - f_i^{k,eq})/\tau = 0 \tag{4.26}$$

$$O(\epsilon^1) : (\partial_{t_1} + e_{i\alpha}\partial_\alpha)f_i^{k,(0)} + \frac{1}{\tau}f_i^{k,(1)} = 0 \tag{4.27}$$

$$O(\epsilon^2) : \partial_{t_2}f_i^{k,(0)} + \left(1 - \frac{1}{2\tau}\right)(\partial_{t_1} + e_{i\alpha}\partial_\alpha)f_i^{k,(1)} + \frac{1}{\tau}f_i^{k,(2)} = 0 \tag{4.28}$$

To obtain the continuity equation and the N–S equations, we start from Eqs. (4.26)–(4.28).

Continuity equation
Summing on i, Eq. (4.27),

$$\partial_{t_1}\left(\sum_i f_i^{k,(0)}\right) + \partial_\alpha\left(\sum_i e_{i\alpha}f_i^{k,(0)}\right) + \frac{1}{\tau}\sum_i f_i^{k,(1)} = 0. \tag{4.29}$$

From equations $\sum_i f_i^{k,(0)} = \rho_k$, $\sum_i f_i^{k,(1)} = 0$, and $\sum_i f_i^{k,(2)} = 0$ (because $\sum_i f_i^k = \rho_k$, and $f_i^k = f_i^{k,(0)} + \Delta t f_i^{k,(1)} + \Delta t^2 f_i^{k,(2)}$) we know that

$$\partial_{t_1}\rho_k + \partial_\alpha(\rho_k u_\alpha) = 0. \tag{4.30}$$

From Eq. (4.27) we know that $f_i^{k,(1)} = -\tau(\partial_{t_1} + e_{i\alpha}\partial_\alpha)f_i^{k,(0)}$ and substituting it into Eq. (4.28) we have

$$\partial_{t_2}f_i^{k,(0)} - \left(\tau - \frac{1}{2}\right)(\partial_{t_1} + e_{i\alpha}\partial_\alpha)(\partial_{t_1} + e_{i\beta}\partial_\beta)f_i^{k,(0)} + \frac{1}{\tau}f_i^{k,(2)} = 0. \tag{4.31}$$

Summing on i, Eq. (4.28) yields

$$\partial_{t_2}\rho_k - \left(\tau - \frac{1}{2}\right)\left\{\partial_{t_1}^2\rho_k + 2\partial_{t_1}\partial_\alpha(\rho_k u_\alpha) + \partial_\alpha\partial_\beta\sum_i f_i^{k,(0)}e_{i\alpha}e_{i\beta}\right\} = 0. \tag{4.32}$$

Combining Eqs (4.30) and (4.32) we have

$$\partial_t\rho_k + \partial_\alpha(\rho_k u_\alpha) - \Delta t\left(\tau - \frac{1}{2}\right)\left\{\partial_{t_1}^2\rho_k + 2\partial_{t_1}\partial_\alpha(\rho_k u_\alpha) + \partial_\alpha\partial_\beta\sum_i f_i^{k,(0)}e_{i\alpha}e_{i\beta}\right\}$$
$$+O((\Delta t)^2) = 0, \tag{4.33}$$

where $\partial_t\rho_k = \partial_{t_1}\rho_k + \Delta t\partial_{t_2}\rho_k$.

Multiplying Eq. (4.27) by $e_{i\beta}$ and summing over i gives

$$\partial_{t_1}(\rho_k u_\beta) + \partial_\alpha\sum_i f_i^{k,(0)}e_{i\alpha}e_{i\beta} = 0. \tag{4.34}$$

Substituting Eqs (4.30) and (4.34) into Eq. (4.33), immediately we find that terms $\left\{\partial_t^2\rho_k + 2\partial_t\partial_\alpha(\rho_k u_\alpha) + \partial_\alpha\partial_\beta\sum_i f_i^{k,(0)}e_{i\alpha}e_{i\beta}\right\}$ vanish and we have

$$\partial_t\rho_k + \partial_\alpha(\rho_k u_\alpha) = 0 \tag{4.35}$$

to the order of $O((\Delta t)^2)$. This is the continuity equation of the flow system.

Momentum equations (N–S)

Using the definition of the equilibrium distribution function in Eq. (4.5) and $f_i^{k,(0)} = f_i^{k,eq}$ (Eq. (4.26)),

$$\partial_\alpha \left(\sum_i f_i^{k,(0)} e_{i\alpha} e_{i\beta} \right) = \partial_\alpha (p_k \delta_{\alpha\beta} + \rho_k u_\alpha u_\beta). \tag{4.36}$$

Substituting Eq. (4.36) into Eq. (4.34) we obtain the Euler equation:

$$\partial_{t_1}(\rho_k u_\alpha) = -\partial_\beta (\rho_k u_\alpha u_\beta + p_k \delta_{\alpha\beta}), \tag{4.37}$$

where

$$p_k = (c_s^k)^2 \rho_k \tag{4.38}$$

and

$$(c_s^k)^2 = \sum_i C_i e_{ix} e_{ix} = \frac{3}{5}(1 - \alpha_k). \tag{4.39}$$

Multiplying Eq. (4.28) by $e_{i\beta}$ and summing over i gives

$$\partial_{t_2}(\rho_k u_\beta) + \left(1 - \frac{1}{2\tau}\right) \left[\partial_{t_1} \left(\sum_i f_i^{k,(1)} e_{i\beta} \right) + \partial_\alpha \sum_i f_i^{k,(1)} e_{i\alpha} e_{i\beta} \right] = 0. \tag{4.40}$$

From the definition of the equilibrium distribution function we have

$$\sum_i f_i^{k,(0)} e_{i\beta} = \rho_k u_\beta. \tag{4.41}$$

From Eq. (4.8) we know $\sum_i f_i^k e_{i\beta} \approx \rho u_\beta$ is valid in the component k-rich region, which is far from the interface. Hence, $\sum_i f_i^{k,(1)} e_{i\beta} = 0$.

To derive the N–S equations we have to know the first-order momentum flux tensor, i.e.,

$$\Pi_{\alpha\beta}^{k,(1)} = \sum_i f_i^{k,(1)} e_{i\alpha} e_{i\beta}$$

$$= -\tau_k \sum_i e_{i\alpha} e_{i\beta} D_{t_1} f_i^{k,eq}$$

$$= -\tau_k \partial_{t_1}(\rho_k u_\alpha u_\beta + p_k \delta_{\alpha\beta}) - \tau_k \partial_\gamma \left[\frac{1}{3} \rho_k c^2 (u_\alpha \delta_{\beta\gamma} + u_\beta \delta_{\gamma\alpha} + u_\gamma \delta_{\alpha\beta}) \right], \tag{4.42}$$

where $D_{t_1} \equiv \partial_{t_1} + e_{i\gamma} \partial_\gamma$.

Then we substitute the following continuity equation (Eq. (4.30)) and the Euler equation (Eq. (4.37)) into the above equation (Eq. (4.42)). If omitting the terms of $O(u^3)$, such as $u_\beta \partial_\gamma(\rho_k u_\alpha u_\gamma)$ and so on, we have

$$\Pi_{\alpha\beta}^{k,(1)} = -\tau_k u_\beta \partial_{t_1}(\rho_k u_\alpha) + u_\alpha \partial_{t_1}(\rho_k u_\beta) - u_\alpha u_\beta \partial_{t_1} \rho_k + (c_s^k)^2 (\partial_{t_1} \rho_k \delta_{\alpha\beta})$$

$$- \tau_k \partial_\gamma \left[\frac{1}{3} \rho_k c^2 (u_\alpha \delta_{\beta\gamma} + u_\beta \delta_{\gamma\alpha} + u_\gamma \delta_{\alpha\beta}) \right]$$

$$= \tau_k (c_s^k)^2 [u_\beta \partial_\alpha(\rho_k) + u_\alpha \partial_\beta(\rho_k) + \partial_\gamma(\rho_k u_\gamma) \delta_{\alpha\beta}]$$

$$- \frac{1}{3} c^2 \tau_k [\partial_\beta(\rho_k u_\alpha) + \partial_\alpha(\rho_k u_\beta) + \partial_\gamma(\rho_k u_\gamma) \delta_{\alpha\beta}] + O(u^3). \tag{4.43}$$

At this step Liu et al. (2012) obtained

$$\Pi_{\alpha\beta}^{k,(1)} = -\tau_k \left[\frac{1}{3}c^2 - (c_s^k)^2 \right] [u_\beta \partial_\alpha(\rho_k) + u_\alpha \partial_\beta(\rho_k) + \partial_\gamma(\rho_k u_\gamma)\delta_{\alpha\beta}]$$

$$- \frac{1}{3}c^2 \tau_k \rho_k [\partial_\beta u_\alpha + \partial_\alpha u_\beta]. \tag{4.44}$$

Finally, the macroscopic momentum equation recovered from the RK model is

$$\partial_t(\rho_k u_\alpha) + \partial_\beta(\rho_k u_\alpha u_\beta) = -\partial_\alpha p + \rho_k \nu_k \partial_\beta(\partial_\alpha u_\beta + \partial_\beta u_\alpha)$$

$$+ \left(\tau_k - \frac{1}{2} \right) \left[\frac{1}{3}c^2 - (c_s^k)^2 \right] \Delta t \partial_\beta[u_\beta \partial_\alpha(\rho_k) + u_\alpha \partial_\beta(\rho_k) + \partial_\gamma(\rho_k u_\gamma)\delta_{\alpha\beta}]. \tag{4.45}$$

The term $\rho_k \nu_k \partial_\beta(\partial_\alpha u_\beta + \partial_\beta u_\alpha)$ is the viscosity term that appears in the N–S equations and the recovered kinematic viscosity is $\nu_k = c_s^2(\tau_k - 0.5)\Delta t$.

Eliminating the unwanted extra term

Compared to the N–S equations, the last term in Eq. (4.45), i.e.,

$$U_\alpha^k = \left(\tau_k - \frac{1}{2} \right) \left[\frac{1}{3}c^2 - (c_s^k)^2 \right] \Delta t \partial_\beta[u_\beta \partial_\alpha(\rho_k) + u_\alpha \partial_\beta(\rho_k) + \partial_\gamma(\rho_k u_\gamma)\delta_{\alpha\beta}], \tag{4.46}$$

is an unwanted extra term that appears in the momentum equations. In Liu et al. (2012) it was assumed that the term is $O(u^3)$. In single-phase flows, the density gradient is small and the assumption may be true. However, for two-phase flows with different densities, $\frac{1}{3}c^2 \neq (c_s^k)^2$, and near the interface the density gradient $u_\beta \partial_\alpha(\rho_k)$ or $u_\alpha \partial_\beta(\rho_k)$ may be significant. Hence the last term in Eq. (4.45) may be important and should not be neglected.

On the other hand, for two-phase flows with identical densities usually $C_i = w_i$ ($\alpha_k = \frac{4}{9}$, $k = 1, 2$) is adopted and $\frac{1}{3}c^2 = (c_s^k)^2$ is always valid. In this way the undesired term would disappear automatically. Hence, the N–S equation is correctly recovered and the RK model is always correct for two-phase flows with identical densities.

Huang et al. (2013) proposed a scheme to eliminate the unwanted extra terms. The terms can be regarded as forcing terms in the N–S equations. To eliminate the forcing terms in a recovered macroscopic equation we can introduce a source term to the Lattice Boltzmann equation, i.e., S_i^k in Eq. (4.3).

$$S_i^k = -w_i U_\alpha^k \Delta t e_{i\alpha} \frac{1}{c_s^2}, \tag{4.47}$$

with $\sum_i S_i^k = 0$ and $\sum_i e_{i\beta} S_i^k = -U_\beta^k \Delta t$. We know that when this source term is added into the Lattice Boltzmann equation the correct N–S equation for kth component is recovered. We also note that the forcing term in Eq. (4.47) is calculated explicitly through the density and velocity at the last time step. Such a treatment cannot eliminate completely the error term in Eq. (4.45) for an unsteady flow problem.

To add the source term into the Lattice Boltzmann equation one has to evaluate the density gradient and relevant derivatives in Eq. (4.46). For example, to evaluate $\partial_y \rho_r$ and $\partial_y(u_x \partial_y \rho_r)$ at a lattice node we adopt the central finite difference method:

$$(\partial_y \rho_r)_{(i,j)} = \frac{1}{2\Delta y}[(\rho_r)_{(i,j+1)} - (\rho_r)_{(i,j-1)}] \tag{4.48}$$

and

$$[\partial_y(u_x \partial_y \rho_r)]_{(i,j)} = \frac{1}{2\Delta y}[(u_x)_{(i,j+1)}(\partial_y \rho_r)_{(i,j+1)} - (u_x)_{(i,j-1)}(\partial_y \rho_r)_{(i,j-1)}], \tag{4.49}$$

where $\Delta y = 1$ lu and the subscript (i, j) denotes the column and row indices of a lattice node in the computational domain.

In the following, cases of parallel two-phase flows in a channel are simulated. We can see the extra term is important; neglecting it may lead to incorrect results.

4.3.1 Discussion of above formulae

Grunau et al. (1993) showed that $\nu_k = c_s^2(\tau_k - 0.5)\Delta t$, where $c_s^2 = \frac{1}{4}c^2$ for the D2Q7 velocity model that they used. Analogously, for the D2Q9 velocity model $c_s^2 = \frac{1}{3}c^2$ and $\nu_k = c_s^2(\tau_k - 0.5)\Delta t$. However, it is unclear why here c_s^2 instead of $(c_s)_k^2$ is used. Besides that it is unknown how $(c_s)_k^2$ is related to c_s^2 in the RK model (Grunau et al. 1993, Reis and Phillips 2007).

Our revision of the RK model (Huang et al. 2013) is not fully successful. For cases with not only different density ratio but also different kinematic viscosity ratio, the results are still not as satisfactory compared with the analytical solutions. This issue is discussed at the end of Section 4.4.

4.4 Layered two-phase flow in a 2D channel

The numerical results of Huang et al. (2013) demonstrate that the RK model is able to simulate the cases of stationary droplets with high density ratios correctly in terms of interfacial tension. This observation is highly consistent with the results in the literature (Grunau et al. 1993; Leclaire et al. 2011; Reis and Phillips 2007; Tölke et al. 2001; Tölke 2002). However, it does not necessary mean the RK model is able to handle general high-density contrast two-phase flows correctly. In this section we investigate the performance of the RK model for parallel two-phase flows in a channel.

In our description lattice units are used. The units of some parameters are listed in Table 4.1. Some other parameters, such as C_i, ω_i, α, β, and A, are non-dimensional parameters.

In the simulation, as illustrated in Figure 2.9, periodic boundary conditions are applied on the left and right boundaries. For the lattice nodes in upper and lower plates, only simple bounce back is implemented and the collision steps are not implemented. It is noted that when simple bounce back is used to mimic the no-slip boundary condition, the wall is actually located halfway between a flow node and a bounce-back node (He et al. 1997).

In this flow, the vertical velocity u_y is assumed to be zero everywhere inside the computational domain. Because of the periodic boundary condition in the x

Table 4.1 Units of parameters.

Variable	Unit	Variable	Unit
f_i, ρ	mu/lu^3	p	mu/(lu ts^2)
$\mathbf{u}, C_s$	lu/ts	$\mathcal{G}, U_\alpha^k$	mu/(lu^2 ts^2)
S_i^k	mu/lu^3	σ	mu/ts^2

direction, the derivatives in the x direction are zero, i.e., $\partial_x\phi = 0$, where ϕ denotes density, velocity, and pressure. Hence, in a steady flow, for the k component, the N–S equations (i.e., Eq. (4.45)) can be simplified to

$$\rho_k v_k \partial_y^2 u_x + \mathcal{G} + (\tau_k - 0.5)\Delta t \left[\frac{c^2}{3} - (c_s^k)^2\right]\partial_y(u_x\partial_y\rho_k) = 0, \quad (4.50)$$

where $\mathcal{G}$ denotes the body force, which has units identical to $\rho\partial_t u_x$ (refer to Table 4.1).

In our simulations, the non-wetting phase flows in the central region $0 < |y| < a$, while the wetting phase flows in the region $a < |y| < b$ (refer to Figure 2.9). Assuming the flow in the channel is Poiseuille-type, the analytical solution for the velocity profile between the parallel plates can be obtained (Huang and Lu 2009, Rannou 2008).

The computational domain is 10×100. Because the periodic boundary condition is used on the left and right boundaries, the number of mesh nodes used in the x direction can be much smaller. The main parameters in the RK model are $A = 10^{-4}$ and $\beta = 0.5$. The error between numerical and analytical solutions is defined as

$$E(t) = \frac{\sum_j |u(j,t) - u_0(j)|}{\sum_j |u_0(j)|}, \quad (4.51)$$

where the summation is over the lattice nodes j in the slice $x = 5$ and u_0 is the analytical solution. The convergence criterion is $|\frac{E(t)-E(t-10^4\Delta t)}{E(t-10^4\Delta t)}| < 10^{-4}$.

4.4.1 Cases of two fluids with identical densities
Figure 4.1 shows the velocity profile across the middle vertical section of the channel for different kinematic viscosity ratios

$$M = \frac{v_n}{v_w} = \frac{v_1}{v_2}, \quad (4.52)$$

where v_n and v_w are the kinematic viscosities of the non-wetting and wetting fluids, respectively. The kinematic viscosity of component k is calculated with $v_k = c_s^2(\tau_k - 0.5)\Delta t$. In the figure, velocity profiles in (a) and (b) are obtained through applying a body force $\mathcal{G} = 1.5 \times 10^{-8}$ to both fluids. From Figure 4.1 we can see that the numerical solutions agree well with the analytical ones. The errors in Figure 4.1(a) and (b) are 2.89% and 5.16%, respectively.

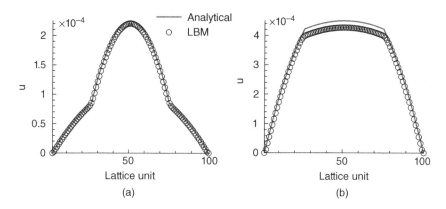

Figure 4.1 Velocity profiles for cases with identical densities and $\mathcal{G} = 1.5 \times 10^{-8}$ applied to both fluids. (a) $M = \frac{1}{5}$, $\tau_1 = 0.6$, and $\tau_2 = 1.0$; (b) $M = 5$, $\tau_1 = 1.0$, and $\tau_2 = 0.6$.

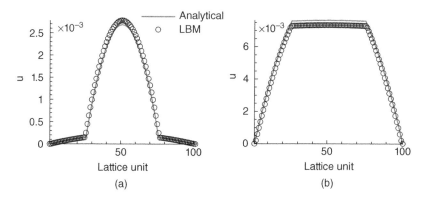

Figure 4.2 Velocity profiles for cases with identical densities and $\mathcal{G} = 1.5 \times 10^{-8}$ applied to both fluids. The original RK approach is (a) $M = \frac{1}{50}$, $\tau_1 = 0.51$, and $\tau_2 = 1.0$; (b) $M = 50$, $\tau_1 = 1.0$, and $\tau_2 = 0.51$.

Figure 4.2(a) and (b) shows the velocity profiles for cases with identical densities and $M = \frac{1}{50}$ and $M = 50$, respectively. The LBM results agree well with the analytical ones. The errors in Figure 4.2(a) and (b) are 5.25% and 4.24%, respectively.

Figure 4.3(a) and (b) shows the velocity profiles for cases with identical densities and $M = \frac{1}{500}$ and $M = 500$, respectively. The errors in Figure 4.2(a) and (b) are 6.0% and 4.27%, respectively.

Applying the RK model (Grunau et al. 1993, Latva-Kokko and Rothman 2005a), some studies of this flow (Leclaire et al. 2012, Rannou 2008) have shown that the numerical error would increase with viscosity ratio contrast but still agree well with the analytical solution. From Figures 4.1, 4.2, and 4.3 we also observed this trend.

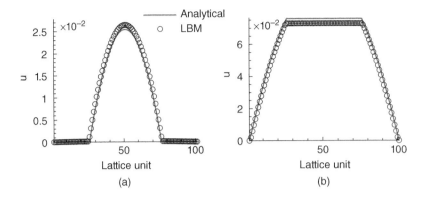

Figure 4.3 Velocity profiles for cases with identical densities and $\mathcal{G} = 1.5 \times 10^{-8}$ applied to both fluids. (a) $M = \frac{1}{500}$, $\tau_1 = 0.501$, and $\tau_2 = 1.0$; (b) $M = 500$, $\tau_1 = 1.0$, and $\tau_2 = 0.501$.

4.4.2   Cases of two fluids with different densities

Rannou (2008) reported that when the densities of the two fluids are different, the numerical results obtained through the RK model for this flow do not agree with the analytical solution. Here we further confirmed this conclusion. Many simulations were performed in Huang et al. (2013). Here one more example is given.

In present tests, the kinematic viscosity ratios are kept at unity but the two fluids have different densities. The main parameters in our simulations are shown in Table 4.2. It is noted that in some cases the source term (i.e., Eq. (4.47)) is added into the Lattice Boltzmann equation to cancel the unwanted terms.

Table 4.2   Parameters for the RK model simulations of parallel flows with different densities ($A = 10^{-4}$, $\tau_1 = \tau_2 = 1$, $\mathcal{G} = 1.5 \times 10^{-8}$ is applied to both fluids).

Case	α_1	α_2	$\frac{\rho_1}{\rho_2} = \frac{1-\alpha_2}{1-\alpha_1}$	β	Eliminate unwanted terms?
A	0.4	0.8	$\frac{1}{3}$	0.5	No
B	0.8	0.4	3	0.5	No
C	0.4	0.8	$\frac{1}{3}$	0.5	Yes
D	0.8	0.4	3	0.5	Yes
E	0.9	0.2	8	0.2	No
F	0.9	0.2	8	0.2	Yes
G	0.2	0.9	$\frac{1}{8}$	0.2	No
H	0.2	0.9	$\frac{1}{8}$	0.2	Yes

The numerical and analytical solutions are shown in Figure 4.4. We can see that neither numerical solutions of $\frac{\rho_1}{\rho_2} = \frac{1}{3}$ (Figure 4.4(a)) nor those of $\frac{\rho_1}{\rho_2} = 3$ (Figure 4.4(b)) are consistent with the analytical solutions. The incorrect numerical result in Figure 4.4(b) is similar to that obtained in Figure 9 in Rannou (2008). Hence, here we corroborate the simulations in Rannou (2008).

In Figure 4.4(a) and (b) we see that the numerical velocities jump near the interface vicinity and they are not continuous, as the analytical solutions are. The undesired term $u_x \partial_y(\rho_1)$, which is not negligible, contributes to this incorrect result.

Above we mentioned that using the source term, i.e., Eq. (4.47), the unwanted terms in Eq. (4.45) may be eliminated and the correct N–S equations for each component recovered. To show this point, Cases C and D (see Table. 4.2) were simulated and the results are shown in Figure 4.4(c) and (d). We see that the results agree well with the analytical solutions except for very small velocity jumps in the interfacial region. The errors between the LBM results and the analytical ones are 2.67% and 1.12% for Cases C and D, respectively. Hence, the unwanted terms play an important role for cases with different densities.

We further simulated cases with higher density contrast (i.e., Cases E to H). The parameters of Cases E to H are listed in Table 4.2. It is noted that in the simulations of Cases F and H the unwanted terms are eliminated. Here $\beta = 0.2$ is used because a larger β (e.g., $\beta = 0.3$) would lead to numerical instability (the effect of β on simulation is shown in Section 4.5). Comparisons between the cases with and without the added source term are shown in Figure 4.4(e)–(h). The results of Cases F and H are more consistent with the analytical solutions. They are much better than the results of the corresponding Cases E and G, which are simulated without adding the source term. The errors (Eq. (4.51)) in Cases F and H are 4.09% and 14.05%, respectively. Hence, if the unwanted terms can be canceled, the RK model is able to simulate the cases with density difference significantly better.

It is also noted that the error in cases of higher density contrast (e.g., 14.05% in Case H) is larger than those of lower density ratio cases (refer to Figure 4.4(c) and (d)). Using a larger computational domain is also helpful to decrease the numerical error. For Case H, when the width of the channel is represented by 200 lu, 300 lu, and 400 lu, the numerical errors are 9.14%, 7.03%, and 5.88%, respectively. Hence, the error decreases with grid refinement.

Smaller β, which means a thicker interface, may improve the accuracy of evaluation of the source term when the finite difference scheme is used. However, the numerical error E does not decrease when the interface becomes thicker. A possible reason is that physically a thicker interface means diffusion between the two components becomes stronger. In the RK model, which is applied to simulate immiscible two-phase flows, strong diffusion would induce a large numerical error. We conclude that to get accurate results a thinner interface (a larger β) is preferred.

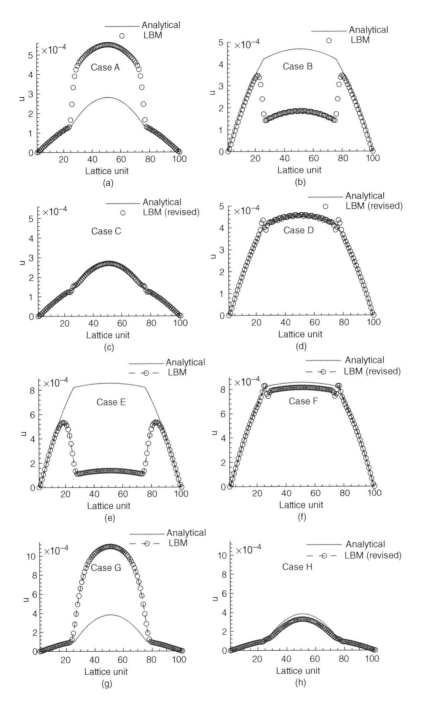

Figure 4.4 Comparison of the velocity profiles obtained from Cases A to H and the corresponding analytical solutions. The parameters of Cases A to H are illustrated in Table 4.2. 'LBM' and 'LBM (revised)' denote the original RK model and the RK model eliminating unwanted terms, respectively. Source: Huang et al. 2013. Reproduced with permission of World Scientific Publishing Company.

Table 4.3 Parameters for the RK model simulations of parallel flows with different densities ($A = 10^{-4}$, $\mathcal{G} = 1.5 \times 10^{-8}$ is applied to both fluids).

Case	α_1	α_2	$\frac{\rho_1}{\rho_2} = \frac{1-\alpha_2}{1-\alpha_1}$	β	Eliminate unwanted terms?	τ_1	τ_2
I	0.8	0.4	3	0.5	Yes	1.0	0.8
J	0.8	0.4	3	0.5	Yes	0.8	1.0

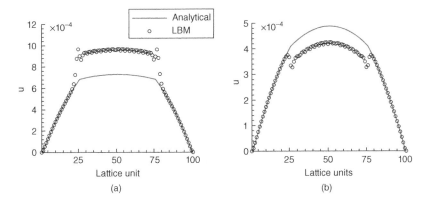

Figure 4.5 Comparison of the velocity profiles obtained from Cases I and J and the corresponding analytical solutions. (a) Case I: $\tau_1 = 1.0$, $\tau_2 = 0.8$; (b) Case J: $\tau_1 = 0.8$, $\tau_2 = 1.0$. In these simulations $\frac{\rho_1}{\rho_2} = 3$ and the unwanted terms are eliminated.

To evaluate the difference between the simulations with and without the source term a residual is defined in Huang et al. (2013) and confirmed our theoretical analysis. Refer to Huang et al. (2013) for more details.

To further test the performance of the revised RK model the τ effect is also evaluated for different density ratios. Table 4.3 shows the main parameters in Cases I and J. In these cases the density ratio is $\frac{\rho_1}{\rho_2} = \frac{1-\alpha_2}{1-\alpha_1} = \frac{0.6}{0.2} = 3$ and τ_1 and τ_2 are different. Figure 4.5 shows the results of the simulations. Although the unwanted terms are eliminated in the simulations, the LBM results have some discrepancy with the analytical solutions. In Figure 4.5 we can see that the interfacial region affects the numerical result significantly. Because the second collision (Eq. (4.14)) and the separation of the two fluids in the RK model (Eqs (4.17) and (4.18)) is too complex, detailed theoretical analysis is not available at present stage.

We conclude that the RK model of Latva-Kokko and Rothman (2005a) is valid for two-phase flows with identical density and different kinematic viscosities. Our revised RK model based on Grunau et al. (1993) and Reis and Phillips (2007) is valid for two-phase flows with different densities but identical kinematic viscosity. For simulations of two-phase flows with not only different densities but

Table 4.4 Interfacial tension as a function of A for different viscosity ratios.

A	τ_1	τ_2	$\frac{\sigma}{A}$
$10^{-6} \sim 10^{-2}$	1.0	1.0	2.69
$10^{-6} \sim 10^{-2}$	1.5	0.55	2.79
$10^{-6} \sim 10^{-2}$	1.5	0.502	2.74
$10^{-6} \sim 10^{-2}$	1.0	0.505	2.05
$10^{-6} \sim 10^{-2}$	1.0	0.501	1.96
$10^{-6} \sim 10^{-2}$	0.501	1.0	1.96

Source: Huang et al. 2014b. Reproduced with permission of Elsevier.

also different kinematic viscosities, the revised RK model (Huang et al. 2013) is still not satisfactory.

4.5 Interfacial tension and isotropy of the RK model

4.5.1 Interfacial tension
Huang et al. (2014b) showed the interfacial tension and isotropy of the RK model. Here it is introduced briefly. In this section cases of a droplet immersed in another fluid are simulated.

The interfacial tension σ as a function of A for the RK simulations with viscosity ratio $M = 1$ can be determined analytically (Reis and Phillips 2007). However, how to analytically determine the interfacial tension for $M \neq 1$ is an open question. Here σ is determined using the Laplace law along with numerical simulations of a droplet. σ as a function of A for different M is listed in Table 4.4. Here we can see that over a wide range if M is fixed, $\frac{\sigma}{A}$ is almost a constant, which means σ changes linearly with the parameter A.

4.5.2 Isotropy
In the study of Hou et al. (1997), the isotropy of the RK model was investigated. However, in their work the "recolor" step utilized an outdated approach (Gunstensen et al. 1991, Rothman and Keller 1988) and the parameters in their study (Hou et al. 1997) are fixed to a narrow range. The combination effect of the MRT and "recolor" step proposed by Latva-Kokko and Rothman (2005a) on isotropy is studied by Huang et al. (2014b). It is introduced here.

Figure 4.6 shows the isotropy, which is defined as $\varepsilon = \frac{r_{max} - r_{min}}{r_{min}}$, and the magnitude of the maximum spurious current $|\mathbf{u}|_{max}$ as functions of β when $A = 10^{-4}$. r_{max} and r_{min} are the maximum and minimum radii in eight directions, which are aligned with the vectors $\mathbf{e}_i, i = 1, \ldots, 8$. We can see that both ε and $|\mathbf{u}|_{max}$ increase with β. Hence, smaller β is helpful to achieve better isotropy of the simulated droplet. However, smaller β is not a good choice because when β is small, the interface becomes thick. For example, the interfacial thicknesses are approximately 7 lu, 5 lu, 4 lu, and 3 lu for $\beta = 0.3, 0.4, 0.5$, and 0.7, respectively. Thick

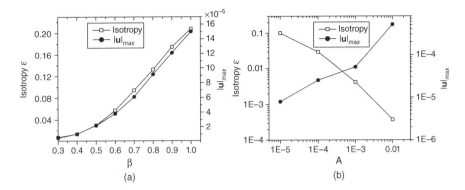

Figure 4.6 (a) Isotropy and the magnitude of maximum spurious current as functions of β with $A = 10^{-4}$. (b) Isotropy and the magnitude of maximum spurious current as functions of A with $\beta = 0.5$. Source: Huang et al. 2014b. Reproduced with permission of Elsevier.

interfaces are not desirable in simulations of two-phase flow in porous media. Usually $\beta = 0.5$ or 0.4 is used in our simulations. Note that β does not change the interfacial tension but affects the interface thickness, isotropy, and the magnitude of spurious current.

4.6 Drainage and capillary filling

In this section two dynamic multiphase flows are simulated. One case is the injection of a non-wetting gas (Fluid 1) into two parallel capillary tubes (Liu et al. 2013). The other is the capillary filling dynamic (Pooley et al. 2009), i.e., liquid (Fluid 2) filling a capillary tube that initially contains gas (Fluid 1). Here the component with smaller dynamic viscosity is regarded as a gas (Pooley et al. 2009).

For the first validation, as shown in Figure 4.7(a), the computational domain is 80×160. The upper and lower boundaries are the inlet and outlet boundaries, respectively. Initially Fluid 1 (non-wetting fluid) is put in the upper section of the domain and does not enter the tubes. The widths of the left and right tubes are $r_L = 24$ lu and $r_R = 32$ lu, respectively. The important parameters in our simulations are $A = 10^{-4}$, $\beta = 0.5$, $\tau_1 = 0.51$, and $\tau_2 = 1.5$. That means the kinematic viscosity ratio between the non-wetting and wetting fluid is $\frac{\tau_1 - 0.5}{\tau_2 - 0.5} = \frac{1}{100}$. According to Table 4.4, the interfacial tension in these simulations is $\sigma = 2.7 \times 10^{-4}$. In our simulations the equilibrium contact angle θ^{eq} is 45°. The contact angle is set through the scheme described in Section 4.8. Hence the corresponding capillary pressures for the left and right tubes are

$$P_{cL} = \frac{2\sigma \cos \theta^{eq}}{r_L} = 1.59 \times 10^{-5} \quad \text{and} \quad P_{cR} = \frac{2\sigma \cos \theta^{eq}}{r_R} = 1.19 \times 10^{-5}, \quad (4.53)$$

respectively.

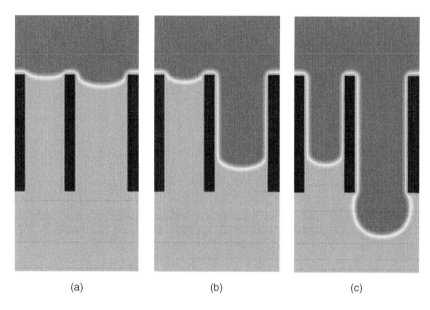

(a)   (b)   (c)

Figure 4.7 (Color online) Snapshots of the injection of a non-wetting fluid into two parallel capillary tubes with different ΔP, which is the pressure difference between the inlet and outlet. The widths of the left and right tubes are 24 lu and 32 lu, respectively. (a) $\Delta P = 1.0 \times 10^{-5} < P_{cR}$, $t = 4 \times 10^6$ ts, (b) $\Delta P = 1.4 \times 10^{-5}$, i.e., $P_{cR} < \Delta P < P_{cL}$, $t = 3.6 \times 10^6$ ts, and (c) $\Delta P = 3.3 \times 10^{-5} > P_{cL}$, $t = 10^6$ ts. The black, dark grey and light grey represent the solid, non-wetting fluid, and wetting fluid respectively, $\theta^{eq} = 45°$. Source: Huang et al. 2014b. Reproduced with permission of Elsevier.

From Figure 4.7(a) we can see that for the smallest pressure difference between the inlet and outlet, $\Delta P = 1.0 \times 10^{-5} < P_{cR}$, the non-wetting fluid (Fluid 1) is not able to enter into either tube. In Figure 4.7(b), when $P_{cR} < \Delta P < P_{cL}$, the non-wetting fluid enters into the wider tube (the right one) but it is unable to percolate into the narrower tube. In Figure 4.7(c), when $\Delta P = 3.3 \times 10^{-5}$, which is larger than P_{cL} and P_{cR}, the non-wetting fluid passes through both tubes. The result is consistent with the basic principle of pore-network simulations.

For the simulations of the capillary filling dynamic, the computational domain is 12×400. As shown in Figure 4.8(a), the capillary tube length is half of the computational height (200 lu), which ranges from $y = 100$ lu to $y = 300$ lu, and hence the left and right boundaries in the middle of the computational domain are solid nodes (black). The periodic boundary condition is applied on both the upper and lower boundaries, as well as the left and right boundaries without solid nodes. In the simulation $\tau_1 = 0.51$, $\tau_2 = 1.5$, the kinematic viscosities of the gas (Fluid 1, non-wetting component) and liquid (Fluid 2) are $\nu_1 = c_s^2(\tau_1 - 0.5)\Delta t = \frac{0.01}{3}$, and $\nu_2 = \frac{1}{3}$. The dynamic viscosity ratio is $\frac{\nu_2}{\nu_1} = 100$, which is sufficiently high that viscous dissipation in the gas phase can be ignored (Pooley et al. 2009). Ignoring the viscous dissipation is a basic

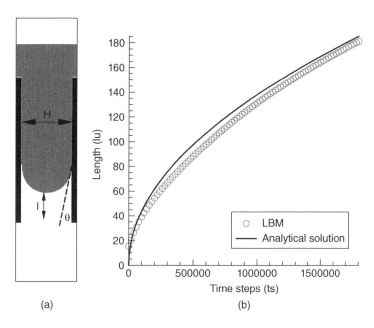

Figure 4.8 (a) Schematic diagram for liquid filling a capillary tube that initially contains gas. H is the width of the capillary tube and l is the length of the liquid column that has penetrated the capillary. The light grey and black are the non-wetting component (gas, Fluid 1) and the solid, respectively. (b) The length of the liquid column (lu) that has penetrated the capillary as a function of time. The initial penetration is about 15 lu at $t = 0$. Source: Huang et al. 2014b. Reproduced with permission of Elsevier.

assumption in the derivation of the following analytical solution (Eq. (4.54)). Here the gas (grey in the figure) is non-wetting and the equilibrium contact angle is set to be $\theta = 45°$. In the simulation $A = 10^{-3}$, $\sigma = 2.7 \times 10^{-3}$, and the width of the capillary tube $H = 10$ lu. The analytical solution for the length of tube filled with liquid l as a function of time is (Pooley et al. 2009)

$$l = \left(\frac{\sigma H cos\theta}{3\rho v_L} \right)^{\frac{1}{2}} \sqrt{t}, \qquad (4.54)$$

where $\rho = 1.0$. Initially the gas column ranges from $y = 115$ lu to $y = 315$ lu.

According to the parameters in our simulation the analytical solution is $l = \sqrt{0.01909}\sqrt{t}$. From Figure 4.8 it is seen that our numerical simulation is in reasonable agreement with the analytical solution.

4.7 MRT RK model

The difference between the MRT and BGK RK models is the collision term. The collision term $(\Omega_i)^1$ in Eq. (4.20) should be replaced by the MRT collision model

(Lallemand and Luo 2000), that is,

$$(\Omega_i)^1 = -M^{-1}\hat{\mathbf{S}}[|m(\mathbf{x},\mathbf{t})\rangle - |m^{(eq)}(\mathbf{x},\mathbf{t})\rangle], \qquad (4.55)$$

where the Dirac notation of ket $|\cdot\rangle$ vectors symbolize column vectors. The collision matrix $\hat{\mathbf{S}} = M \cdot S \cdot M^{-1}$ is diagonal with $\hat{\mathbf{S}} = diag(s_0, s_1, \dots, s_b)$ and $|m^{(eq)}\rangle$ is the equilibrium value of the moment $|m\rangle$. The matrix M illustrated in the appendix is a linear transformation which is used to map a vector $|f\rangle$ in discrete velocity space to a vector $|m\rangle$ in moment space, i.e., $|m\rangle = M \cdot |f\rangle$, $|f\rangle = M^{-1} \cdot |m\rangle$.

The momenta $j_\zeta = \rho u_\zeta$ are obtained from

$$j_\zeta = \sum_i f_i e_{i\zeta}, \qquad (4.56)$$

where ζ denotes x or y coordinates. The collision process is executed in moment space (Lallemand and Luo 2000). For the D2Q9 model

$$|m\rangle = (\rho, e, \epsilon, j_x, q_x, j_y, q_y, p_{xx}, p_{xy})^T, \qquad (4.57)$$

where e, ϵ, and q_ζ are the energy, the energy square, and the heat flux, respectively. The equilibrium value of the moment is

$$|m^{(eq)}\rangle = (\rho, e^{eq}, \epsilon^{eq}, j_x^{eq}, q_x^{eq}, j_y^{eq}, q_y^{eq}, p_{xx}^{eq}, p_{xy}^{eq})^T, \qquad (4.58)$$

where

$$e^{eq} = -2\rho + 3(j_x^2 + j_y^2)/\rho,$$

$$\epsilon^{eq} = \rho - 3(j_x^2 + j_y^2)/\rho,$$

$$q_x^{eq} = -j_x,$$

$$q_y^{eq} = -j_y,$$

$$p_{xx}^{eq} = (j_x^2 - j_y^2)/\rho,$$

$$p_{xy}^{eq} = j_x j_y/\rho. \qquad (4.59)$$

The diagonal collision matrix $\hat{\mathbf{S}}$ is (Lallemand and Luo 2000) $\hat{\mathbf{S}} \equiv diag(s_0, s_1, s_2, s_3, s_4, s_5, s_6, s_7, s_8)$. The parameters are chosen as $s_0 = s_3 = s_5 = 1.0$, $s_1 = 1.64$, $s_2 = 1.54$, $s_4 = s_6 = 1.9$, and $s_7 = s_8 = \frac{1}{\tau}$. In the following, one of the advantages of the MRT RK model will be shown: smaller spurious currents.

4.8 Contact angle

Here wetting phenomena are simulated using the MRT RK model. In our 2D simulations the computational domain is 200×100, the top and bottom of the domain are bounded by walls, and periodic boundary conditions are applied on the left and right boundaries. We note that both the densities of the majority and minority components should be specified at each lattice node inside the computational domain. In our simulations a circle with radius $r = 25$ lu is initialized just

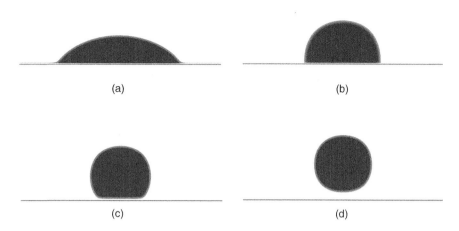

(a)   (b)

(c)   (d)

Figure 4.9 Equilibrium contact angles (measured from Fluid 1, dark grey one) for the present MRT RK model ($\tau_1 = 1$, $\tau_2 = 0.501$). (a) $\rho_{w1} = 0.7$, $\rho_{w2} = 0.0$, $\theta = 46.9°$, (b) $\rho_{w1} = \rho_{w2} = 0.5$, $\theta = 90°$, (c) $\rho_{w1} = 0.0$, $\rho_{w2} = 0.7$, $\theta = 136.7°$, and (d) $\rho_{w1} = 0.0$, $\rho_{w2} = 1.0$, $\theta = 180°$.

above the bottom wall as Fluid 1 ($\rho_1 = \rho_i$ and the minority component $\rho_2 = 0$) and the remaining area is initialized as Fluid 2 ($\rho_2 = \rho_i$ and $\rho_1 = 0$), where $\rho_i = 1.0$ is an initial density.

The contact angles obtained from the present RK model are illustrated in Figure 4.9. Through setting ρ_1 and ρ_2 values on the wall nodes, i.e., ρ_{w1} and ρ_{w2}, different contact angles can be obtained.

The contact angle θ can be analytically determined by Latva-Kokko and Rothman (2005a)

$$\theta = arccos\left(\frac{\rho_{w1} - \rho_{w2}}{\rho_i}\right),\tag{4.60}$$

where θ is measured from Fluid 1 and ρ_i is the initial density of the majority component. From the above equation we can see that the difference between ρ_{w1} and ρ_{w2} determines the contact angle.

In Figure 4.10 the simulated angles are compared with Eq. (4.60). Cases a, b, c, and d in Figure 4.9 are also labeled in this figure. From the figure we can see that for $M = 1, 25, \frac{1}{500}$, and 500, the contact angles obtained from the present MRT RK model agree well with those obtained from Eq. (4.60). Hence, the validity of Eq. (4.60) is independent of the viscosity contrast.

4.8.1 Spurious currents

The spurious currents in the simulations are investigated in order to compare the performance of the MRT and BGK models. Simulations of different contact angles with various viscosity ratios were carried out. The magnitudes of the spurious currents are shown in Table 4.5. We can see that when $M = 0.002$ or $M = 500$, the spurious current in the MRT simulations is approximately one order of magnitude less than that obtained from the BGK model. Hence, compared to the BGK RK model, the MRT RK model decreases the spurious current significantly at high viscosity contrast.

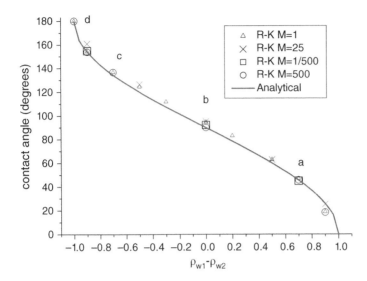

Figure 4.10 Contact angles obtained from the present MRT RK model (measured from Fluid 1) compared with the analytical solution. Viscosity ratios $M = 1$, $M = 25$, $M = \frac{1}{500}$, and $M = 500$ are simulated. Source: Huang et al. 2014b. Reproduced with permission of Elsevier.

Table 4.5 Spurious current in the MRT and BGK RK simulations (cases of contact angle with different viscosity ratios ($A = 10^{-4}, \beta = 0.5$).

BGK or MRT	τ_1	τ_2	$\rho_{w1} - \rho_{w2}$	$|\mathbf{u}|_{max}$
BGK	1.0	1.0	0.0	5.75×10^{-5}
MRT	1.0	1.0	0.0	5.46×10^{-5}
BGK	1.0	0.6	−0.7	8.15×10^{-5}
MRT	1.0	0.6	−0.7	6.77×10^{-5}
BGK	0.501	1.0	0.0	2.28×10^{-3}
MRT	0.501	1.0	0.0	9.97×10^{-5}
BGK	0.501	1.0	−0.9	1.43×10^{-3}
MRT	0.501	1.0	−0.9	7.03×10^{-5}
BGK	1.0	0.501	−1.0	9.49×10^{-4}
MRT	1.0	0.501	−1.0	1.85×10^{-4}
BGK	1.0	0.501	0.7	2.01×10^{-3}
MRT	1.0	0.501	0.7	2.09×10^{-4}
BGK	1.0	0.501	0.9	1.42×10^{-3}
MRT	1.0	0.501	0.9	1.75×10^{-4}

Source: Huang et al. 2014b. Reproduced with permission of Elsevier.

4.9   Tests of inlet/outlet boundary conditions

The inlet/outlet boundary conditions discussed in Section 3.7.1 are used in the following few cases. One of the advantages of the boundary conditions is that after the outlet pressure is specified, the outflow of each fluid is determined automatically. The outflow is determined only by the geometry of the porous media and the pressure difference between the inlet and outlet or by the inlet velocity boundary condition. That can mimic the two-phase displacement in porous media in experiments. In these experiments usually the inlet pressure or flux (velocity) of each phase is specified and the outlet is exposed to the atmosphere (a constant pressure).

Here, two simple cases are simulated. The computational domain is 100×100. The no-slip wall boundary is applied on the left and right boundaries (see Figure 4.11). On the top boundary, two short vertical plates (length 20 lu) are used to separate the two fluids (Fluid 2 is in the center region). Fluids 1 and 2 with an identical density are injected into the vertical channel. There is no gravity. The inlet velocities of the fluids are specified. As shown in Figure 4.11, at the ultimate equilibrium state the fluid with large flux would occupy a large fraction of the channel, which also means the corresponding saturation is higher. In Figure 4.11(a) the inlet velocities for Fluids 1 and 2 are 0.04 lu/ts and 0.01 lu/ts, respectively. As expected, the saturation of Fluid 1 is larger. The streamlines in the inlet converge first and diverge after the fluids leave the inlet section. The interface of the two fluids is tangent to the streamline. That also demonstrates

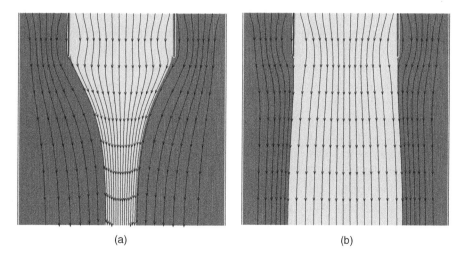

(a)	(b)

Figure 4.11   Equilibrium of the coflow. Fluid 1 (dark grey, non-wetting) and Fluid 2 (light grey, wetting) are injected from the upper boundary (two fluids have an identical kinematic viscosity because $\tau_1 = \tau_2 = 1$). (a) Inlet velocities for Fluids 1 and 2 are 0.04 and 0.01, respectively. (b) Inlet velocities for Fluids 1 and 2 are 0.01 and 0.03, respectively.

that the flow reaches the equilibrium state. Mass and momentum exchanges between the fluids reach an equilibrium state because no streamlines penetrate the interface. In Figure 4.11(b), it can be seen that the flow also reaches the equilibrium state. Because $u_{in} = 0.01$ for Fluid 1 is smaller than $u_{in} = 0.03$ for Fluid 2, the saturation of Fluid 1 is smaller than that in Figure 4.11(a).

4.10   Immiscible displacements in porous media

In the study of Lenormand et al. (1988), it was found that in porous media the capillary number and dynamic viscosity ratio are the two most important factors that affect the displacement behaviors. Generally speaking, there are three typical displacement patterns: capillary fingering, viscous fingering, and stable displacement (Lenormand et al. 1988). For stable displacement, the front between the two fluids is almost flat with some irregularities at the scale of a few pores. Very little of the displaced fluid is trapped behind the front. For viscous fingering, tree-like fingers spread across the whole porous medium and present no loops (Lenormand et al. 1988). For capillary fingering, "the fingers grow in all directions, even backward (toward the entrance)" (Lenormand et al. 1988). The displaced fluid may be trapped due to the loops developed by the fingers. The size of trapped clusters varies from pore size to the size of the whole porous medium (Lenormand et al. 1988).

In our simulations the porous medium is generated by placing circles with different radii in a 900×900 domain. Circles are not allowed to overlap and the minimum gap between any two circles is set to be 4 lu. The radii of the randomly distributed circles range from 24 lu to 5 lu. The porosity of the domain is 0.661 and the intrinsic permeability of the porous medium is 15.3 lu^2, which is determined by single-phase Lattice Boltzmann simulations. The left and right sides of the domain are periodic. In the displacement simulations, the domain was originally occupied by the wetting fluid (Fluid 2, which is white in the figures, $\rho_2 = 1.0$; the initial density of Fluid 1 is $\rho_1 = 10^{-8}$). The wetting fluid is displaced vertically by the non-wetting fluid (Fluid 1, light grey in the figure, contact angle 180° and $\rho_1 = 1.0$, $\rho_2 = 10^{-8}$) from top to bottom. In our study, the densities of the two fluids on the wall nodes are set to be $\rho_{w1} = 0.0$ and $\rho_{w2} = 1.0$.

Our simulations are run until "breakthrough" occurs, which means the injected fluid reaches the lower boundary. The non-wetting saturations (S_n) are measured when breakthrough occurs. From the above definitions of the three patterns we can see that S_n of stable displacement is the highest and S_n for viscous fingering is the lowest.

The capillary number is defined as

$$Ca = \frac{\rho_1 \nu_1 u_i}{\sigma},\qquad(4.61)$$

where ρ_1 is the initial density of Fluid 1 and u_i is the inlet velocity of Fluid 1.

Table 4.6 Displacement in porous medium cases with different viscosity ratios M and capillary numbers Ca.

Case	u_i	A	log (Ca)	τ_2	τ_1	log (M)	Saturation (S_n)
1	1.e-2	1.e-6	2.79	1	1	0	0.433
2	5.e-3	1.e-3	−0.51	1	1	0	0.451
3	5.e-4	5.e-3	−2.21	1	1	0	0.461
4	1.e-2	1.e-3	−1.92	1.5	0.51	−2	0.310
5	1.e-2	1.e-2	−2.92	1.5	0.51	−2	0.321
6	1.e-3	1.e-2	−2.92	1.5	0.6	−1	0.330
7	1.e-3	1.e-6	2.08	0.502	1.5	2.7	0.862
8	1.e-3	1.e-5	1.08	0.502	1.5	2.7	0.862
9	1.e-3	1.e-4	0.08	0.502	1.5	2.7	0.832
10	5.e-3	1.e-6	2.78	0.55	1.5	1.3	0.815
11	1.e-2	1.e-3	0.08	0.55	1.5	1.3	0.813
12	2.5e-3	1.e-3	−0.519	0.52	1.5	1.7	0.833
13	5.e-4	5.e-3	−1.92	0.502	1.5	2.7	0.721
14	5.e-4	5.e-3	−1.92	0.51	1.5	2	0.680
15	5.e-4	5.e-3	−1.92	0.6	1.5	1	0.677
16	6.e-4	2.e-2	−2.44	0.51	1.5	2	0.637
17	6.e-4	2.e-2	−2.44	0.8	1.5	0.5	0.563

Source: Huang et al. 2014b. Reproduced with permission of Elsevier.

We investigated the numerical stability of our simulations. It is found that some cases can be simulated by the MRT model but they become unstable if the BGK model is used, e.g., the case with $u_i = 0.01$, $A = 0.01$, log (Ca) = −2.92, $\tau_1 = 1.5$, and $\tau_2 = 0.51$. Hence, the MRT model has better numerical stability than the BGK model.

Seventeen cases with different viscosity ratios and capillary numbers were simulated and the parameters are shown in Table 4.6. The non-wetting saturations (S_n) are listed in the last column. Figure 4.12 shows flow pattern transitions from stable displacement to capillary fingering when M is fixed to be 500. When the capillary number is large (log(Ca) = 2.08, Case 7), the flow pattern is stable displacement and it has a large non-wetting saturation ($S_n \approx 0.83$). When log(Ca) = −1.92, the flow pattern becomes capillary fingering and $S_n = 0.721$ (Case 13). Figure 4.12(b) shows an intermediate case with non-wetting saturation about 0.78.

From Figure 4.13(a) we can see that for Case 2 with log $M = 0$ and an intermediate Ca, the displacement pattern is viscous fingering. When M is fixed and Ca increases to log (Ca) = 2.79 (Figure 4.13(b)), the flow pattern does not change. When both M and Ca decrease the flow pattern still remains unchanged but the fingers become thicker and S_n slightly decreases. Here we can see that in this flow pattern S_n approximately ranges from 0.31 to 0.46 in Figure 4.13 (a), (b), and (c).

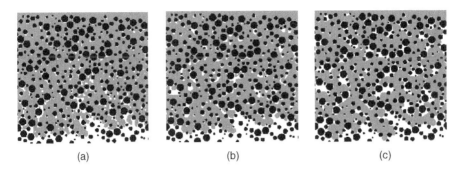

Figure 4.12 From stable displacement to capillary fingering ($M = 500$). (a) Case 7 (log (Ca) = 2.08), (b) a case of $u_{in} = 5. \times 10^{-4}$, $A = 10^{-3}$, log (Ca) = −1.217, $\tau_1 = 1.5$, $\tau_2 = 0.502$, and (c) Case 13 (log(Ca) = −1.92). The grey and black represent the non-wetting fluid (Fluid 1) and solid, respectively. The white area is occupied by wetting fluid (Fluid 2). Source: Huang et al. 2014b. Reproduced with permission of Elsevier.

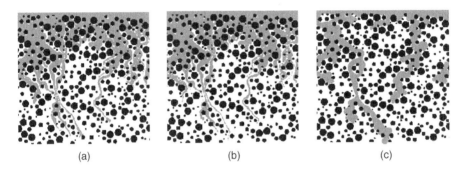

Figure 4.13 Viscous fingering obtained from typical combination of different M and Ca. (a) Case 2 (log $M = 0$, log (Ca) = −0.51), (b) Case 1 (log $M = 0$, log (Ca) = 2.79), (c) Case 6 (log $M = −1$, log(Ca) = −2.92). The grey and black represent the non-wetting fluid (Fluid 1) and solid, respectively. The white area is occupied by wetting fluid (Fluid 2). Source: Huang et al. 2014b. Reproduced with permission of Elsevier.

According to the flow patterns, the 17 cases illustrated in Table 4.6 are classified into three groups. The first six cases are viscous fingering, which approximately have a saturation $S_n \in [0.31, 0.46]$. The 7th to the 12th cases are stable displacement, which have the highest saturations ($S_n \approx 0.83$) among the three patterns. The last five cases are capillary fingering with $S_n \in [0.56, 0.72]$. From Table 4.6 we can see that the saturations in each group are slightly different. In the following discussion the saturation inside each group is regarded as "constant".

Figure 4.14 shows displacement pattern distributions in the M–Ca plane. The above three groups roughly form three domains and individual cases in each domain have similar flow patterns. The boundaries that separate the domains are approximately drawn and are represented by the thick gray dashed lines. Our phase diagram for the 900×900 porous medium is consistent with that obtained

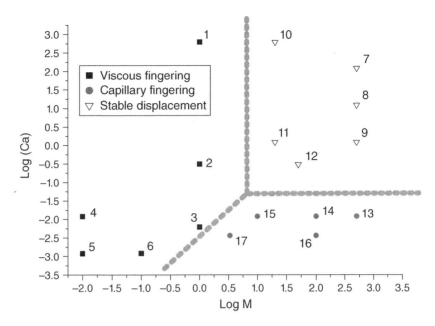

Figure 4.14 Plot of the constant saturation domains for the simulations of the 900 × 900 porous medium with different viscosity ratios and capillary numbers. The thick gray dashed lines represent the boundaries approximately drawn to separate the three domains. Source: Huang et al. 2014b. Reproduced with permission of Elsevier.

through experiments (Lenormand et al. 1988). The general shape of the boundaries of the three domains is almost the same as that illustrated in Lenormand et al. (1988). The difference between our results and those of Lenormand et al. (1988) is some translation of the boundaries of the domains. This translation might be attributed to differences in pore size distributions (Lenormand et al. 1988) and the overall size of the porous medium.

4.11   Appendix A

The transformation matrix M for 2D is (Lallemand and Luo 2000):

$$
\begin{pmatrix}
1 & 1 & 1 & 1 & 1 & 1 & 1 & 1 & 1 \\
-4 & -1 & -1 & -1 & -1 & 2 & 2 & 2 & 2 \\
4 & -2 & -2 & -2 & -2 & 1 & 1 & 1 & 1 \\
0 & 1 & 0 & -1 & 0 & 1 & -1 & -1 & 1 \\
0 & -2 & 0 & 2 & 0 & 1 & -1 & -1 & 1 \\
0 & 0 & 1 & 0 & -1 & 1 & 1 & -1 & -1 \\
0 & 0 & -2 & 0 & 2 & 1 & 1 & -1 & -1 \\
0 & 1 & -1 & 1 & -1 & 0 & 0 & 0 & 0 \\
0 & 0 & 0 & 0 & 0 & 1 & -1 & 1 & -1
\end{pmatrix}
$$

4.12 Appendix B

```
NOTES:
! 1. This programme is a 2D RK model code using D2Q9 model to
!    simulate layered two phase flow in a channel.
! 2. The whole project is composed of head.inc, main.for,
!    streamcollision.for, redistribute.for, and output.for.
! 3. head.inc is a header file, where some common blocks are defined.
! 4. Before running this project, please create
!    a new sub-folder 'out' under the working directory.
! 5. The output file "coflow_ana.dat", the first to fourth
!    columns denote the lattice nodes index, the analytical velocity
!    profile when a body force acts on both fluid,
!    on fluid 1, and on fluid 2,
!    respectively.

c FILE head.inc
c=================================================================
! Computational domain can be changed here.
       integer lx,ly
       PARAMETER(lx=2,ly=101)

       common/AA/ density,G,tau,ex(0:8),ey(0:8),rsq(0:8),Bi(0:8)
       real*8 density,G,tau,ex,ey, rsq, Bi

       common/b/ error,xc(0:8),yc(0:8),t_k(0:8)
       real*8 error,xc,yc,t_k

       common/rb/ rho_b(lx,ly), rho_r(lx,ly), rho_ri, rho_bi
       common/rb2/ f_r(0:8,lx,ly),f_b(0:8,lx,ly)
       real*8 rho_b, rho_r, f_r, f_b, rho_ri, rho_bi

       common/vel/ c_squ,cc,  beta, aa1, df, taur, taub
       real*8 c_squ,cc,  beta, aa1, df, taur, taub
       common/gg/   A, c_squ_b, c_squ_r, err_pre, u_ana(ly)
       real*8   A, c_squ_b, c_squ_r, u_ana, err_pre
       common/obs/ obst(lx,ly)
       integer obst

       common/oo/  Nwri, t_max, t_0, t_1, t_2,nu1, nu2, conv,ana_t
       integer Nwri, t_max
       real*8  t_0, t_1, t_2, nu1, nu2, conv, ana_t

       common/app/  alfa_r, alfa_b
       real*8   alfa_r, alfa_b


c FILE  main.for
c=================================================================
! Rothman-Keller Lattice Boltzmann model
! for layered two phase flow in a 2D channel.
      program Rothman_D2Q9
c --------------------------
c Author: Haibo Huang, huanghb@ustc.edu.cn
```

```
c -----------------------------
      implicit none
      include "head.inc"
      integer time,k, m, n
      integer*4 now(3)

      integer x,y
      real*8  u_x(lx,ly),u_y(lx,ly),rho(lx,ly)
      real*8  ff(0:8,lx,ly),Fx(lx,ly),Fy(lx,ly), p(lx,ly)

     data xc/0.d0,1.d0,0.d0, -1.d0, 0.d0, 1.d0, -1.d0, -1.d0, 1.d0/,
    &     yc/0.d0,0.d0,1.d0, 0.d0, -1.d0, 1.d0, 1.d0, -1.d0, -1.d0/,
    &     ex/0, 1, 0, -1,  0, 1, -1, -1, 1/,
    &     ey/0, 0, 1,  0, -1, 1,  1, -1, -1/

      character filename*16, B2*2, C*3, D*2

c------------------------------------------------------
! Lattice speed
      cc = 1.d0
! Square of sound speed
      c_squ = cc *cc / 3.d0

! Initialize the weighting coefficient in equilibrium
!  distribution function.
! B(k) is the parameters in the second collision term.
! rsq(k) is length of the velocities in the D2Q9 velocity model.
      t_0 =  4.d0 / 9.d0
      t_1 =  1.d0 / 9.d0
      t_2 =  1.d0 / 36.d0

      t_k(0) = t_0
      Bi(0) =-4.d0/27.d0
      rsq(0) = 0.d0
      do 1 k =1,4
      t_k(k) = t_1
      Bi(k) =2.d0/27.d0
      rsq(k) = 1.d0
  1 continue
      do 2 k =5,8
      t_k(k) = t_2
      Bi(k) = 5.d0/108.d0
      rsq(k) = dsqrt(2.d0)
  2 continue

! Dump data every 'Nwri' time steps.
      Nwri= 1000

      write (6,*)
      write (6,*) '@@@ Rothman D2Q9 starting ... @@@'
! Initialization, the parameters can be changed directly in the
!  subroutine
!  read_parameters().
      call read_parametrs()
      call read_obstacles()
      call init_density(u_x ,u_y ,rho , ff, Fx, Fy)
```

```
      call itime(now)
      write(*,"(i2.2, ':', i2.2, ':', i2.2)") now

! Get the analytical solution of the layered two phase flow
      call analytical_sol2()
      open(42, file='./out/residue.dat')

!     Begin iteration
      do 100 time = 0, t_max

      if ( mod(time, Nwri) .eq. 0) call relative_error
     & (rho, u_x, u_y, n)
c------------ Output
      if ( mod(time, 50*Nwri) .eq. 0) then
          write(*,*) time
          call itime(now)
c         write(*,"(i2.2, ':', i2.2, ':', i2.2)") now
             call write_results_ASCII(rho,u_x,u_y,time)
          call write_slice(rho, u_x, u_y, time)
          write(42, '(i8, 2f15.8)')
     &  time, error/ana_t, conv
          write(42,"(i2.2, ':', i2.2, ':', i2.2)") now
c-----------
      end if

! Streaming step (propagation)
          call stream(f_r )
          call stream(f_b )
! Get macrovariables
      call getuv(u_x ,u_y, rho, ff )

! Collision step
      call collision(u_x,u_y,rho ,ff, time )
! Redistribution step
      call redistribute(ff, rho, Fx, Fy)
! Handle the wall, bounce back for the wall nodes
      call bounceback(f_r)
      call bounceback(f_b)

  100 continue
!     End iteration
          write (6,*) '------------  end  ---------------'
      close(42)
      end

c
c-------------------------------------------------
      subroutine read_parametrs()
      implicit none
      include "head.inc"
      real*8 Re1, V1

! A critical parameter to adjust the thickness of the interface
      beta = 0.5d0
```

```
! The critical parameter to adjust the strength of the interface
!  tension
      A = 0.0001d0
      aa1 =  25.d0  ! The width of the inner fluid
! The body force applied to the fluid
      df = 0.000000015d0
! alfa_r and alfa_b are the parameters in the equilibrium
!  distribution function,
! and the values are relevant to the density of the fluids.
      alfa_r = 0.8  !4.d0/9.d0
      alfa_b = 0.4  !4.d0/9.d0
      rho_ri = 1.d0 - alfa_b
      rho_bi = 1.d0 - alfa_r

      c_squ_r = 3.d0*(1.d0-alfa_r)/5.d0
      c_squ_b = 3.d0*(1.d0-alfa_b)/5.d0
! Maximum time steps in the simulation.
      t_max  = 450000
! Our model may only work well for \tau_r=\tau_b
      taur = 1.0d0
      taub = 1.0d0

! Dump some important parameters in the simulation to a data file.
      open(41, file = './out/parameter.dat')
      write(41,'("alfar=",f12.5, 3X, "alfab=", f12.5)') alfa_r, alfa_b
      write(41,'("taur=",f12.5, 3X, "taub=", f12.5)') taur, taub
      write(41,'("rhor=",f12.5, 3X, "rhob=", f12.5)') rho_ri, rho_bi
      write(41,'("df=",f12.5, 3X, "aa1=", f12.5)') df, aa1
      write(41,'("A=",f12.8, 3X, "beta=", f12.5)') A, beta
      close(41)

      end

c-------------------------------------------------
! Initialize the node type: obst=0 is fluid node; obst=1 is
!  solid node
      subroutine read_obstacles()
      implicit none
      include "head.inc"
      integer  x,y

        do 10 y = 1, ly
          do 40 x = 1, lx
          obst(x,y) =  0
          if(y .eq. 1) obst(x,y) = 1
          if(y .eq. ly) obst(x,y) = 1
   40     continue
   10     continue
      end
c-------------------------------------------------

! Initialize the whole fluid field
      subroutine init_density(u_x,u_y,rho,ff, Fx, Fy)
      implicit none
      include "head.inc"
      integer i,j,x,y,k,n
```

```
      real*8  u_squ, xx, u_n(0:8),u_x(lx,ly),u_y(lx,ly),
     & rho(lx,ly),ff(0:8,lx,ly),ff1(0:8), f_r1(0:8),f_b1(0:8)
      real*8 Fx(lx,ly),Fy(lx,ly)

! Initialize the macroscopic variables in the flow field
      do 10 y = 1, ly
        do 10 x = 1, lx
         u_x(x,y) = 0.d0
         u_y(x,y) = 0.d0
         Fx(x,y) = 0.d0
         Fy(x,y) = 0.d0
         rho_r(x,y) = 0.d0
         rho_b(x,y) = rho_bi
  10 continue

      do 30 y = 1, ly
        do 20 x = 1, lx
! Initialize the fluid in the center of the channel
         if(abs(y-(1+ly)/2) .le. int(aa1)) then
         rho_r(x,y) = rho_ri
         rho_b(x,y) = 0.d0
         endif

      rho(x,y) = rho_r(x,y) +rho_b(x,y)

      call get_feq(rho_r(x,y),u_x(x,y),u_y(x,y),alfa_r, f_r1)
      call get_feq(rho_b(x,y),u_x(x,y),u_y(x,y),alfa_b, f_b1)

      do k = 0, 8
      f_r(k,x,y) = f_r1(k)
      f_b(k,x,y) = f_b1(k)
      ff(k,x,y)  = f_r1(k) + f_b1(k)
      enddo

c--------------------
! Here the 'light grey' and 'dark grey' fluids are set to
! be nonwetting and wettting fluid
      if(obst(x,y) .eq. 1) then
         rho_r(x,y) = 0.d0
         rho_b(x,y) = 1.d0
      endif
c-------------------

  20   continue
  30 continue
      return
      end


c FILE  streamcollision.for
c===================================================================
! Streaming step (propagation step in the LBM)
      subroutine stream(ff)
      implicit none
      include "head.inc"
      integer  k
```

```
      real*8 ff(0:8,lx,ly),f_hlp(0:8,lx,ly)
      integer  x,y,x_e,x_w,y_n,y_s,l,m,n

      do 10 y = 1, ly
         do 11 x = 1, lx
         y_n = mod(y,ly) + 1
         x_e = mod(x,lx) + 1
         y_s = ly - mod(ly + 1 - y, ly)
         x_w = lx - mod(lx + 1 - x, lx)
c
         f_hlp(1,x_e,y  ) = ff(1,x,y)
         f_hlp(2,x  ,y_n) = ff(2,x,y)
         f_hlp(3,x_w,y  ) = ff(3,x,y)
         f_hlp(4,x  ,y_s) = ff(4,x,y)
         f_hlp(5,x_e,y_n) = ff(5,x,y)
         f_hlp(6,x_w,y_n) = ff(6,x,y)
         f_hlp(7,x_w,y_s) = ff(7,x,y)
         f_hlp(8,x_e,y_s) = ff(8,x,y)
 11 continue
 10 continue
c
      do 20 y = 1, ly
         do 21 x = 1, lx
         do k = 1, 8
         ff(k,x,y)=f_hlp(k,x,y)
      enddo
 21 continue
 20 continue
      return
      end

c----------------------------------------------------
! Get macro variables
      subroutine getuv(u_x,u_y,rho,ff)
      include "head.inc"
      integer x,y
      real*8  u_x(lx,ly),u_y(lx,ly),rho(lx,ly),
     & ff(0:8,lx,ly)

      do 10 y = 1, ly
       do 20 x = 1, lx

      if(obst(x,y) .eq. 0) then

      rho_r(x,y) = f_r(0,x,y) +f_r(1,x,y) +f_r(2,x,y)
     &             + f_r(3,x,y) +f_r(4,x,y) +f_r(5,x,y)
     &             + f_r(6,x,y) +f_r(7,x,y) +f_r(8,x,y)

      rho_b(x,y) = f_b(0,x,y) +f_b(1,x,y) +f_b(2,x,y)
     &             + f_b(3,x,y) +f_b(4,x,y) +f_b(5,x,y)
     &             + f_b(6,x,y) +f_b(7,x,y) +f_b(8,x,y)

         rho(x,y) = rho_r(x,y) +rho_b(x,y)

      u_x(x,y) = (f_b(1,x,y) + f_b(5,x,y) + f_b(8,x,y)
     &           -(f_b(3,x,y) + f_b(6,x,y) + f_b(7,x,y))
```

```
      &                 + f_r(1,x,y) + f_r(5,x,y) + f_r(8,x,y)
      &                 -(f_r(3,x,y) + f_r(6,x,y) + f_r(7,x,y))   ) /rho(x,y)

        u_y(x,y) = (f_b(2,x,y) + f_b(5,x,y) + f_b(6,x,y)
      &                 -(f_b(4,x,y) + f_b(7,x,y) + f_b(8,x,y))
      &                 + f_r(2,x,y) + f_r(5,x,y) + f_r(6,x,y)
      &                 -(f_r(4,x,y) + f_r(7,x,y) + f_r(8,x,y))   ) /rho(x,y)

        else
! To set wetting boundary condition for the dark grey fluid
        rho_r(x,y) = 0.0d0
        rho_b(x,y) = rho_bi
        endif
  20    continue
  10    continue
        end


c-------------------------------------------------------
      subroutine collision(u_x,u_y,rho,ff, time)

      implicit none
      include "head.inc"
      integer  l, time
      real*8   u_x(lx,ly),u_y(lx,ly),ff(0:8,lx,ly),rho(lx,ly)
      integer  x,y,k,ip,jp
      real*8   u_n(0:8),fequ1(0:8),fequ2(0:8),u_squ,temp, invtau
      real*8 Fx(lx,ly), Fy(lx,ly), fm, cosfai, fei, invtaur, invtaub
      real*8 alfa1, beta1, kappa1, delta1, eta1, kxi1
      real*8 pr(ly),pb(ly), cr(ly), cb(ly), pu2(ly), nur(ly),nub(ly),
      & par(ly), pab(ly), par1(ly),  G0(ly)

! The following parameters are chosen to make the
! relaxation parameter \tau change smoothly across the interface.
      delta1 = 0.98d0
      alfa1 = 2.d0*taur *taub/(taur +taub)
      beta1 = 2.d0*(taur -alfa1)/delta1
      kappa1 =  -beta1/(2.d0* delta1)
      eta1 = 2.d0*(alfa1- taub)/delta1
      kxi1 =      eta1/(2.d0* delta1)

! Get the unwanted terms in the revised RK model.
      do y = 1, ly
      if(y.gt.1 .and. y.lt.ly) then
      pr(y) = 0.5d0 *(rho_r(1,y+1)-rho_r(1,y-1))
      pb(y) = 0.5d0 *(rho_b(1,y+1)-rho_b(1,y-1))
      endif
      enddo

      do 3 y = 1, ly
      if(y.gt.1 .and. y.lt.ly) then
      par(y) =   0.5d0 *(pr(y+1)*u_x(1,y+1)
      &  -pr(y-1)*u_x(1,y-1))
      pab(y) =   0.5d0 *(pb(y+1)*u_x(1,y+1)
      &  -pb(y-1)*u_x(1,y-1))
      endif
  3   continue
```

```
      do 5 y = 1, ly
       do 5 x = 1, lx

      if(obst(x,y) .eq. 0 )   then

c-------------------------
       call get_feq(rho_r(x,y), u_x(x,y), u_y(x,y), alfa_r,fequ1)
       call get_feq(rho_b(x,y), u_x(x,y), u_y(x,y), alfa_b,fequ2)

! Get the relaxation time as a function of position
       fei = (rho_r(x,y) - rho_b(x,y))/(rho_r(x,y) + rho_b(x,y))

      if(fei .gt. delta1) then
      tau = taur
      else if (fei .gt. 0.d0 .and. fei .le. delta1) then
      tau = alfa1+ beta1 *fei + kappa1* fei*fei
      else if (fei .le. 0.d0 .and. fei . ge. -delta1) then
      tau = alfa1+ eta1 *fei  + kxi1 * fei *fei
      else if (fei .lt. -delta1) then
      tau = taub
      endif

! Adding the body force into the LB simulation.
c      if( abs(y-(1+ly)/2) .le. aa1) then
      G0(y) = df
c      endif
c---------------------------
       cr(y) = (taur-0.5)*(1./3.-c_squ_r )*par(y)
       cb(y) = (taub-0.5)*(1./3.-c_squ_b )*pab(y)
! Collision step (the unwanted terms are eliminated)
      do 12 k = 0, 8
      f_r(k,x,y) = fequ1(k) + (1.d0-1.d0/tau)*( f_r(k,x,y) -fequ1(k) )
     &  - cr(y)* t_k(k)*xc(k)/c_squ     ! the unwanted term

      f_b(k,x,y) = fequ2(k) + (1.d0-1.d0/tau)*( f_b(k,x,y) -fequ2(k) )
     &  -  cb(y)* t_k(k)*xc(k)/c_squ     ! the unwanted term

! Body force is added as a source term
      ff(k,x,y) = f_r(k,x,y) + f_b(k,x,y) + t_k(k)* xc(k)*G0(y)/c_squ

  12   continue
c---------------------------

      endif
  5  continue

      end


c FILE  redistribute.for
c================================================================
! Get the equilibrium distribution functions for both
! the 'light grey' and 'dark grey' fluids
      subroutine get_feq(rho, ux, uy, alfa, fequ)
      include "head.inc"
```

```
      real*8 rho, ux, uy, fequ(0:8), u_n(0:8), tmp1, u_squ, alfa
      integer k
            u_squ = ux * ux + uy * uy

            u_n(0) =    0.d0
       do k = 1, 8
            u_n(k) =    xc(k)*ux +yc(k)*uy
            enddo
            tmp1 =  u_squ *1.5d0
! equilibrium distribution functions (refer to Eq. (4.9))
          fequ(0) = t_k(0)* rho * ( u_n(0) / c_squ
     &                + u_n(0) * u_n(0) *4.5d0
     &                - tmp1) + rho*alfa
       do k = 1, 4
           fequ(k) = t_k(k)* rho * ( u_n(k) *3.d0
     &                + u_n(k) * u_n(k) *4.5d0
     &                - tmp1) + rho*(1.d0-alfa)/5.d0
       enddo

       do k = 5, 8
           fequ(k) = t_k(k)* rho * ( u_n(k) *3.d0
     &                + u_n(k) * u_n(k) *4.5d0
     &                - tmp1) + rho*(1.d0-alfa)/20.d0
       enddo


       end
c-------------------------------------------------------
      subroutine redistribute(ff, rho, Fx, Fy)
      implicit none
      include 'head.inc'
      integer x,y, k
      real*8 ff(0:8,lx,ly), rho(lx,ly), feq(0:8), cosfai(8)
      real*8 Fx(lx,ly) , Fy(lx,ly), fm, temp
      integer  x_e,x_w,y_n,y_s

      do 100 y = 1, ly
       do 200 x = 1, lx

       if(obst(x,y) .eq. 0) then
          y_n = mod(y,ly) + 1
          x_e = mod(x,lx) + 1
          y_s = ly - mod(ly + 1 - y, ly)
          x_w = lx - mod(lx + 1 - x, lx)
! Calculate the color-gradient (a vector having x and y
!   components in 2D cases)
       Fx(x,y) = xc(1) *(rho_r(x_e,y  ) - rho_b(x_e,y  ))
     &      + xc(5) *(rho_r(x_e,y_n) - rho_b(x_e,y_n))
     &      + xc(8) *(rho_r(x_e,y_s) - rho_b(x_e,y_s))
     &      + xc(3) *(rho_r(x_w,y  ) - rho_b(x_w,y  ))
     &      + xc(6) *(rho_r(x_w,y_n) - rho_b(x_w,y_n))
     &      + xc(7) *(rho_r(x_w,y_s) - rho_b(x_w,y_s))

       Fy(x,y) = yc(2) *(rho_r(x  ,y_n) - rho_b(x  ,y_n))
     &      + yc(5) *(rho_r(x_e,y_n) - rho_b(x_e,y_n))
     &      + yc(6) *(rho_r(x_w,y_n) - rho_b(x_w,y_n))
     &      + yc(4) *(rho_r(x  ,y_s) - rho_b(x  ,y_s))
```

```
      &          + yc(7) *(rho_r(x_w,y_s) - rho_b(x_w,y_s))
      &          + yc(8) *(rho_r(x_e,y_s) - rho_b(x_e,y_s))

         fm = dsqrt(Fx(x,y)*Fx(x,y) + Fy(x,y)*Fy(x,y))

C---------------
         if(fm .lt. 0.00000001d0) then
! Specify how to handle cases of "denominator =0"

         do 10 k = 0, 8

         f_r(k,x,y) = rho_r(x,y)/rho(x,y)*ff(k,x,y)
         f_b(k,x,y) = rho_b(x,y)/rho(x,y)*ff(k,x,y)

   10 continue
C--------------
         else
C--------------
! The second collision term is added here
!  (Reis and Philips, 2007, J. Phys. A)
         k = 0
         ff(k,x,y) = ff(k,x,y) + A * fm *(  - Bi(k) )

          do 12 k = 1, 8
            cosfai(k) = ( xc(k) *Fx(x,y) + yc(k) *Fy(x,y) )/rsq(k)/fm

            ff(k,x,y) = ff(k,x,y)
      &       + A * fm *( t_k(k) *cosfai(k)*cosfai(k)*rsq(k)*rsq(k)- Bi(k))
   12       continue

         temp = rho_b(x,y)*rho_r(x,y)/rho(x,y)/rho(x,y)

! Separation of the distributions of the two fluids.
         f_r(0,x,y) = rho_r(x,y)/rho(x,y)*ff(0,x,y)
         f_b(0,x,y) = rho_b(x,y)/rho(x,y)*ff(0,x,y)

         do 20 k = 1, 8

         feq(k) = t_k(k) *rho(x,y)

         f_r(k,x,y) = rho_r(x,y)/rho(x,y)*ff(k,x,y)
      &          + beta* temp * feq(k) *cosfai(k)

         f_b(k,x,y) = rho_b(x,y)/rho(x,y)*ff(k,x,y)
      &          - beta* temp * feq(k) *cosfai(k)
   20 continue

      endif
C-----------------
      endif
  200    continue
  100    continue

      end

C----------------------------------------------
```

```
! Simple bounce back is used to handle the no-slip wall boundary
!  condition
      subroutine bounceback(ff)
      implicit none
      include 'head.inc'
      integer x, y
      real*8 temp, ff(0:8,lx,ly)

      do x = 1, lx
      do y = 1, ly
       if(obst(x,y) .eq. 1) then
          temp    = ff(1,x,y)
          ff(1,x,y) = ff(3,x,y)
          ff(3,x,y) = temp

          temp    = ff(2,x,y)
          ff(2,x,y) = ff(4,x,y)
          ff(4,x,y) = temp

          temp    = ff(5,x,y )
          ff(5,x,y) = ff(7,x,y)
          ff(7,x,y) = temp

          temp    = ff(6,x,y)
          ff(6,x,y) = ff(8,x,y)
          ff(8,x,y) = temp
        endif
       enddo
      enddo
c------------------------
      end


c FILE  output.for
c=================================================================
! Analytical solution
      subroutine analytical_sol2()
      include "head.inc"
      integer y, ycen
      real*8   u(ly), L, rho1, rho2, A1, A2, B2, C1, C2,
     &   u2g20(ly), u2g10(ly),  Mo,  u1g20(ly), u1g10(ly)
     &     , u_1(ly), u_2(ly), u_ho(ly)

      open(48, file ="./out/coflow_ana.dat")
      write(48,'("_y,____u,_____g20,___g10_")')

      nu1 = 1.d0/3.d0*(taur-0.5)*rho_ri
      nu2 = 1.d0/3.d0*(taub-0.5)*rho_bi ! Fluid near to wall
      L = dble(ly-2)/2.d0
      ycen = (ly+1)/2

      rho1 = rho_ri
      rho2 = rho_bi  ! Fluid near to wall
      Mo = nu1/nu2   ! Dynamic viscosity ratio

    ! Please refer to the analytical solution expression.
```

```
      A1 = -df/2.d0/nu1
      A2 = -df/2.d0/nu2

      B2 = -2.d0*A2*aa1 +2.d0*Mo*A1*aa1

      C1 = (A2-A1)*aa1*aa1- B2*(L-aa1) - A2*L*L
      C2 = -A2*L*L- B2*L
C----------------------------
      do y = 1, ly
        y1 = dabs(dble(y-ycen))

        if(abs(y-ycen) .le. int(aa1)) then
        u1g20(y) = A1*y1*y1 -A1*aa1*aa1 -2.d0*Mo*A1*aa1*(L-aa1)
        u1g10(y) = A2*aa1*aa1+ 2.d0*A2*aa1*(L-aa1) -A2*L*L
        u(y) = -A2 *( L* L - aa1 *aa1) +
     &         -A1 *(aa1 *aa1- y1*y1 )
        else
        u(y) = -A2 *(L *L- y1*y1 )

        u2g20(y) = 2.d0* A1*aa1*(y1-L)*Mo
        u2g10(y) = A2*y1*y1 - 2.d0* A2*aa1*y1- A2*L*L+ 2.d0*A2*aa1*L
        endif
C  -----------------------
        if(y1 .le. aa1) then
        u_1(y) = A1*y1*y1 +C1
        else
        u_2(y) = A2*y1*y1 + B2*y1 + C2
        endif
! 'u_ho' means the analytical velocity profile
! when the body force 'df' is applied to both fluid.
        u_ho(y) = u_1(y) + u_2(y)
C----------------------------
        u_ana(y) = u_ho(y) !Alternatively it can be written
!        as 'u2g10(y)+u1g10(y)'

! 'u1g20(y)+u2g20(y)' denotes the analytical velocity profile
! when the body force 'df' is only applied to fluid 1
! (light grey fluid in center of the channel).

! 'u1g10(y)+u2g10(y)' denotes the analytical velocity profile
! when the body force 'df' is only applied to fluid 2
! (dark grey fluid near to the wall).
        write(48,'(i5, 3f15.7)') y, u_ho(y), u1g20(y)+u2g20(y),
     *     u2g10(y)+u1g10(y)
      enddo

      close(48)
      return
      end

C-----------------------------------------------------
! Calculate the relative error to justify whether the
! convergence criterion is reached.
      subroutine  relative_error(rho, u_x, u_y, n)
      implicit none
      include "head.inc"
```

```
      integer  x,y,i,j,n,obsval,k1,k
      real*8  rho1,rho(lx,ly), u_x(lx,ly), u_y(lx,ly)
     & , p(lx,ly), rho10, rho_u, rho_d, Q_h, Q_l
      character filename*29,  C*7

      ana_t = 0.d0
      error = 0.d0
C-------------------------
      x = 1
      do 10 y = 2, ly-1
C----------------------
      ana_t =  ana_t +dabs( u_ana(y) )
      error = error + dabs(u_ana(y) - u_x(x,y))
! u1g20(y)+u2g20(y)   ! u(y)
C-------------------
   10 continue

      conv = (error- err_pre)/err_pre

      if(n.gt. 10000 .and. dabs(conv) .lt. 0.00001d0) stop
      err_pre = error
      end


C---------------------------------------------------
! Dump the macro-variables on a slice
! The velocity profile can be compared with the data that
!  dumped by subroutine
! 'analytical_sol2()'
      subroutine  write_slice(rho, u_x, u_y, n)
      implicit none
      include "head.inc"
      integer  x,y,i,j,n,obsval,k1,k
      real*8  rho1,rho(lx,ly), u_x(lx,ly), u_y(lx,ly)
     & , p(lx,ly), rho10, rho_u, rho_d, Q_h, Q_l
      character filename*29,  C*9

C-------------------------
      write(C,'(i9.9)') n
      filename='./out/slice_'//C//'.plt'

        open(41,file=filename)

        write(41,*) '_y,_u,_v,_rho1,_rho2'

      x = 1
      do 10 y = 1, ly
         write(41,'(i5, 4f15.8 )') y, u_x(x,y), u_y(x,y),
     &    rho_r(x,y), rho_b(x,y)
   10 continue

      close(41)
      end


C---------------------------------------------------
! Dump the tecplot file
      subroutine  write_results_ASCII(rho, u_x, u_y, n)
```

```fortran
      implicit none
      include "head.inc"
      integer  x,y,i,j,n,obsval,k1,k
      real*8  rho1,rho(lx,ly), u_x(lx,ly), u_y(lx,ly)
     & , p(lx,ly), rho10, rho_u, rho_d, Q_h, Q_l
      character filename*20,  C*6

      write(C,'(i6.6)') n
      filename='./out/2D_'//C//'.plt'

      open(41,file=filename)

      write(41,*) 'variables = x, y, u, v, rho, rho1, rho2,  p, obst'
      write(41,*) 'zone i=', lx, ', j=', ly, ', f=point'

      do 10 y = 1, ly
      do 10 x = 1, lx
       p(x,y) = 1.d0/3.d0 *rho(x,y)
          write(41,'(i4,i5, 6f15.8, i4 )') x, y, u_x(x,y), u_y(x,y),
     &    rho(x,y), rho_r(x,y), rho_b(x,y), p(x,y), obst(x,y)
   10 continue
      close(41)
c----------------------------------------------------
      end
```

CHAPTER 5

Free-energy-based multiphase Lattice Boltzmann model

5.1 Swift free-energy based single-component multiphase LBM

The original Swift free-energy-based SCMP LBM was described in Swift et al. (1996). Here detailed derivations are given to make its development easy to follow. In the original paper (Swift et al. 1996) the D2Q7 model (Chapter 1) is used. Here we give the derivation according to the D2Q7 and D2Q9 velocity models. This model is developed from the free-energy model, which is Galilean invariant when both fluids are ideal gases (He and Doolen 2002). Holdych et al. (1998) added density gradient terms to reduce the lack of Galilean invariance to order $O(u^2)$ when non-ideal fluids are considered. One of the advantages of this model over the SC model is that the surface tension is much more easily adjusted. The model has been applied to simulate bubble rise (Frank et al. 2005; Takada et al. 2000), droplet formation and movement in microchannels (Hao and Cheng 2009; Kusumaatmaja et al. 2006, 2008; Van der Graaf et al. 2006; Van der Sman and Van der Graaf 2008; Varnik et al. 2008; Vrancken et al. 2009; Zhang 2011), spinodal decomposition (Gonnella et al. 1997; Kendon et al. 2001; Sofonea and Mecke 1999), phase separation (Suppa et al. 2002; Wagner and Yeomans 1999, 1998; Xu et al. 2003, 2004), migration of a bubble (Holdych et al. 2001), vapor-liquid flow in porous media (Angelopoulos et al. 1998, Hao and Cheng 2010), micron-scale droplets and jetting (Leopoldes et al. 2003), surfactant adsorption into interfaces (Van der Sman and Van der Graaf 2006), thermal two-phase flow (Palmer and Rector 2000), colloid suspension (Stratford et al. 2005), supercritical CO_2 injection into porous media containing water (Suekane et al. 2005), components mixing in microchannels (Kuksenok et al. 2002), etc. It has also been applied to the study of more complex flow systems, such as liquid crystal hydrodynamics (Denniston et al. 2001) and ternary fluid mixtures

Multiphase Lattice Boltzmann Methods: Theory and Application, First Edition.
Haibo Huang, Michael C. Sukop and Xi-Yun Lu.
© 2015 John Wiley & Sons, Ltd. Published 2015 by John Wiley & Sons, Ltd.
Companion Website: www.wiley.com/go/huang/boltzmann

(Lamura et al. 1999). In the appendix to this chapter code implementing this model in the D2Q9 framework is provided.

Zheng et al. (2006) proposed a Galilean-invariant free-energy-based LB model. This model is simpler but only valid for density-matched cases (Fakhari and Rahimian 2010). Niu et al. (2007) applied the model to investigate water-gas transport processes in the gas-diffusion layer of a proton exchange membrane (PEM) fuel cell. Recently, to handle high-density contrast cases, Shao et al. (2014) revised the model of Zheng et al. (2006) through considering the effect of local density variation in the LBE as a forcing term. Based on the free-energy model of Swift et al. (1996), Tiribocchi et al. (2009) also proposed a hybrid scheme for binary fluid mixtures, i.e., a Lattice Boltzmann scheme with a forcing term for solving the N–S equation and a finite-difference scheme for the Cahn–Hilliard equation.

The Lattice Boltzmann equation in Swift's SCMP model is the same as used for other models:

$$f_i(\mathbf{x} + \mathbf{e}_i \Delta t, t + \Delta t) - f_i(\mathbf{x}, t) = -\frac{1}{\tau}(f_i - f_i^{(eq)}), \tag{5.1}$$

where Δt is the time step and τ the relaxation parameter. $f_i^{(eq)}$ is an equilibrium distribution function. As usual, the density of the fluid and the fluid momentum can be calculated through

$$\rho = \sum_i f_i \quad \text{and} \quad \rho u_\alpha = \sum_i f_i e_{i\alpha}. \tag{5.2}$$

The thermodynamic aspects of this model take effect through the pressure tensor $P_{\alpha\beta}$ (Swift et al. 1996). In Swift et al. (1996) the typical non-ideal system represented by the van der Waals equation of state is chosen. The corresponding free energy of the system is (Cahn and Hilliard 1958, Rowlinson and Widom 1982)

$$\Psi = \int d\mathbf{r} \left(\psi(T, \rho) + \frac{\kappa}{2}(\nabla\rho)^2 \right), \tag{5.3}$$

where $\mathbf{r}$ represents a tiny volume fraction, k is a constant, and $\psi(T, \rho)$ is the bulk free-energy density at a temperature T. For the van der Waals system the bulk free-energy density is

$$\psi(T, \rho) = \rho T ln \left(\frac{\rho}{1 - \rho b} \right) - a\rho^2. \tag{5.4}$$

The pressure tensor is related to the free energy in this way (Swift et al. 1996)

$$P_{\alpha\beta} = p\delta_{\alpha\beta} + \kappa \frac{\partial\rho}{\partial x_\alpha} \frac{\partial\rho}{\partial x_\beta}, \tag{5.5}$$

with

$$p = p_0 - \kappa\rho\nabla^2\rho - \frac{\kappa}{2}|\nabla\rho|^2, \tag{5.6}$$

where

$$p_0 = \rho \frac{\partial \psi(\rho)}{\partial \rho} - \psi(\rho) = \frac{\rho T}{1 - \rho b} - a\rho^2 \qquad (5.7)$$

is the EOS of the van der Waals fluid. The van der Waals parameters a and b are the same as described in Chapter 2.

For single-phase flow, the equilibrium distribution function is

$$f_i^{eq}(\mathbf{x}, t) = w_i \rho \left[1 + \frac{e_{i\beta} u_\beta}{c_s^2} + \frac{e_{i\beta} u_\beta e_{i\alpha} u_\alpha}{2c_s^4} - \frac{u_\beta u_\beta}{2c_s^2} \right], \qquad (5.8)$$

however, to incorporate the thermodynamic aspects mentioned above a more general equilibrium distribution function is suggested (Swift et al. 1996),

$$f_i^{eq} = A + B u_\beta e_{i\beta} + C u^2 + D u_\alpha u_\beta e_{i\alpha} e_{i\beta} + G_{\alpha\beta} e_{i\alpha} e_{i\beta}, \qquad (5.9)$$

for the velocity components $\mathbf{e}_i$ $(i \neq 0)$ and

$$f_0^{eq} = A_0 + C_0 u^2 \qquad (5.10)$$

for the rest component $\mathbf{e}_0 = (0, 0)$. The coefficients A, B, ... can be determined through the procedure given in the following section.

As usual the zeroth, first, and second moments of f_i^{eq} should satisfy

$$\sum_i f_i^{eq} = \rho, \qquad (5.11)$$

$$\sum_i f_i^{eq} e_{i\alpha} = \rho u_\alpha, \qquad (5.12)$$

$$\sum_i f_i^{eq} e_{i\alpha} e_{i\beta} = P_{\alpha\beta} + \rho u_\alpha u_\beta. \qquad (5.13)$$

5.1.1 Derivation of the coefficients in the equilibrium distribution function

The derivations for the coefficients in the equilibrium distribution functions of the D2Q7 and D2Q9 models are shown in the following subsections.

D2Q7 model

In the D2Q7 model (Chapter 1, Figure 1.1(b)) the non-zero velocity components are $\mathbf{e}_1 = (1, 0)c$, $\mathbf{e}_2 = (\frac{1}{2}, \frac{\sqrt{3}}{2})c$, $\mathbf{e}_3 = (-\frac{1}{2}, \frac{\sqrt{3}}{2})c$, $\mathbf{e}_4 = (-1, 0)c$, $\mathbf{e}_5 = (-\frac{1}{2}, -\frac{\sqrt{3}}{2})c$, and $\mathbf{e}_6 = (\frac{1}{2}, -\frac{\sqrt{3}}{2})c$.

According to the definition of the D2Q7 velocity model the following lattice-tensor formulae can be obtained:

$$\sum_i e_{i\alpha} = 0, \quad \sum_i e_{i\alpha} e_{i\beta} = 3c^2 \delta_{\alpha\beta},$$

$$\sum_i e_{i\alpha} e_{i\beta} e_{i\gamma} = 0, \quad \text{and}$$

$$\sum_i e_{i\alpha} e_{i\beta} e_{i\gamma} e_{i\delta} = \frac{3c^2}{4}(\delta_{\alpha\beta}\delta_{\gamma\delta} + \delta_{\gamma\beta}\delta_{\alpha\delta} + \delta_{\delta\beta}\delta_{\gamma\alpha}). \tag{5.14}$$

For the D2Q7 velocity model, from the constraint Eq. (5.11), we have

$$\sum_i f_i^{eq} = A_0 + C_0 u^2 + 6(A + Cu^2) + (Du_\alpha u_\beta + G_{\alpha\beta})(3c^2\delta_{\alpha\beta}))$$

$$= (A_0 + 6A) + (C_0 + 6C + 3c^2 D)u^2 + 3c^2 G_{\alpha\alpha} = \rho. \tag{5.15}$$

Because there is only ρ and no terms involving u^2 or c^2 on the right-hand side of Eq. (5.15), the left-hand side cannot depend on those variables either and hence

$$\begin{cases} A_0 + 6A = \rho, \\ C_0 + 6C + 3c^2 D = 0, \quad \text{and} \\ G_{\alpha\alpha} = 0. \end{cases} \tag{5.16}$$

From the constraint Eq. (5.12) we have

$$\sum_i f_i^{eq} e_{i\alpha} = \sum_i B u_\beta e_{i\beta} e_{i\alpha} = 3c^2 B u_\alpha = \rho u_\alpha, \tag{5.17}$$

or simply

$$3c^2 B = \rho. \tag{5.18}$$

From the constraint Eq. (5.13) we have

$$\sum_i f_i^{eq} e_{i\alpha} e_{i\beta} = (A + Cu^2) \sum_i e_{i\alpha} e_{i\beta} + (Du_\gamma u_\delta + G_{\gamma\delta}) \sum_i e_{i\alpha} e_{i\beta} e_{i\gamma} e_{i\delta}$$

$$= 3c^2\delta_{\alpha\beta}(A + Cu^2) + \frac{3c^4}{4}(Du_\gamma u_\delta + G_{\gamma\delta})(\delta_{\alpha\beta}\delta_{\gamma\delta} + \delta_{\gamma\beta}\delta_{\alpha\delta} + \delta_{\delta\beta}\delta_{\gamma\alpha})$$

$$= 3c^2\delta_{\alpha\beta}(A + Cu^2) + \frac{3c^4}{4}(Du^2\delta_{\alpha\beta} + 2Du_\alpha u_\beta + G_{\gamma\gamma}\delta_{\alpha\beta} + 2G_{\alpha\beta})$$

$$= 3c^2 A\delta_{\alpha\beta} + 3c^2(C + \frac{c^2}{4}D)u^2\delta_{\alpha\beta} + \frac{3c^4}{2}Du_\alpha u_\beta + \frac{3c^2}{4}(G_{\gamma\gamma}\delta_{\alpha\beta} + 2G_{\alpha\beta})$$

$$= \rho u_\alpha u_\beta + (p_0 - \kappa\rho\nabla^2\rho - \frac{\kappa}{2}|\nabla\rho|^2)\delta_{\alpha\beta} + \kappa\frac{\partial\rho}{\partial x_\alpha}\frac{\partial\rho}{\partial x_\beta}. \tag{5.19}$$

We can make any of the following choices to satisfy the above constraint (Eq. (5.19)) (Briant et al. 2004):

$$\begin{cases} 3c^2 A = p_0 - \kappa\rho\nabla^2\rho, \\ C + \frac{c^2}{4}D = 0, \\ \frac{3c^4}{2}D = \rho, \\ \frac{3c^4}{4}(G_{\gamma\gamma}\delta_{\alpha\beta} + 2G_{\alpha\beta}) = -\frac{k}{2}|\nabla\rho|^2\delta_{\alpha\beta} + \kappa\frac{\partial\rho}{\partial x_\alpha}\frac{\partial\rho}{\partial x_\beta}. \end{cases} \tag{5.20}$$

The last choice in Eq. (5.20) is consistent with the formula $G_{\alpha\alpha} = 0$ because when $\alpha = \beta = x$ we have

$$\frac{3c^2}{4}(G_{\gamma\gamma}\delta_{\alpha\beta} + 2G_{\alpha\beta}) = -\frac{\kappa}{2}((\partial_x\rho)^2 + (\partial_y\rho)^2) + \kappa(\partial_x\rho)^2 = \frac{\kappa}{2}((\partial_x\rho)^2 - (\partial_y\rho)^2), \quad (5.21)$$

and when $\alpha = \beta = y$ we have

$$\frac{3c^2}{4}(G_{\gamma\gamma}\delta_{\alpha\beta} + 2G_{\alpha\beta}) = -\frac{\kappa}{2}((\partial_x\rho)^2 + (\partial_y\rho)^2) + \kappa(\partial_y\rho)^2 = \frac{\kappa}{2}((\partial_y\rho)^2 - (\partial_x\rho)^2), \quad (5.22)$$

so that $G_{\alpha\alpha} = G_{xx} + G_{yy} = 0$.

In Eqs (5.16), (5.18), and (5.20) there are seven coefficients (A, B, C, D, A_0, C_0, $G_{\alpha\beta}$) and seven effective equations (the last equation in Eq. (5.16) and the last one in Eq. (5.20) are consistent). The coefficients can therefore be solved for. The solution is

$$A_0 = n - 6A = n - \frac{2}{c^2}(p_0 - \kappa\rho\nabla^2\rho), \quad A = \frac{1}{3c^2}(p_0 - \kappa\rho\nabla^2\rho),$$

$$B = \frac{\rho}{3c^2}, \quad C_0 = -6C - 3c^2D = -\frac{\rho}{c^2}, \quad C = -\frac{c^2}{4}D = -\frac{\rho}{6c^2},$$

$$D = \frac{2\rho}{3c^4}, \quad G_{\alpha\beta} = \frac{2}{3c^4}\left(-\frac{k}{2}|\nabla\rho|^2\delta_{\alpha\beta} + \kappa\frac{\partial\rho}{\partial x_\alpha}\frac{\partial\rho}{\partial x_\beta}\right) \quad (5.23)$$

$G_{\alpha\beta}$ can be written in terms of the following components, which are convenient for inclusion in LBM code:

$$G_{xx} = \frac{k}{3c^4}\left[\left(\frac{\partial\rho}{\partial x}\right)^2 - \left(\frac{\partial\rho}{\partial y}\right)^2\right],$$

$$G_{yy} = \frac{k}{3c^4}\left[\left(\frac{\partial\rho}{\partial y}\right)^2 - \left(\frac{\partial\rho}{\partial x}\right)^2\right],$$

$$G_{xy} = \frac{2k}{3c^4}\frac{\partial\rho}{\partial x}\frac{\partial\rho}{\partial y}. \quad (5.24)$$

Because the D2Q9 velocity model is more commonly used, in the following section we also derive the coefficients for the D2Q9 model.

D2Q9 model

According to the definition of the D2Q9 velocity model, the following formulae can be obtained:

$$\sum_i e_{i\alpha} = 0, \quad \sum_{i=1}^{4} e_{i\alpha}e_{i\beta} = 2c^2\delta_{\alpha\beta}, \quad \sum_{i=5}^{8} e_{i\alpha}e_{i\beta} = 4c^2\delta_{\alpha\beta},$$

$$\sum_i e_{i\alpha}e_{i\beta}e_{i\gamma} = 0, \quad \sum_{i=1}^{4} e_{i\alpha}e_{i\beta}e_{i\gamma}e_{i\delta} = 2c^4\delta_{\alpha\beta\gamma\delta},$$

$$\sum_{i=5}^{8} e_{i\alpha}e_{i\beta}e_{i\gamma}e_{i\delta} = 4c^4(\delta_{\alpha\beta}\delta_{\gamma\delta} + \delta_{\gamma\beta}\delta_{\alpha\delta} + \delta_{\delta\beta}\delta_{\gamma\alpha}) - 8c^4\delta_{\alpha\beta\gamma\delta}. \quad (5.25)$$

Here we adopt the assumption that due to symmetry each of the f_i^{eq} ($i = 1, 2, 3, 4$) have identical form, as follows:

$$f_i^{eq} = A_1 + B_1 u_\beta e_{i\beta} + C_1 u^2 + D_1 u_\alpha u_\beta e_{i\alpha} e_{i\beta} + G_{\alpha\beta1} e_{i\alpha} e_{i\beta}. \tag{5.26}$$

It is also reasonable to assume that f_i^{eq} ($i = 5, 6, 7, 8$) have identical form:

$$f_i^{eq} = A_2 + B_2 u_\beta e_{i\beta} + C_2 u^2 + D_2 u_\alpha u_\beta e_{i\alpha} e_{i\beta} + G_{\alpha\beta2} e_{i\alpha} e_{i\beta}. \tag{5.27}$$

For the D2Q9 velocity model, from the constraint Eq. (5.11), we have

$$\sum_i f_i^{eq} = A_0 + C_0 u^2 + 4(A_1 + C_1 u^2) + D_1 u_\alpha u_\beta (2c^2 \delta_{\alpha\beta}) + G_{\alpha\beta1}(2c^2 \delta_{\alpha\beta})$$

$$+ 4(A_2 + C_2 u^2) + D_2 u_\alpha u_\beta (4c^2 \delta_{\alpha\beta}) + G_{\alpha\beta2}(4c^2 \delta_{\alpha\beta})$$

$$= [A_0 + 4(A_1 + A_2)] + [C_0 + 4(C_1 + C_2) + 2c^2(D_1 + 2D_2)]u^2 + 2c^2(G_{\alpha\alpha1} + 2G_{\alpha\alpha2})$$

$$= \rho. \tag{5.28}$$

As in the D2Q7 case we have

$$\begin{cases} A_0 + 4(A_1 + A_2) = \rho, \\ C_0 + 4(C_1 + C_2) + 2c^2(D_1 + 2D_2) = 0, \\ G_{\alpha\alpha1} + 2G_{\alpha\alpha2} = 0. \end{cases} \tag{5.29}$$

From the constraint Eq. (5.12) we have

$$\sum_i f_i^{eq} e_{i\alpha} = \sum_{i=1}^{4} B_1 u_\beta e_{i\beta} e_{i\alpha} + \sum_{i=5}^{8} B_2 u_\beta e_{i\beta} e_{i\alpha} = 2c^2(B_1 + 2B_2)u_\alpha = \rho u_\alpha, \tag{5.30}$$

i.e.,

$$2c^2(B_1 + 2B_2) = \rho. \tag{5.31}$$

From the constraint Eq. (5.13) we have

$$\sum_i f_i^{eq} e_{i\alpha} e_{i\beta} = (A_1 + C_1 u^2) \sum_{i=1}^{4} e_{i\alpha} e_{i\beta} + (D_1 u_\gamma u_\delta + G_{\gamma\delta1}) \sum_{i=1}^{4} e_{i\alpha} e_{i\beta} e_{i\gamma} e_{i\delta}$$

$$+ (A_2 + C_2 u^2) \sum_{i=5}^{8} e_{i\alpha} e_{i\beta} + (D_2 u_\gamma u_\delta + G_{\gamma\delta2}) \sum_{i=5}^{8} e_{i\alpha} e_{i\beta} e_{i\gamma} e_{i\delta}$$

$$= 2c^2 \delta_{\alpha\beta}(A_1 + C_1 u^2) + 2c^4 \delta_{\alpha\beta\gamma\delta}(D_1 u_\gamma u_\delta + G_{\gamma\delta1}) + 4c^2 \delta_{\alpha\beta}(A_2 + C_2 u^2)$$

$$+ [4c^4(\delta_{\alpha\beta}\delta_{\gamma\delta} + \delta_{\gamma\beta}\delta_{\alpha\delta} + \delta_{\delta\beta}\delta_{\gamma\alpha}) - 8c^4 \delta_{\alpha\beta\gamma\delta}](D_2 u_\gamma u_\delta + G_{\gamma\delta2})$$

$$= 2c^2(A_1 + 2A_2)\delta_{\alpha\beta} + u^2(2c^2)\delta_{\alpha\beta}(C_1 + 2C_2 + 2c^2 D_2) + 8c^4 D_2 u_\alpha u_\beta$$

$$+ 2c^4 \delta_{\alpha\beta\gamma\delta}[(D_1 - 4D_2)u_\gamma u_\delta + (G_{\gamma\delta1} - 4G_{\gamma\delta2})] + 4c^4 G_{\gamma\gamma2}\delta_{\alpha\beta} + 8c^4 G_{\alpha\beta2}$$

$$= \rho u_\alpha u_\beta + (p_0 - \kappa\rho\nabla^2\rho - \frac{\kappa}{2}|\nabla\rho|)\delta_{\alpha\beta} + \kappa\frac{\partial\rho}{\partial x_\alpha}\frac{\partial\rho}{\partial x_\beta}. \tag{5.32}$$

We can make the following choices to satisfy the above constraint:

$$\begin{cases} 2c^2(A_1 + 2A_2) = p_0 - \kappa\rho\nabla^2\rho, \\ G_{\gamma\gamma2} = 0, \\ C_1 + 2C_2 + 2c^2 D_2 = 0, \\ D_1 - 4D_2 = 0, \\ G_{\gamma\delta1} - 4G_{\gamma\delta2} = 0, \\ 8c^4 D_2 = \rho, \\ 8c^4 G_{\alpha\beta2} = -\dfrac{k}{2}|\nabla\rho|^2\delta_{\alpha\beta} + \kappa\dfrac{\partial\rho}{\partial x_\alpha}\dfrac{\partial\rho}{\partial x_\beta}. \end{cases} \qquad (5.33)$$

As mentioned in the derivation of the D2Q7 case, it is easy to check that the last choice in Eq. (5.33) is consistent with the formula $G_{\gamma\gamma} = 0$.

From Eqs (5.29), (5.31), and (5.33) it is seen that there are 12 coefficients $(A_0, C_0, A_1, B_1, C_1, D_1, A_2, B_2, C_2, D_2, G_{\alpha\beta1}, G_{\alpha\beta2})$ and nine effective equations (the last equation in Eq. (5.29) and the second, the fifth, and the last ones in Eq. (5.33) are not independent). Three extra equations need to be added to solve the whole equation system.

Here it is assumed that (Briant and Yeomans 2004)

$$A_1 = 4A_2, \quad B_1 = 4B_2, \quad C_1 = 4C_2. \qquad (5.34)$$

We would like to demonstrate the validity of this assumption. In the equilibrium distribution function (D2Q9 model), the weighting coefficients $w_0 = \frac{4}{9}$, $w_i = \frac{1}{9}$ for $i = 1, 2, 3, 4$, and $w_i = \frac{1}{36}$ for $i = 5, 6, 7, 8$. We can see that $w_0 = 4w_i$, where $i = 1, 2, 3, 4$ and $w_i = 4w_j$, where $i = 1, 2, 3, 4$ and $j = 5, 6, 7, 8$. In Eq. (5.34) the formulae are analogous to the above relationships between the w_i.

Then the solution is

$$A_0 = n - 4(A_1 + A_2) = n - \frac{5}{3c^2}(p_0 - \kappa\rho\nabla^2\rho), \quad A_1 = 4A_2 = \frac{1}{3c^2}(p_0 - \kappa\rho\nabla^2\rho),$$

$$A_2 = \frac{1}{3c^2}(p_0 - \kappa\rho\nabla^2\rho), \quad B_1 = 4B_2 = \frac{\rho}{3c^2}, \quad B_2 = \frac{\rho}{12c^2}$$

$$C_0 = -4(C_1 + C_2) - \frac{3\rho}{2c^2} = -\frac{2\rho}{3c^2}, \quad C_1 = 4C_2 = -\frac{\rho}{6c^2}, \quad C_2 = -\frac{\rho}{24c^2}$$

$$D_1 = \frac{\rho}{2c^4}, \quad D_2 = \frac{\rho}{8c^4} \quad G_{\alpha\beta1} = 4G_{\alpha\beta2} = \frac{1}{2c^4}\left(-\frac{k}{2}|\nabla\rho|^2\delta_{\alpha\beta} + \kappa\frac{\partial\rho}{\partial x_\alpha}\frac{\partial\rho}{\partial x_\beta}\right)$$

$$G_{\alpha\beta2} = \frac{1}{8c^4}\left(-\frac{k}{2}|\nabla\rho|^2\delta_{\alpha\beta} + \kappa\frac{\partial\rho}{\partial x_\alpha}\frac{\partial\rho}{\partial x_\beta}\right) \qquad (5.35)$$

Similarly, $G_{\alpha\beta}$ and $G_{\alpha\beta}$ can be written in terms of their components G_{xx}, G_{yy}, and G_{xy}, which are convenient for LBM code. In the following, examples for G_{xx1},

G_{yy1}, and G_{xy1} are given:

$$G_{xx1} = -G_{yy1} = \frac{k}{4c^4}[(\partial_x\rho)^2 - (\partial_y\rho)^2],$$

$$G_{xy1} = \frac{k}{2c^4}\partial_x\rho\partial_y\rho. \tag{5.36}$$

Here we can see that in this scheme we have to determine the density derivatives and $\nabla^2\rho$, and $\partial_x\rho\partial_y\rho$. The finite difference method can be used to handle that. The derivatives are homogeneous derivatives that use the surrounding six points in the point view of the finite difference. In the LBM this collapses to the following formulae for the D2Q7 model (Swift et al. 1996):

$$\partial_\alpha\rho \approx \frac{1}{3c\Delta x}\sum_i \rho(\mathbf{x} + \mathbf{e}_i\Delta t)e_{i\alpha} \tag{5.37}$$

and

$$\nabla^2\rho \approx \frac{2}{3(\Delta x)^2}\left[\sum_i \rho(\mathbf{x} + \mathbf{e}_i\Delta t) - 6\rho(\mathbf{x})\right]. \tag{5.38}$$

For the D2Q9 model the corresponding formulae are:

$$\partial_\alpha\rho \approx \frac{1}{c_s^2\Delta t}\sum_i \omega_i\rho(\mathbf{x} + \mathbf{e}_i\Delta t)e_{i\alpha} \tag{5.39}$$

and

$$\nabla^2\rho \approx \frac{2}{c_s^2(\Delta t)^2}\sum_i \omega_i[\rho(\mathbf{x} + \mathbf{e}_i\Delta t) - \rho(\mathbf{x})]. \tag{5.40}$$

5.2   Chapman–Enskog expansion

Through the Taylor expansion, retaining terms to $O((\Delta t)^2)$, the Lattice Boltzmann equation (Eq. (5.1)) becomes

$$\Delta t\left(\frac{\partial}{\partial t} + e_{i\alpha}\frac{\partial}{\partial x_\alpha}\right)f_i + \frac{\Delta t^2}{2}\left(\frac{\partial}{\partial t} + e_{i\alpha}\frac{\partial}{\partial x_\alpha}\right)^2 f_i + O((\Delta t)^2)$$
$$= -\frac{1}{\tau}(f_i - f_i^{(eq)}). \tag{5.41}$$

Because we are interested in the expansion to second order, we perform the expansion as $f_i = f_i^{(0)} + \Delta t f_i^{(1)} + (\Delta t)^2 f_i^{(2)}$ and $\partial_t = \partial_{t_1} + \Delta t\partial_{t_2}$, and we have

$$O(\varepsilon^0) : (f_i^{(0)} - f_i^{eq})/\tau = 0 \tag{5.42}$$

$$O(\varepsilon^1) : (\partial_{t_1} + e_{i\alpha}\partial_\alpha)f_i^{(0)} + \frac{1}{\tau}f_i^{(1)} = 0 \tag{5.43}$$

$$O(\varepsilon^2) : \partial_{t_2}f_i^{(0)} + \left(1 - \frac{1}{2\tau}\right)(\partial_{t_1} + e_{i\alpha}\partial_\alpha)f_i^{(1)} + \frac{1}{\tau}f_i^{(2)} = 0. \tag{5.44}$$

ε has the same order as Δt. To obtain the continuity equation and the N–S equations, we start from Eqs (5.42)–(5.44).

Continuity equation

Summing on i, Eq. (5.43) yields

$$\partial_{t_1}\left(\sum f_i^{(0)}\right) + \partial_\alpha\left(\sum e_{i\alpha} f^{(0)}\right) + \frac{1}{\tau}\sum f_i^{(1)} = 0. \tag{5.45}$$

From equations $\sum_i f_i^{(0)} = \rho$, $\sum_i f_i^{(1)} = 0$, and $\sum_i f_i^{(2)} = 0$ (because $\sum_i f_i = \rho$ and $f_i = f_i^{(0)} + \Delta t f_i^{(1)} + (\Delta t)^2 f_i^{(2)}$) we know that

$$\partial_{t_1}\rho + \partial_\alpha(\rho u_\alpha) = 0. \tag{5.46}$$

From Eq. (5.43) we know that $f_i^{(1)} = -\tau(\partial_{t_1} + e_{i\alpha}\partial_\alpha)f_i^{(0)}$, and substituting it into Eq. (5.44) we have

$$\partial_{t_2}f_i^{(0)} - \left(\tau - \frac{1}{2}\right)(\partial_{t_1} + e_{i\alpha}\partial_\alpha)(\partial_{t_1} + e_{i\beta}\partial_\beta)f_i^{(0)} + \frac{1}{\tau}f_i^{(2)} = 0. \tag{5.47}$$

Summing on i, Eq. (5.44) yields

$$\partial_{t_2}\rho - \left(\tau - \frac{1}{2}\right)\left\{\partial_{t_1}^2\rho + 2\partial_{t_1}\partial_\alpha(\rho u_\alpha) + \partial_\alpha\partial_\beta\Pi_{\alpha\beta}^0\right\} = 0, \tag{5.48}$$

where

$$\Pi_{\alpha\beta}^0 = \sum_i f_i^{(0)} e_{i\alpha} e_{i\beta}. \tag{5.49}$$

Combining Eqs (5.46) and (5.48) we have

$$\partial_t\rho + \partial_\alpha(\rho u_\alpha) - \Delta t\left(\tau - \frac{1}{2}\right)\left\{\partial_{t_1}^2\rho + 2\partial_{t_1}\partial_\alpha(\rho u_\alpha) + \partial_\alpha\partial_\beta\Pi_{\alpha\beta}^0\right\}$$
$$+ O((\Delta t)^2) = 0, \tag{5.50}$$

where $\partial_t\rho = \partial_{t_1}\rho + \Delta t\partial_{t_2}\rho$.

Multiplying Eq. (5.43) by $e_{i\beta}$ and summing over i gives

$$\partial_{t_1}(\rho u_\beta) + \partial_\alpha\Pi_{\alpha\beta}^0 = 0. \tag{5.51}$$

Substituting Eqs (5.46) and (5.51) into Eq. (5.50), immediately we find that terms $\left\{\partial_{t_1}^2\rho + 2\partial_{t_1}\partial_\alpha(\rho u_\alpha) + \partial_\alpha\partial_\beta\Pi_{\alpha\beta}^0\right\}$ vanish and we have

$$\partial_t\rho + \partial_\alpha(\rho u_\alpha) = 0, \tag{5.52}$$

to the order of $O((\Delta t)^2)$. This is the continuity equation of the flow system.

Momentum equations (N–S)

Multiplying Eq. (5.44) by $e_{i\beta}$ and summing over i gives

$$\partial_{t_2}(\rho u_\beta) - (\tau - \frac{1}{2})\Delta t \left[\partial_{t_1}^2(\rho u_\beta) + 2\partial_{t_1}\partial_\alpha \Pi_{\alpha\beta}^0 + \partial_\alpha\partial_\gamma \sum_i f_i^{(0)} e_{i\alpha} e_{i\beta} e_{i\gamma}\right]$$

$$= 0 \tag{5.53}$$

From Eqs (5.51) and (5.53) we know that

$$\partial_t(\rho u_\beta) = -\partial_\alpha(\Pi_{\alpha\beta}^0) + O(\Delta t), \tag{5.54}$$

where $\partial_t(\rho u_\beta) = \partial_{t_1}(\rho u_\beta) + \Delta t \partial_{t_2}(\rho u_\beta)$. This is the Euler equation.

Using the definition of the second moment of f_i^0 in Eq. (5.13),

$$\partial_\alpha \Pi_{\alpha\beta}^0 = \partial_\alpha(P_{\alpha\beta} + \rho u_\alpha u_\beta) \approx \partial_\beta p_0 + \partial_\alpha(\rho u_\alpha u_\beta). \tag{5.55}$$

Here Eq. (5.5) is used and the high-order derivatives are neglected. Similarly, we have

$$\partial_{t_1}\partial_\alpha \Pi_{\alpha\beta}^0 = \partial_{t_1}\partial_\alpha(P_{\alpha\beta} + \rho u_\alpha u_\beta)$$

$$\approx \partial_\beta \partial_{t_1} p_0 + \partial_\alpha[u_\alpha \partial_{t_1}(\rho u_\beta) + u_\beta \partial_{t_1}(\rho u_\alpha) - u_\alpha u_\beta \partial_{t_1}\rho]$$

$$\approx -\partial_\beta\left[\frac{dp_0}{d\rho}\partial_\gamma(\rho u_\gamma)\right] - \partial_\alpha(u_\alpha\partial_\beta p_0 + u_\beta\partial_\alpha p_0). \tag{5.56}$$

In the above derivation, $\partial_t p_0 = \frac{\partial p_0}{\partial\rho}\partial_t\rho = \frac{\partial p_0}{\partial\rho}(-\partial_\gamma(\rho u_\gamma))$. From Eqs (5.54) and (5.55) it is noted that $\partial_{t_1}(\rho u_\alpha) \approx -\partial_\alpha p_0 + \partial_\beta(\rho u_\alpha u_\beta)$; from Eq. (5.52) we have $\partial_{t_1}\rho = -\partial_\alpha(\rho u_\alpha)$. Substituting these formula into the above derivation and omitting the higher order terms such as $O(u^3)$, we finally have Eq. (5.56).

Using the definition of $f_i^{(0)}$, and Eq. (5.25), the final term inside the square bracket in Eq. (5.53) can be written as

$$\partial_\alpha\partial_\gamma\left(\sum_i f_i^{(0)} e_{i\alpha} e_{i\beta} e_{i\gamma}\right) = \frac{c^2}{4}\partial_\alpha\partial_\gamma(\rho u_\gamma\delta_{\alpha\beta} + \rho u_\beta\delta_{\alpha\gamma} + \rho u_\alpha\delta_{\beta\gamma})$$

$$= \frac{c^2}{4}[2\partial_\beta\partial_\gamma(\rho u_\gamma) + \partial_\gamma\partial_\gamma\rho u_\beta]. \tag{5.57}$$

Substituting Eqs. (5.55)–(5.57) into Eqs (5.51) and (5.53), and then combining Eqs (5.51) and (5.53) together, we have

$$\partial_t(\rho u_\beta) + \partial_\alpha(\rho u_\alpha u_\beta) = -\partial_\beta p_0 + \partial_\alpha(\nu\partial_\alpha\rho u_\beta)$$

$$+\partial_\beta(\lambda(\rho)\partial_\alpha\rho u_\alpha) - \left(\tau - \frac{1}{2}\right)\Delta t \partial_\alpha\left(\frac{dp_0}{d\rho}(u_\beta\partial_\alpha\rho + u_\alpha\partial_\beta\rho)\right) \tag{5.58}$$

where

$$\lambda(\rho) = \left(\tau - \frac{1}{2}\right)\left(\frac{c^2}{2} - \frac{dp_0}{d\rho}\right)\Delta t, \quad \nu = \left(\tau - \frac{1}{2}\right)\frac{c^2}{4}\Delta t. \tag{5.59}$$

In the derivation of Eq. (5.58) the following formula is used:

$$\partial_{t_1}^2(\rho u_\beta) + 2\partial_{t_1}\partial_\alpha\Pi_{\alpha\beta}^0 + \partial_\alpha\partial_\gamma\sum_i f_i^{(0)}e_{i\alpha}e_{i\beta}e_{i\gamma}$$

$$= -\partial_{t_1}(\partial_\alpha\Pi_{\alpha\beta}^0) + 2\partial_{t_1}\partial_\alpha\Pi_{\alpha\beta}^0 + \partial_\alpha\partial_\gamma\sum_i f_i^{(0)}e_{i\alpha}e_{i\beta}e_{i\gamma}$$

$$= \partial_{t_1}\partial_\alpha\Pi_{\alpha\beta}^0 + \partial_\alpha\partial_\gamma\sum_i f_i^{(0)}e_{i\alpha}e_{i\beta}e_{i\gamma}$$

$$= -\partial_\beta\left[\frac{dp_0}{d\rho}\partial_\gamma(\rho u_\gamma)\right] - \partial_\alpha(u_\alpha\partial_\beta p_0 + u_\beta\partial_\alpha p_0) + \frac{c^2}{4}[2\partial_\beta\partial_\gamma(\rho u_\gamma) + \partial_\gamma^2(\rho u_\beta)]. \quad (5.60)$$

In the derivation of the Eq. (5.60) the chain rule is used:

$$\partial_\beta p_0 = \frac{\partial p_0}{\partial\rho}\frac{\partial\rho}{\partial x_\beta} = \frac{dp_0}{d\rho}\frac{\partial\rho}{\partial x_\beta} = \frac{dp_0}{d\rho}\partial_\beta\rho. \quad (5.61)$$

Note that in Eq. (5.61) $\frac{\partial p_0}{\partial\rho} = \frac{dp_0}{d\rho}$ because from Eq. (5.7) we know that p_0 is only a function of density ρ for the isothermal case and here only isothermal cases are studied.

If we define

$$\zeta = \left(\tau - \frac{1}{2}\right)\frac{dp_0}{d\rho}\Delta t, \quad (5.62)$$

the N–S equation (Eq. (5.58)) becomes

$$\partial_t(\rho u_\beta) + \partial_\alpha(\rho u_\alpha u_\beta) = -\partial_\beta p_0 + \partial_\alpha(\nu\partial_\alpha\rho u_\beta)$$

$$+\partial_\beta(\lambda\partial_\alpha\rho u_\alpha) - \partial_\alpha(\zeta(u_\beta\partial_\alpha\rho + u_\alpha\partial_\beta\rho)). \quad (5.63)$$

From the definitions in Eq. (5.59) it is noted that

$$\zeta = \left(\tau - \frac{1}{2}\right)\frac{dp_0}{d\rho}\Delta t = \left(\tau - \frac{1}{2}\right)\frac{c^2}{2}\Delta t - \lambda = 2\nu - \lambda. \quad (5.64)$$

5.3 Issue of Galilean invariance

Galilean invariance means that a function $A(u)$, which is a function of velocity u, is equal to $A(u + U)$, where U is a constant velocity, i.e., $A(u) = A(u + U)$.

In Eq. (5.58) the viscous term $\partial_\beta(\lambda\partial_\alpha(\rho u_\alpha))$ is not Galilean invariant when density gradients $\partial_\alpha\rho \neq 0$ because

$$\partial_\alpha(\rho(u_\alpha + U)) = \rho\partial_\alpha(u_\alpha + U) + (u_\alpha + U)\partial_\alpha\rho = \rho\partial_\alpha u_\alpha + (u_\alpha + U)\partial_\alpha\rho \neq \partial_\alpha(\rho u_\alpha). \quad (5.65)$$

It is also noted that here the Galilean invariance is always associated with the density gradient. Without density gradient, i.e., when $\partial_\alpha\rho = 0$, the term is Galilean invariant since $\partial_\alpha(\rho(u_\alpha + U)) = \partial_\alpha(\rho u_\alpha)$.

Some non-Galilean invariant terms can be removed by adding terms to the pressure tensor (Eq. (5.13)) (Swift et al. 1996):

$$\sum_i f_i^{eq} e_{i\alpha} e_{i\beta} = P_{\alpha\beta} + \rho u_\alpha u_\beta + \xi_1 (u_\beta \partial_\alpha \rho + u_\alpha \partial_\beta \rho) + \xi_2 u_\gamma \partial_\gamma \rho \qquad (5.66)$$

In this way, the N–S equation (Eq. (5.63)) can be written as

$$\partial_t(\rho u_\beta) + \partial_\alpha(\rho u_\alpha u_\beta) = -\partial_\beta p_0 + \partial_\alpha(\nu \rho \partial_\alpha u_\beta) + \partial_\beta(\lambda \rho \partial_\alpha u_\alpha)$$
$$+\partial_\alpha[(\nu - \xi_1 - \zeta)u_\beta \partial_\alpha \rho] + \partial_\beta[(\lambda - \xi_2)u_\alpha \partial_\alpha \rho] + \partial_\alpha[(-\zeta - \xi_1)u_\alpha \partial_\beta \rho].$$

$$(5.67)$$

In this derivation the following formulae should be noted:

$$\partial_\alpha(\nu \partial_\alpha \rho u_\beta) = \partial_\alpha(\nu \rho \partial_\alpha u_\beta) + \partial_\alpha(\nu u_\beta \partial_\alpha \rho) \qquad (5.68)$$

and

$$\partial_\beta(\lambda \partial_\alpha \rho u_\alpha) = \partial_\beta(\lambda \rho \partial_\alpha u_\alpha) + \partial_\beta(\lambda u_\alpha \partial_\alpha \rho). \qquad (5.69)$$

In Eq. (5.67), if the equations $\nu - \xi_1 - \zeta = 0$, $\lambda - \xi_2 = 0$, and $-\zeta - \xi_1 = 0$ are all satisfied, the N–S equation is fully Galilean invariant. However, it is not possible to choose ξ_1 and ξ_2 to satisfy this condition. According to Swift et al. (1996), "This occurs because the second moment of $f_i^{(0)}$ is symmetric with respect to interchange of its indices, but the viscosity terms do not have the symmetry." Swift et al. (1996) suggested choosing

$$\xi_1 = -\zeta, \quad \xi_2 = \lambda, \qquad (5.70)$$

to improve the Galilean invariance of the model.

Holdych et al. (1998) improved the Galilean invariance property for the free-energy model of Swift et al. (1996). Through choosing the following form for the equilibrium momentum tensor, " ... the leading error terms of the momentum equation are eliminated at the small cost of introducing new lower order terms ... " (Holdych et al. 1998).

In Holdych et al. (1998) the term $\partial_{t_1} \Pi_{\alpha\beta}^0$ is written as

$$\partial_{t_1} \Pi_{\alpha\beta}^0 = \partial_{t_1} P_{\alpha\beta} + \partial_{t_1}(\rho u_\alpha u_\beta)$$
$$= -\partial_\rho P_{\alpha\beta} \partial_\gamma(\rho u_\gamma) - [u_\alpha \partial_\gamma P_{\beta\gamma} + u_\beta \partial_\gamma P_{\alpha\gamma} + \partial_\gamma(\rho u_\alpha u_\beta u_\gamma)]. \qquad (5.71)$$

In the derivation the following formulae, the continuity equation (i.e., Eq. (5.52)), and the Euler equation (i.e., Eq. (5.54)) are used:

$$\partial_{t_1} P_{\alpha\beta} = \frac{\partial P_{\alpha\beta}}{\partial \rho}\left(\frac{\partial \rho}{\partial t_1}\right) = -(\partial_\rho P_{\alpha\beta})\partial_\gamma(\rho u_\gamma) \qquad (5.72)$$

$$\partial_{t_1}(\rho u_\alpha u_\beta) = u_\alpha \partial_{t_1}(\rho u_\beta) + u_\beta \partial_{t_1}(\rho u_\alpha) - u_\alpha u_\beta \partial_{t_1}\rho$$

$$= -u_\alpha[\partial_\gamma P_{\beta\gamma} + \partial_\gamma(\rho u_\beta u_\gamma)] - u_\beta[\partial_\gamma P_{\alpha\gamma} + \partial_\gamma(\rho u_\alpha u_\gamma)] + u_\alpha u_\beta \partial_\gamma(\rho u_\gamma)$$

$$= -u_\alpha \partial_\gamma P_{\beta\gamma} - u_\beta \partial_\gamma P_{\alpha\gamma} - \partial_\gamma(\rho u_\alpha u_\beta u_\gamma). \tag{5.73}$$

In the above derivative it is also noted that

$$u_\alpha \partial_\gamma(\rho u_\beta u_\gamma) + u_\beta \partial_\gamma(\rho u_\alpha u_\gamma) - u_\alpha u_\beta \partial_\gamma(\rho u_\gamma) = \partial_\gamma(\rho u_\alpha u_\beta u_\gamma). \tag{5.74}$$

Through the Chapman–Enskog expansion the N–S equation is (Holdych et al. 1998)

$$\partial_t(\rho u_\alpha) + \partial_\beta(\rho u_\alpha u_\beta) = -\partial_\beta P_{\alpha\beta} + \nu[\partial_\beta \rho(\partial_\beta u_\alpha + \partial_\alpha u_\beta + \partial_\gamma u_\gamma \delta_{\alpha\beta})]$$

$$+ \nu[\partial_\beta(u_\alpha \partial_\beta \rho + u_\beta \partial_\alpha \rho + u_\gamma \partial_\gamma \rho \delta_{\alpha\beta})]$$

$$- \frac{3\nu}{c^2}\partial_\beta[u_\alpha \partial_\gamma P_{\beta\gamma} + u_\beta \partial_\gamma P_{\alpha\gamma} + \partial_\gamma(\rho u_\alpha u_\beta u_\gamma)]$$

$$- \frac{3\nu}{c^2}\partial_\beta[(\partial_\rho P_{\alpha\beta})\partial_\gamma(\rho u_\gamma)] + O(\delta^2), \tag{5.75}$$

where the kinematic viscosity $\nu = c_s^2(\tau - 0.5)\Delta t = \frac{1}{3}c^2(\tau - 0.5)\Delta t$. Assuming the gradient of pressure to be at most of order (u), and ν to be order one, the errors of the third and fourth lines of the above equation are at most of order $O(u^2)$. In two-phase flows, the density gradients can be of order one, and the terms on the second line (relevant to non-Galilean invariance) of the above equation are of the same order as the viscous term and convection terms of the N–S equation (Holdych et al. 1998). Hence, the terms on the second line should be eliminated. It is noted that there is a small typographical error in the last line of Eq. (22) in Holdych et al. (1998). It should be $\partial_\rho P_{\alpha\beta}$ instead of $\partial_P P_{\alpha\beta}$.

To eliminate the terms relevant to the non-Galilean invariance, Holdych et al. (1998) redefined the equilibrium stress tensor as (Holdych et al. 1998)

$$\Pi_{\alpha\beta}^{0+} = P_{\alpha\beta} + \rho u_\alpha u_\beta + \nu(u_\alpha \partial_\beta \rho + u_\beta \partial_\alpha \rho + u_\gamma \partial_\gamma \rho \delta_{\alpha\beta}). \tag{5.76}$$

In this way, the terms that do not satisfy Galilean invariance, i.e.,

$$\nu[\partial_\beta(u_\alpha \partial_\beta \rho + u_\beta \partial_\alpha \rho + u_\gamma \partial_\gamma \rho \delta_{\alpha\beta})], \tag{5.77}$$

in the recovered N–S equation are eliminated (Holdych et al. 1998).

Redefining the equilibrium stress tensor will introduce new lower error terms. For example, due to this new definition, Eq. (5.71) becomes

$$\partial_{t_1}\Pi_{\alpha\beta}^{0+} = \partial_{t_1}P_{\alpha\beta} + \partial_{t_1}(\rho u_\alpha u_\beta) + \partial_{t_1}[\nu(u_\alpha \partial_\beta \rho + u_\beta \partial_\alpha \rho + u_\gamma \partial_\gamma \rho \delta_{\alpha\beta})]$$

$$= -\partial_\rho P_{\alpha\beta}\partial_\gamma(\rho u_\gamma) - [u_\alpha \partial_\gamma P_{\beta\gamma} + u_\beta \partial_\gamma P_{\alpha\gamma} + \partial_\gamma(\rho u_\alpha u_\beta u_\gamma)]$$

$$- u_\alpha \nu(u_\beta \partial_\gamma \rho + u_\gamma \partial_\beta \rho + u_\lambda \partial_\lambda \rho \delta_{\alpha\beta})$$

$$- u_\beta \nu(u_\alpha \partial_\gamma \rho + u_\gamma \partial_\alpha \rho + u_\lambda \partial_\lambda \rho \delta_{\alpha\beta})$$

$$+ \nu \partial_{t_1}(u_\alpha \partial_\beta \rho + u_\beta \partial_\alpha \rho + u_\gamma \partial_\gamma \rho \delta_{\alpha\beta}). \tag{5.78}$$

Because of the new definition, the corresponding Eq. (5.73) becomes

$$\partial_{t_1}(\rho u_\alpha u_\beta) = u_\alpha \partial_{t_1}(\rho u_\beta) + u_\beta \partial_{t_1}(\rho u_\alpha) - u_\alpha u_\beta \partial_{t_1}\rho$$

$$= -u_\alpha[\partial_\gamma P_{\beta\gamma} + \partial_\gamma(\rho u_\beta u_\gamma) + \nu(u_\beta \partial_\gamma \rho + u_\gamma \partial_\beta \rho + u_\lambda \partial_\lambda \rho \delta_{\alpha\beta})]$$

$$-u_\beta[\partial_\gamma P_{\alpha\gamma} + \partial_\gamma(\rho u_\alpha u_\gamma) + \nu(u_\alpha \partial_\gamma \rho + u_\gamma \partial_\alpha \rho + u_\lambda \partial_\lambda \rho \delta_{\alpha\beta})]$$

$$+u_\alpha u_\beta \partial_\gamma(\rho u_\gamma)$$

$$= -u_\alpha \partial_\gamma P_{\beta\gamma} - u_\beta \partial_\gamma P_{\alpha\gamma} - \partial_\gamma(\rho u_\alpha u_\beta u_\gamma)$$

$$-u_\alpha \nu(u_\beta \partial_\gamma \rho + u_\gamma \partial_\beta \rho + u_\lambda \partial_\lambda \rho \delta_{\alpha\beta})$$

$$-u_\beta \nu(u_\alpha \partial_\gamma \rho + u_\gamma \partial_\alpha \rho + u_\lambda \partial_\lambda \rho \delta_{\alpha\beta}). \tag{5.79}$$

Finally, the recovered N–S equation is (Holdych et al. 1998)

$$\partial_t(\rho u_\alpha) + \partial_\beta(\rho u_\alpha u_\beta) = -\partial_\beta P_{\alpha\beta} + \nu[\partial_\beta \rho(\partial_\beta u_\alpha + \partial_\alpha u_\beta + \partial_\gamma u_\gamma \delta_{\alpha\beta})]$$

$$-\frac{3\nu}{c^2}\partial_\beta[u_\alpha \partial_\gamma P_{\beta\gamma} + u_\beta \partial_\gamma P_{\alpha\gamma} + \partial_\gamma(\rho u_\alpha u_\beta u_\gamma)]$$

$$-\frac{3\nu}{c^2}\partial_\beta[(\partial_\rho P_{\alpha\beta})\partial_\gamma(\rho u_\gamma)]$$

$$-\frac{3\nu^2}{c^2}\partial_\beta[u_\alpha \partial_\gamma(u_\beta \partial_\gamma \rho + u_\gamma \partial_\beta \rho + u_\lambda \partial_\lambda \rho \delta_{\alpha\beta})]$$

$$-\frac{3\nu^2}{c^2}\partial_\beta[u_\beta \partial_\gamma(u_\alpha \partial_\gamma \rho + u_\gamma \partial_\alpha \rho + u_\lambda \partial_\lambda \rho \delta_{\alpha\beta})]$$

$$+\frac{3\nu^2}{c^2}\partial_\beta[\partial_{t_1}(u_\alpha \partial_\beta \rho + u_\beta \partial_\alpha \rho + u_\gamma \partial_\gamma \rho \delta_{\alpha\beta})] + O(\delta^2). \tag{5.80}$$

We can see that compared to Eq. (5.75) the terms in the last three lines of the above equation are new error terms and the terms relevant to non-Galilean invariance on the second line of Eq. (5.75) are eliminated. The new error terms are at most of order $O(u^2)$ (Holdych et al. 1998). In Holdych et al. (1998) three benchmark flows, including Couette flow, pure shear of a droplet, and advection of a droplet, are used to demonstrate the improvement of the free-energy two-phase model.

Because the equilibrium stress tensor is different from that in Swift et al. (1996), the coefficients in the equilibrium distribution function should be slightly different from those we obtained in Section 5.1.1. If readers are interested in this model, they can easily derive their own formulae following the procedure in Section 5.1.1. Similar derivations can also be found in Kalarakis et al. (2002) and Inamuro et al. (2000).

5.4 Phase separation

In this section numerical examples of phase separation and coexistence of the fluid and gas are provided. The parameters chosen in Swift et al. (1996) are $a = \frac{9}{49}$

and $b = \frac{2}{21}$. Here we adopt the same values. The coexistence curve due to the EOS of Eq. (5.7) and the Maxwell construction are shown in Figure 5.2.

To check whether the equilibrium coexistence densities of the fluid and the gas are consistent with the analytical ones, test cases of a droplet immersed in gas in the absence of gravity are simulated. In our simulations, the computational domain is 100×100 and periodic boundary conditions are applied on all boundaries. A circular area in the center of the domain is initialized as the liquid area (higher density) while the other region is initialized with gas at lower density. After the simulations reach the equilibrium state, we check the equilibrium densities of the liquid and gas. If not specified, in this section $\tau = 1$. If $\kappa = 0.01$, the corresponding coexisting densities are $\rho_l = 4.523$ mu/lu^3 and $\rho_g = 2.552$ mu/lu^3 for $T = 0.56$. The convergence criterion can be set based on the radius of the droplet or the spurious current reaching a constant value, for example $E_{\mathbf{u}} < 0.5\%$ over 1000 ts, where

$$E_{\mathbf{u}} = \sqrt{\frac{\sum_{\mathbf{x}}(|\mathbf{u}(\mathbf{x}, t) - \mathbf{u}(\mathbf{x}, t - 1000\Delta t)|^2)}{\sum(|\mathbf{u}(\mathbf{x}, t)|^2)}}. \qquad (5.81)$$

A typical simulation result is illustrated in Figure 5.1(a). Initially, the density inside a circular drop with a radius of $r_0 = 30$ lu is set to be $\rho_l = 4.5$ and the other area is set to $\rho_g = 2.3$. Figure 5.1(a) shows the equilibrium state and the spurious current. The droplet becomes smaller (the equilibrium radius of the droplet

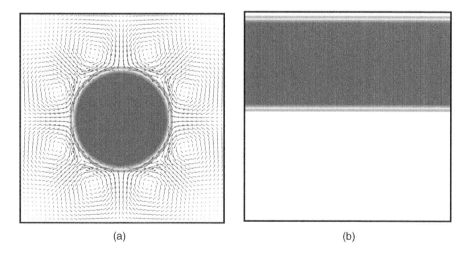

(a)	(b)

Figure 5.1   (a) Equilibrium state of a droplet immersed in gas. In the simulation, $T = 0.56$ and $\kappa = 0.01$. The equilibrium densities of liquid and gas are $\rho_l = 4.523$ and $\rho_g = 2.552$, respectively. The maximum magnitude of the spurious current is 8.8×10^{-5} lu/ts. (b) Equilibrium snapshot of the coexisting liquid and gas with flat interfaces for $T = 0.56$. Densities above and below the middle horizontal line are initialized as $\rho_l = 4.5$ and $\rho_g = 2.3$, respectively. The equilibrium densities are $\rho_l = 4.51$ and $\rho_g = 2.54$.

is about 24 lu). This is correct because the whole mass should be conserved. We can do a simple approximation as follows. Initially the mass in the system is $\rho_l \pi r_0^2 + \rho_g(L^2 - \pi r_0^2) = 4.5\pi \times 30^2 + 2.3(100^2 - \pi \times 30^2) \approx 29,200$ mu. After it reaches the equilibrium state, the radius of the droplet is about 24 lu, and the whole mass in the system is $\rho_l \pi r^2 + \rho_g(L^2 - \pi r^2) = 4.523\pi \times 24^2 + 2.552(100^2 - \pi \times 24^2) \approx 29100$ mu. The very small discrepancy between the initial and final mass is due to minor measurement errors in the radius of the droplet. Hence, because the initial gas density $\rho_g = 2.3$ is lower than the thermodynamic equilibrium one $\rho_g = 2.552$, some mass of initial liquid was transformed into gas. If the densities set initially are thermodynamically consistent (i.e., coexisting densities at the specified T), the diameter of the droplet would remain constant.

The surface tension can be calculated numerically. In the above case, the pressures inside and outside of the droplet are $p_{in} = 0.692118$ mu/(lu ts^2) and $p_{out} = 0.691805$ mu/(lu ts^2), respectively. The radius of the droplet is $r = 23.75$ lu. Hence the surface tension is $\sigma = (p_{in} - p_{out})r = 0.00743$ mu/ ts^2.

For simulations focused on measuring equilibrium coexistence, cases with a horizontal interface are appropriate to eliminate interface curvature-induced pressure (density) variation. The result for this model is shown in Figure 5.1(b). In the simulation, periodic boundary conditions are applied on all boundaries. Initially, the upper and lower half planes are occupied by liquid and gas, respectively. Here we can see again that because the initial density values are not the equilibrium coexisting ones, the area occupied by the liquid (light grey area) at equilibrium is slightly less than half. It is also noted that for this horizontal interface there are no spurious currents near the interface. The spurious currents that appear in numerical simulations are associated with the curvature of the interface Shan (2006).

Through simulating many cases of flat interfaces at different T, and measuring the densities, we can recover the coexistence curve. Figure 5.2 shows the LBM result for $T = 0.52, 0.54, 0.55, \ldots$ compared with the analytical coexisting densities. It demonstrates that the LBM result is consistent with the analytical one. The analytical coexisting densities can be obtained from the Maxwell equal-area construction as described in Chapter 2.

A typical van der Waals EOS $p = \frac{T}{1/\rho - b} - \frac{a}{1/\rho^2}$ with $T = 0.56$, $a = \frac{9}{49}$, and $b = \frac{2}{21}$ is shown in Figure 5.3. Here, for $T = 0.56$, the equilibrium $\rho_l \approx 4.50$ and $\rho_g \approx 2.55$.

If the fluid system is initially specified as having a homogeneous density ρ and the density is in the unstable part of the EOS (section between points 1 and 2 in Figure 5.3, i.e., $\rho_2 < \rho < \rho_1$), then due to perturbation and thermodynamic instability two phases with densities ρ_l and ρ_g would separate. In Figure 5.3, the ρ values at points 1 and 2 are 4.08 and 2.92, respectively. Here, a case of phase separation is tested. Initially the density field is set to be

$$\rho(\mathbf{x}) = 3.00 + 0.01\theta, \tag{5.82}$$

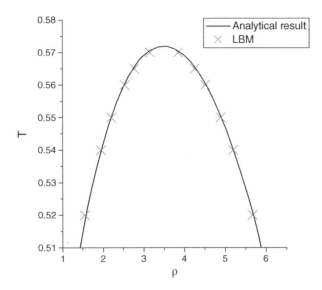

Figure 5.2 Coexistence curve of the van der Waals fluid. In the EOS (i.e., Eq. (5.7)), $a = \frac{9}{49}$ and $b = \frac{2}{21}$. The corresponding critical density and temperature are $\rho_c = \frac{7}{2}$ and $T_c = \frac{4}{7}$, respectively.

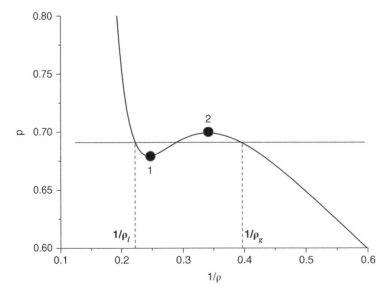

Figure 5.3 EOS $p = \frac{T}{1/\rho - b} - \frac{a}{1/\rho^2}$ with $T = 0.56$, $a = \frac{9}{49}$, and $b = \frac{2}{21}$. The ρ values at points 1 and 2 are 4.08 and 2.92, respectively, and are at the limits of the non-physical portion of the EOS. In order to simulate spontaneous phase separation, the initial densities should be selected between these values. The equilibrium ρ_l and ρ_g are about 4.50 and 2.55, respectively.

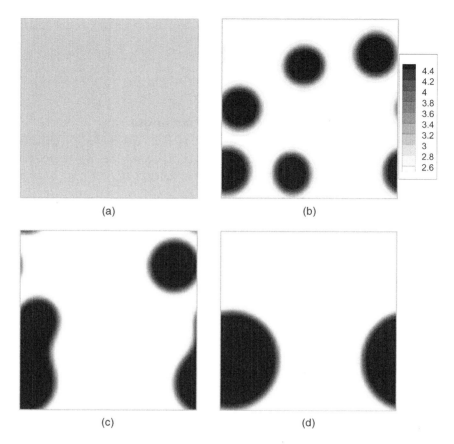

Figure 5.4 Phase separation (a) $t = 0$ ts, (b) $t = 2000$ ts, (c) $t = 8000$ ts, and (d) $t = 20000$ ts. In this case initially there is a small-amplitude perturbation in the density field (Eq. (5.82)), $T = 0.56$, $\kappa = 0.01$. Density contours are shown in (b)–(d). The densities correspond to the greyscales shown in the key and are identical in all images.

where θ is a random number $\theta \in (0, 1)$. Figure 5.4 shows some snapshots of the phase separation. It is seen that at first small droplets are formed. When they collide, they may coalesce to form bigger droplets. In addition, evaporation, condensation, and vapor phase transport occur in SCMP models like this one. Finally, a single large droplet appears at equilibrium. Whether there is a large droplet or a bubble at equilibrium is determined by the initial density we specify. When the initial homogeneous density $\rho \approx 3$ is closer to the equilibrium ρ_g, gas may be dominant in the final equilibrium state. That is why the droplet appears in the system at equilibrium. If the initial homogeneous density is closer to the liquid density ρ_l, for example $\rho = 3.8$, liquid will be dominant and a bubble will appear. Our numerical simulations confirm that behavior.

Here again we can confirm mass conservation. Initially the mass in the system is about $3.05 \times L^2 \approx 3 \times 10^4$ mu. At equilibrium, the droplet radius is $r \approx 26.4$ lu and the mass is $\rho_l \pi r^2 + \rho_g (L^2 - \pi r^2) = 4.52\pi \times 26.4^2 + 2.552(100^2 - \pi \times 26.4^2) \approx 3 \times 10^4$ mu.

5.5   Contact angle

5.5.1   How to specify a desired contact angle

Free-energy-based LBMs, including the Swift model, have become a good tool for the study of wetting phenomena (Briant et al. 2004; Dupuis and Yeomans 2004; Kusumaatmaja and Yeomans 2007, 2009; Pooley et al. 2008). For example, Briant et al. (2004) investigated contact line movement near a sheared interface. Dupuis and Yeomans (2004) investigated the spreading of mesoscopic droplets on homogeneous and heterogeneous surfaces.

In the simulations cited above the van der Waals EOS is used. However, the EOS used in the model of Briant et al. (2004) and Dupuis and Yeomans (2004) is different. The bulk free-energy density is designed to be

$$\psi = p_c(\rho' + 1)^2(\rho'^2 - 2\rho' + 3 - 2\beta T'), \tag{5.83}$$

where $\rho' = \frac{\rho - \rho_c}{\rho_c}$ and $T' = \frac{T_c - T}{T_c}$ are a reduced density and temperature. T_c, p_c, and ρ_c are the critical temperature, pressure, and density, respectively. β is a constant (Briant et al. 2004) and $p_0 = \rho \partial_\rho \psi - \psi$ is the EOS of the fluid. This choice of free-energy density enables calculation of the wall-fluid surface tension in a closed form (Briant et al. 2004).

To specify a contact angle in the free-energy model a surface free-energy density function $\Phi(\rho_s)$, which depends only on the density at the surface ρ_s, should be added into the free-energy calculation (i.e., Eq. (5.3)), that is (Briant et al. 2004)

$$\Psi = \int_V dr \left(\psi(T, \rho) + \frac{\kappa}{2}(\nabla \rho)^2 \right) + \int_S ds \Phi(\rho_s), \tag{5.84}$$

where S is the surface bounding V (Briant et al. 2004).

The equilibrium boundary condition on the wall is (Briant et al. 2004)

$$\kappa \mathbf{n} \cdot \nabla \rho = \frac{d\Phi}{d\rho_s}. \tag{5.85}$$

where $\mathbf{n}$ is the local normal direction of the wall pointing into the fluid.

Following de Gennes (1985), we chose to expand Φ as a power series: a linear term only is sufficient, i.e., $\Phi(\rho_s) = -\omega\rho_s$, where ω is referred to as the wetting potential (related to the surface wetting property). Thus Eq. (5.85) becomes $\kappa \partial_y \rho = -\omega$, where the y direction is supposed to be the normal direction of the wall.

Hence, a natural boundary condition for ρ is (Briant et al. 2004)

$$\mathbf{n} \cdot (\nabla \rho)_s = \partial_y \rho|_{y=1} = -\frac{\omega}{\kappa}. \tag{5.86}$$

In Briant et al. (2004) the contact angle measured in dense fluid ρ_l is (Briant et al. 2004)

$$\cos\theta = \frac{1}{2}[(\sqrt{1-\Omega})^3 - (\sqrt{1+\Omega})^3], \tag{5.87}$$

where

$$\Omega = \frac{\omega}{\beta T'\sqrt{2\kappa p_c}}. \tag{5.88}$$

For a desired contact angle θ, the desired ω can be obtained according to Eq. (5.87).

Although the choice of EOS other than the traditional van der Waals EOS may create some advantage (calculating the wall-fluid surface tension in a closed form) (Briant et al. 2004), to be consistent with the original Swift model here we still choose the van der Waals EOS. One problem is how to specify the desired wetting potential when using the van der Waals EOS. Akin to Eq. (3.33), here we propose a formula,

$$\Omega = \frac{\omega}{(\frac{\rho_l - \rho_d}{2\rho_c})^2 \sqrt{\sigma}}, \tag{5.89}$$

and then Eq. (5.87) is still applicable. In the following subsection we will give verifications for our Eq. (5.89).

5.5.2 Numerical verification

In Section 5.4, from the droplet simulation and the Laplace law, we know that the surface tension $\sigma = 0.00743$ mu/ts^2 when $\kappa = 0.01$. Similarly we get $\sigma = 0.0112$ mu/ts^2 when $\kappa = 0.02$.

In our numerical simulations the computational domain is 200×60. Periodic boundary conditions are applied on the left and right boundaries. Suppose (i, j) are the horizontal and vertical lattice indices, respectively. The upper $(j = ly)$ and lower $(j = 1)$ layers are solid nodes. The bounce-back rule should be implemented for these solid nodes. For simplicity, $\tau = 1$ is adopted.

From Eq. (5.35) we know that both $\nabla\rho$ and $\nabla^2\rho$ have to be specified or calculated at the wall nodes. The first-order partial derivative in the normal direction of the wall should be specified. The second-order derivatives, i.e., $\nabla^2\rho$, at the wall nodes are calculated using a hybrid of the biased and central difference schemes (Huang et al. 2009):

$$\partial_x\rho|_{i,j} = (\rho_{i+1,j} - \rho_{i-1,j})/(2\Delta x), \tag{5.90}$$

$$\partial_y\rho|_{i,j} = -\omega/\kappa, \tag{5.91}$$

$$\partial_{xx}\rho|_{i,j} = (\rho_{i+1,j} - 2\rho_{i,j} + \rho_{i-1,j})/(\Delta x)^2, \tag{5.92}$$

$$\partial_{yy}\rho|_{i,j} = \frac{1}{4\Delta x^2}(6\omega\Delta x/\kappa + \rho_{i,j+2} + 4\rho_{i,j+1} - 5\rho_{i,j}). \tag{5.93}$$

The biased difference scheme used in the derivation of Eq. (5.93) is obtained through (Briant et al. 2002, Huang et al. 2009)

$$\partial_{yy}\rho|_{i,j} = (-3\partial_y\rho|_{i,j} + 4\partial_y\rho|_{i,j+1} - \partial_y\rho|_{i,j+2})/(2\Delta x), \qquad (5.94)$$

where $\partial_y\rho|_{i,j+1}$ and $\partial_y\rho|_{i,j+2}$ are substituted by

$$\partial_y\rho|_{i,j+1} = (\rho_{i,j+2} - \rho_{i,j})/(2\Delta x), \qquad (5.95)$$

$$\partial_y\rho|_{i,j+2} = (3\rho_{i,j+2} - 4\rho_{i,j+1} + \rho_{i,j})/(2\Delta x). \qquad (5.96)$$

For the other fluid nodes inside the computational domain, the first- and second-order partial derivatives can be conveniently obtained through a central difference scheme.

Applying the above boundary condition one can obtain different contact angles, as illustrated in Figure 5.5, through changing the parameter ω. In the simulations the initial shapes of "gray" phase (with $\rho_l = 4.52$) are half circles with radii 25 lu attached to the wall and the final steady states are shown in the figure. From Figure 5.5 it is seen that the simulated contact angles agree well with the theoretical ones (θ^a) calculated from Eq. (5.87). This equation seems to

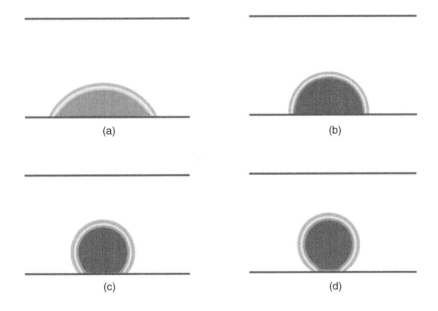

Figure 5.5 Different contact angles obtained through adjusting the parameter ω. The computational domain is 200×60 lu^2, $\tau = 1$, and $\kappa = 0.02$. The ω, measured θ, and calculated θ^a are (a) $\omega = 0.003$, $\theta = 62.1°$, $\theta^a = 57.0°$, (b) $\omega = 0.0$, $\theta = 93.1°$, $\theta^a = 90°$, (c) $\omega = -0.004$, $\theta = 133.1°$, $\theta^a = 136.2°$, and (d) $\omega = -0.0055$, $\theta = 152.3°$, $\theta^a = 169.2°$.

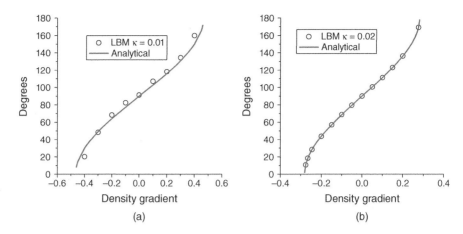

Figure 5.6 The equilibrium contact angle as a function of the density gradient at the wall, i.e., $-\frac{\omega}{\kappa}$. The circles are angles measured from LBM results. The solid lines are calculated from Eq. (5.87): (a) $\kappa = 0.01$ and (b) $\kappa = 0.02$. In the LBM simulation, the computational domain is 200×60 and $\tau = 1$.

have limitations similar to those of the equation for the MCMP SC in Chapter 3, Eq.(3.33). There is a large error (17°) at the high contact angle (Figure 5.5(d)).

The contact angles as a function of the density gradient $(\nabla\rho)_s$ near the wall are shown in Figure 5.6. From the figure we can see that the simulated contact angles (circles) agree relatively well with the theoretical ones (solid lines) calculated from Eq. (5.87).

In the code provided in this chapter we should insert the small segment as follows for calculation of the first derivative of density. Here the array "obst(x,y)" is used to indicate whether the node (x,y) is a wall node ("obst$(x,y) = 1$" or a fluid node "obst$(x,y) = 0$").

```
if(obst(x,y) .eq. 1) then
    Fx(x,y) = 0.
    Fy(x,y) = -omg1/kappa
    endif
```

For the second derivative

```
    if(y.eq. 1) then
    temp(x,y)= 0.25d0*(6.d0* omg1/kappa +rho(x,3) + 4.d0*rho(x,2)
&   -5.d0*rho(x,1))
    endif
    if(y.eq. ly) then
    temp(x,y)= 0.25d0*(6.d0 *omg1/kappa +rho(x,y-2) + 4.d0*rho(x,y-1)
&   -5.d0*rho(x,y))
    endif
```

In the implementation we should notice that the bounce-back scheme should be applied. After the streaming step is performed, the unknown distribution function in the lower and upper layers of the computational domain (wall nodes) can be obtained through the simple bounce-back scheme (He et al. 1997). For example, at layer $y = 1$, the unknowns f_2, f_5, and f_6 are obtained through $f_2 = f_4$, $f_5 = f_7$, and $f_6 = f_8$.

It is noted that ρ on the solid node (ρ_s) should not be specified in the code. It should be calculated as usual through summation of the distribution function, i.e., $\rho_s = \sum_i f_i$, after streaming and the above bounce-back step are implemented.

5.6 Swift free-energy-based multi-component multiphase LBM

We do not review Swift's multi-component multiphase model here. For more information readers should refer to Swift et al. (1996). For specifying the contact angle in this model, Grubert and Yeomans (1999) give some examples. Langaas and Yeomans (2000) used the binary fluid model to study viscous fingering in two dimensions. Takada et al. (2001) applied the model to investigate bubble motion under gravity.

5.7 Appendix

```
NOTES:
! 1. This code can be used to simulate the phase separation case.
! 2. The whole project is composed of head.inc, MAIN.for,
! Getfeq.for, collision.for.
! 3. head.inc is a header file, where some common blocks are defined.
! 4. Before running this project, please create a new sub-folder
!    named 'out' under
! the working directory.
! 5. Here van der Waals EOS is given as an example. The other EOS can
! be included similarly.


c FILE head.inc
C===========================================
        integer lx,ly
c Define the dimension of the computational domain
        PARAMETER(lx=100,ly=100)

c Array xc, yc are $e_{ix}$ and $e_{iy}$, respectively.
c Array t_k is the weighting coefficient in f_{i}^{eq}
        common/b/ xc(0:8),yc(0:8),t_k(0:8)
        real*8 xc,yc,t_k

c 'cc' is the lattice velocity. 'RR' is the droplet's initial radius.
c a, b, TT are parameters in van der Waals EOS.
```

```
      common/vel/ c_squ,cc, RR,df,  a,b, TT
      real*8 c_squ,cc, RR, df,  a,b,  TT

c \tau is the relaxation parameter and 'nu' is the kinematic
!    viscosity.
c \kappa is a key parameter to adjust the surface tension strength
!      common/gg/  kappa, tau, nu
      real*8  kappa, tau, nu

c 'Nwri' is the frequency of output (dump data every 'Nwri'
!    time step).
c t_max is the desired maximum time step.
      common/oo/  Nwri, t_max
      integer Nwri, t_max



c FILE  MAIN.for c===========================================  c In
this program the Swift free-energy model for D2Q9 model is
presented.

      program Swift_D2Q9
      implicit none
      include "head.inc"
      integer time, k, m, n

      integer obst(lx,ly),x,y
c 'obst' denotes the lattice node occupied by fluid (obst=0) or
! solid (obst=1)
c The 9-speed lattice is used here:
c          6    2    5
c           \  |  /
c          3 -  0 - 1
c           /  |  \
c          7    4    8
c Array u_x and u_y are x- and y- components of the velocity,
! respectively.
c 'rho' denotes the density of the fluid.
c-------------------------------------------
c     Author: Haibo Huang, huanghb@ustc.edu.cn
c-------------------------------------------

      real*8  u_x(lx,ly),u_y(lx,ly),rho(lx,ly)

c Array 'ff' is the distribution function. Array 'p' is the pressure.
      real*8  ff(0:8,lx,ly),Fx(lx,ly),Fy(lx,ly),
     &   p(lx,ly)

      data xc/0.d0,1.d0,0.d0, -1.d0, 0.d0, 1.d0, -1.d0, -1.d0, 1.d0/,
     &     yc/0.d0,0.d0,1.d0, 0.d0, -1.d0, 1.d0, 1.d0, -1.d0, -1.d0/

      character filename*16, B2*2, C*3, D*2

c---------------------------------------------------
c   lattice speed cc=1 lu/ts.
      cc = 1.d0
```

```
      c_squ = cc *cc / 3.d0

      Nwri= 2000

      write (6,*) '@@@__Swift_D2Q9_starting_..._____@@@'
      write (6,*) 'Computational domain lx = ,', lx, ' ly = ',ly

c Begin initialization-------
c Read parameters.
      call read_parametrs()
c Initialize the fluid nodes.
      call read_obstacles(obst)
c Initialize the macro variables and distribution functions.
      call init_density(obst,u_x ,u_y ,rho ,p, ff, Fx, Fy)
c Dump initial flow field
        call write_results(obst,rho,u_x,u_y,p, 0)

c Begin iteration ------

      do 100 time = 1, t_max
c Dump flow field data
        if ( mod(time, Nwri) .eq. 0) then
            write(*,*) time
            call write_results(obst,rho,u_x,u_y,p, time)
        end if
c Streaming step.
        call stream(obst,ff )

c Update the macro variables.
        call getuv(obst,u_x ,u_y, rho, ff )

c Collision step.
        call collision(tau,obst,u_x,u_y,rho ,ff, p )

  100 continue
c End iteration ------

      write (6,*) '------   end   ---------'
      end

c-----------------------------------------------
      subroutine read_parametrs()
      implicit none
      include "head.inc"
      real*8 Re1, V1
! Parameters  in the van der Waals EOS.
      TT = 0.56d0
      a = 9.d0/49.d0
      b = 2.d0/21.d0

! Parameter to adjust the surface tension.
       kappa = 0.01d0

! The other parameters control the flow.
      t_max  = 40000
      tau = 1.0d0     ! Relaxation time constant
```

```
         nu = ( tau -0.5d0 )/3.d0  ! Kinematic viscosity
         end


c-------------------------------------------------
         subroutine read_obstacles(obst)
         implicit none
         include "head.inc"
         integer  x,y,obst(lx,ly)
           do 10 y = 1, ly
             do 40 x = 1, lx
   40          obst(x,y) =  0    ! Set all nodes to be fluid nodes
   10      continue
         end


c-------------------------------------------------
         subroutine init_density(obst,u_x,u_y,rho,p, ff, Fx, Fy)
         implicit none
         include "head.inc"
         integer i,j,x,y,k,n,obst(lx,ly)
         real*8  u_squ, xx, u_n(0:8),u_x(lx,ly),u_y(lx,ly),
        & rho(lx,ly),ff(0:8,lx,ly), p(lx,ly)
         real*8 Fx(lx,ly),Fy(lx,ly)

c Initialize the macro variables.
         do 10 y = 1, ly
           do 10 x = 1, lx
           u_x(x,y) = 0.d0
           u_y(x,y) = 0.d0
           Fx(x,y) = 0.d0
           Fy(x,y) = 0.d0
c             rho(x,y) = 2.3d0
      10 continue

         do 20 x = 1, lx
         do 30 y = 1, ly

c----------------------
c Initial a uniform density with small disturbance.
         call random_number (xx) ! Get a random number between [0,1]
             rho(x,y) = 3.8d0 + 0.01d0 * xx
c----------------------
c  Initialize a droplet or a bubble inside the computational domain
c        if(  sqrt(float(x-lx/2)**2+float(y-ly/2)**2) .lt. 30.d0)
c        &  rho(x,y) = 4.5d0

    30 continue
    20 continue

c  Initialize the distribution functions.
         call get_feq(rho,u_x,u_y,ff, p)

         end


c----------------------------------------------------------
         subroutine EOS(rho, tmp)
         include 'head.inc'
```

```fortran
      real*8 rho, tmp
        tmp= rho*TT/(1.d0 -rho*b) -a *rho *rho
      return
      end


c-----------------------------------------------------
      subroutine  write_results(obst,rho,upx,upy,p, n)
      implicit none
      include "head.inc"
      integer  x,y,n,obst(lx,ly)
      real*8  rho1, rho(lx,ly), upx(lx,ly), upy(lx,ly)
     & , p(lx,ly)
      character filename*20,  D*7

      write(D,'(i7.7)') n
      filename='out/swift'//D//'.plt'

      open(41,file=filename)

      write(41,*) 'variables = x, y, u, v, rho1, p, obst'
      write(41,*) 'zone i=', lx, ', j=', ly, ', f=point'

c   Write results to file (an ASCII file)
      do 10 y = 1, ly
      do 10 x = 1, lx
         write(41,9) x, y,
     &        upx(x,y), upy(x,y), rho(x,y), p(x,y), obst(x,y)
   10 continue

    9 format(i4,i5, 3f15.8, f15.8, i4 )

      close(41)
      end
c-----------------------------------------------------




c FILE Getfeq.for
C=========================================

      subroutine get_feq(rho, u_x, u_y, fequ, pre)
      implicit none
      include "head.inc"
      integer x,y,yn,yp,xn,xp
      real*8 Fx(lx,ly),Fy(lx,ly),temp(lx,ly),
     & rho(lx,ly), u_x(lx,ly), u_y(lx,ly),R

      real*8 fequ(0:8,lx,ly),pre(lx,ly),u_n(0:8),rh,u_squ
      real*8 A1, A2, A0, B1, B2, D1, D2, C0, C1, C2, p, Gxx1, Gxx2,
     & Gyy1, Gyy2, Gxy1, Gxy2
      integer k


        do 15 x = 1, lx
          do 15 y = 1, ly
```

```
! Periodic boundary condition
        xp = x+1
        yp = y+1
        xn = x-1
        yn = y-1
        if (xp.gt.lx )  xp = 1
        if (xn.lt.1 )   xn = lx
        if (yp.gt.ly )  yp = 1
        if (yn.lt.1 )   yn = ly
! Calculate Laplacian operator on density
        temp(x,y) =
     *   ( ( rho(xp,y) + rho(xn,y) + rho(x,yp)+ rho(x,yn) )*4.d0/6.d0
     *       +( rho(xp,yp)+ rho(xn,yn)
     *       +  rho(xp,yn)+ rho(xn,yp) )/6.d0
     *       - 20.d0* rho(x,y)/6.d0
     *       )

   15 continue

c----------------------------------------
      do 20 x = 1, lx
        do 20 y = 1, ly
! Periodic boundary condition
        xp = x+1
        yp = y+1
        xn = x-1
        yn = y-1
        if (xp.gt.lx )  xp = 1
        if (xn.lt.1 )   xn = lx
        if (yp.gt.ly )  yp = 1
        if (yn.lt.1 )   yn = ly

! Calculate density gradient. Either simple finite differ-
ence or homogeneous
! finite difference can be used.
        Fx(x,y) = (rho(xp,y)- rho(xn,y) )/2.d0
c     * ( (rho(xp,y)- rho(xn,y) )/3.d0
c     *     +(rho(xp,yp)- rho(xn,yn) )/12.d0
c     *     +(rho(xp,yn)- rho(xn,yp) )/12.d0
c     *     )

        Fy(x,y) =  (rho(x,yp)- rho(x,yn) )/2.d0
c     & ( (rho(x,yp)- rho(x,yn) )/3.d0
c     &     +(rho(xp,yp)- rho(xn,yn) )/12.d0
c     &     +(rho(xn,yp)- rho(xp,yn) )/12.d0  )

   20 continue

      do 30 x = 1, lx
        do 30 y = 1, ly

        rh = rho(x,y)

        u_squ = u_x(x,y)*u_x(x,y) + u_y(x,y)*u_y(x,y)
```

```
! Get thermodynamic pressure from the equation of state
         call EOS(rh, p)
         pre(x,y) = p

! Calculate the coefficients or terms in the equilibrium
! distribution function.

         A1 = 1.d0/3.d0 *(p- kappa*rh *temp(x,y) )
         A2 = A1/4.d0
         A0 = rho(x,y) - 5.d0/3.d0*(p- kappa*rh *temp(x,y))
         B1 = rho(x,y)/3.d0
         B2 = B1/4.d0
         D1 = rho(x,y)/2.d0
         D2 = D1/4.d0
         C1 = -rho(x,y)/6.d0
         C2 = C1/4.d0
         C0 = -rho(x,y)*2.d0/3.d0   !!!important
         Gxx1 = kappa/4.d0*(Fx(x,y)*Fx(x,y)- Fy(x,y)*Fy(x,y))
         Gxx2 = Gxx1/4.d0
         Gyy1 = -Gxx1
         Gyy2 = -Gxx2
         Gxy1 = kappa/2.d0 *Fx(x,y)*Fy(x,y)
         Gxy2 = Gxy1/4.d0

      do 65 k = 0, 8
           u_n(k)  = xc(k)*u_x(x,y) + yc(k)*u_y(x,y)
   65 continue

c   Calculate the equilibrium distribution functions.

         fequ(0,x,y) = A0 + C0* u_squ

         fequ(1,x,y) = A1 + B1*u_n(1) +C1*u_squ +D1*u_n(1)*u_n(1)
   &       + Gxx1
         fequ(2,x,y) = A1 + B1*u_n(2) +C1*u_squ +D1*u_n(2)*u_n(2)
   &       + Gyy1
         fequ(3,x,y) = A1 + B1*u_n(3) +C1*u_squ +D1*u_n(3)*u_n(3)
   &       + Gxx1
         fequ(4,x,y) = A1 + B1*u_n(4) +C1*u_squ +D1*u_n(4)*u_n(4)
   &       + Gyy1

         fequ(5,x,y) = A2 + B2*u_n(5) +C2*u_squ +D2*u_n(5)*u_n(5)
   &       + Gxx2 + Gyy2 + 2.d0 * Gxy2
         fequ(6,x,y) = A2 + B2*u_n(6) +C2*u_squ +D2*u_n(6)*u_n(6)
   &       + Gxx2 + Gyy2 - 2.d0 * Gxy2
         fequ(7,x,y) = A2 + B2*u_n(7) +C2*u_squ +D2*u_n(7)*u_n(7)
   &       + Gxx2 + Gyy2 + 2.d0 * Gxy2
         fequ(8,x,y) = A2 + B2*u_n(8) +C2*u_squ +D2*u_n(8)*u_n(8)
   &       + Gxx2 + Gyy2 - 2.d0 * Gxy2

   30 continue

      end
```

```fortran
c FILE collision.for
C===========================================
      subroutine stream(obst,ff)
      implicit none
      include "head.inc"
      integer   k,obst(lx,ly)
      real*8 ff(0:8,lx,ly),f_hlp(0:8,lx,ly)
      integer   x,y,x_e,x_w,y_n,y_s,l,m,n

      do 10 y = 1, ly
         do 10 x = 1, lx

! Set periodic boundary conditions
            y_n = mod(y,ly) + 1
            x_e = mod(x,lx) + 1
            y_s = ly - mod(ly + 1 - y, ly)
            x_w = lx - mod(lx + 1 - x, lx)

            f_hlp(1,x_e,y  ) = ff(1,x,y)
            f_hlp(2,x  ,y_n) = ff(2,x,y)
            f_hlp(3,x_w,y  ) = ff(3,x,y)
            f_hlp(4,x  ,y_s) = ff(4,x,y)
            f_hlp(5,x_e,y_n) = ff(5,x,y)
            f_hlp(6,x_w,y_n) = ff(6,x,y)
            f_hlp(7,x_w,y_s) = ff(7,x,y)
            f_hlp(8,x_e,y_s) = ff(8,x,y)

   10 continue

      do 20 y = 1, ly
         do 20 x = 1, lx
           do 20 k = 1, 8
           ff(k,x,y) = f_hlp(k,x,y)
   20 continue

      end
C--------------------------------------------------------
      subroutine getuv(obst,u_x,u_y,rho,ff)
      include "head.inc"
      integer x,y,obst(lx,ly)
      real*8  u_x(lx,ly),u_y(lx,ly),rho(lx,ly),
     & ff(0:8,lx,ly)

      do 10 y = 1, ly
         do 10 x = 1, lx

      if(obst(x,y) .eq. 0) then
            rho(x,y) = ff(0,x,y) +ff(1,x,y) +ff(2,x,y)
     &                + ff(3,x,y) +ff(4,x,y) +ff(5,x,y)
     &                + ff(6,x,y) +ff(7,x,y) +ff(8,x,y)
      endif

      if(obst(x,y) .eq. 0) then
          u_x(x,y) = cc* (ff(1,x,y) + ff(5,x,y) + ff(8,x,y)
     &         -(ff(3,x,y) + ff(6,x,y) + ff(7,x,y))) /rho(x,y)
```

```fortran
          u_y(x,y) = cc* (ff(2,x,y) + ff(5,x,y) + ff(6,x,y)
     &          -(ff(4,x,y) + ff(7,x,y) + ff(8,x,y))) /rho(x,y)
      else
      u_x(x,y) = 0.d0
      u_y(x,y) = 0.d0

      endif

  10  continue

      end

c--------------------------------------------------------

c The BGK collision
      subroutine collision(tauc,obst,u_x,u_y,rho,ff, pre)
      implicit none
      include "head.inc"
      integer  l,obst(lx,ly)
      real*8   u_x(lx,ly),u_y(lx,ly),ff(0:8,lx,ly),rho(lx,ly)
      real*8 Fx(lx,ly),Fy(lx,ly), pre(lx,ly)
      integer  x,y,k,ip,jp
      real*8   u_n(0:8),fequ(0:8, lx,ly),u_squ,temp,tauc,ux,uy

        call get_feq(rho, u_x, u_y, fequ, pre)


      do 5 y = 1, ly
        do 5 x = 1, lx
           do 10 k = 0,8
           ff(k,x,y) = fequ(k,x,y)
     &       + (1.d0-1.d0/tauc)*( ff(k,x,y) - fequ(k,x,y) )
  10  continue
  5   continue
      end
```

CHAPTER 6

Inamuro's multiphase Lattice Boltzmann model

6.1 Introduction

The Inamuro model we discuss here also belongs to the class of free-energy models. This model and all models in the following chapters differ from those in the preceding chapters in that while they employ two distribution functions, the purpose of these is solution of the N–S equation and interface tracking governed by the Cahn–Hilliard equation.

This model is able to simulate high-density-ratio two-phase flows but the computational load is extremely heavy. We analyze this model both theoretically and numerically. There have been some applications of this model, for example Gu et al. (2009) investigated droplet impingement on a wall, Ben Salah et al. (2012) investigated two-phase flow in a channel of a polymer electrolyte membrane fuel cell, and Yan and Zu (2007) simulated two-phase flows on partial wetting surface. The model has also been applied in wetting phenomena of chemically heterogeneous surfaces (Iwahara et al. 2003).

6.1.1 Inamuro's method

First we briefly describe the numerical method in accordance with Inamuro et al. (2004). As an example, Inamuro et al. (2004) present their method using the D3Q15 model, which is illustrated in Chapter 2. In Inamuro et al. (2004) the index of the discrete velocity is from 1 to 15. Here, to be consistent with the D3Q15 model in Chapter 2, the index is from 0 to 14. In the method three-particle distribution functions f_i, g_i, and h_i are used. f_i is used to calculate the order parameter, which distinguishes the two phases. g_i is used to predict the velocity field of the flow without a pressure gradient. h_i is used to solve the Poisson equation (velocity field correction using pressure field). The evolution equations for f_i and g_i are (Inamuro et al. 2004)

$$f_i(\mathbf{x} + \mathbf{e}_i \Delta t, t + \Delta t) = f_i(\mathbf{x}, t) - \frac{1}{\tau_f}[f_i(\mathbf{x}, t) - f_i^{eq}(\mathbf{x}, t)] \tag{6.1}$$

Multiphase Lattice Boltzmann Methods: Theory and Application, First Edition.
Haibo Huang, Michael C. Sukop and Xi-Yun Lu.
© 2015 John Wiley & Sons, Ltd. Published 2015 by John Wiley & Sons, Ltd.
Companion Website: www.wiley.com/go/huang/boltzmann

and

$$\bar{g}_i(\mathbf{x} + \mathbf{e}_i \Delta t, t + \Delta t) = \bar{g}_i(\mathbf{x}, t) - \frac{1}{\tau_g}[\bar{g}_i(\mathbf{x}, t) - \bar{g}_i^{eq}(\mathbf{x}, t)] + S_i, \tag{6.2}$$

where the source term S_i is

$$S_i = 3E_i e_{i\alpha} \frac{1}{\rho} \{\partial_\beta[\eta(\partial_\alpha u_\beta + \partial_\beta u_\alpha)]\} \Delta t. \tag{6.3}$$

In the above equations τ_f and τ_g are non-dimensional relaxation parameters. Eq. (6.3) represents the effect of the viscous stress tensor. The order parameter ϕ and the predicted velocity $\mathbf{u}^*$ can calculated through

$$\phi = \sum_{i=0}^{14} f_i \tag{6.4}$$

and

$$\mathbf{u}^* = \sum_{i=0}^{14} \mathbf{e}_i g_i. \tag{6.5}$$

The equilibrium distribution functions f_i^{eq} and g_i^{eq} are (Inamuro et al. 2004)

$$f_i^{eq} = H_i\phi + F_i\left[p_0 - k_f\phi\partial_\alpha^2\phi - \frac{k_f}{6}(\partial_\alpha\phi)^2\right] + 3E_i\phi e_{i\alpha}u_\alpha + E_i k_f G_{\alpha\beta}(\phi)e_{i\alpha}e_{i\beta} \tag{6.6}$$

and

$$g_i^{eq} = E_i\left[1 + 3e_{i\alpha}u_\alpha + \frac{9}{2}e_{i\alpha}u_\alpha e_{i\beta}u_\beta - \frac{3}{2}u_\alpha u_\alpha + \frac{3}{2}\left(\tau_g - 0.5\right)\Delta x(\partial_\alpha u_\beta + \partial_\beta u_\alpha)e_{i\alpha}e_{i\beta}\right]$$
$$+ E_i\frac{k_g}{\rho}G_{\alpha\beta}(\rho)e_{i\alpha}e_{i\beta} - \frac{2}{3}F_i\frac{k_g}{\rho}\left(\frac{\partial\rho}{\partial x_\alpha}\right)^2, \tag{6.7}$$

where

$$G_{\alpha\beta}(\phi) = \frac{9}{2}\partial_\alpha\phi\partial_\beta\phi - \frac{3}{2}\partial_\gamma\phi\partial_\gamma\phi\delta_{\alpha\beta}. \tag{6.8}$$

In the D3Q15 model the coefficients are

$$E_0 = \frac{2}{9}, \quad E_1 = E_2 \dots = E_6 = \frac{1}{9}, \quad E_7 = \dots = E_{14} = \frac{1}{72},$$
$$H_0 = 1, \quad H_1 = H_2 \dots = H_{14} = 0,$$
$$F_0 = -\frac{7}{3}, \quad F_i = 3E_i(i = 1, \dots, 14). \tag{6.9}$$

It is noted that f_i^{eq} (Eq. (6.6)) is the same as that in the model by Swift et al. (1996) except that Eq. (6.6) includes no quadratic term of the flow velocity u_α. The last two terms of Eq. (6.7) represent the effect of interfacial tension (Inamuro et al. 2004).

In the above equations k_f is a constant parameter determining the width of the interface and k_g is a constant parameter determining the strength of the interfacial tension (Inamuro et al. 2004).

In Eq. (6.6) p_0 is calculated by the van der Waals EOS

$$p_0 = \phi \frac{\partial \psi}{\partial \phi} - \psi = \frac{\phi T}{1 - b\phi} - a\phi^2, \tag{6.10}$$

where $\psi(\phi, T) = \phi T ln \left(\frac{\phi}{1-b\phi} \right) - a\phi^2$ and a and b are constants. Two ϕs can coexist at one temperature and pressure, representing a liquid state ϕ_l and a gas state ϕ_g.

For the first and second derivatives that appear in the above equations, the finite difference approximation can be used:

$$\frac{\partial \zeta}{\partial x_\alpha} \approx \frac{1}{10\Delta x} \sum_{i=1}^{14} e_{i\alpha} \zeta(\mathbf{x} + \mathbf{e}_i \Delta t), \tag{6.11}$$

and

$$\frac{\partial^2 \zeta}{\partial x_\alpha^2} \approx \frac{1}{5(\Delta x)^2} \sum_{i=1}^{14} (\zeta(\mathbf{x} + \mathbf{e}_i \Delta t) - 14\zeta(\mathbf{x})), \tag{6.12}$$

where ζ is a macrovariable.

The density in the interface is obtained by using the cut-off values of the order parameter, ϕ_L^* and ϕ_G^*, for the liquid and gas phases with the following relation:

$$\rho = \begin{cases} \rho_G, & \phi < \phi_G^*, \\ \frac{\Delta \rho}{2} \left[\sin \left(\frac{\phi - \bar{\phi}^*}{\Delta \phi^*} \pi \right) + 1 \right] + \rho_G, & \phi_G^* \le \phi \le \phi_L^*, \\ \rho_L, & \phi > \phi_L^*, \end{cases} \tag{6.13}$$

where ρ_G and ρ_L are densities of gas and liquid phases, respectively. $\Delta \rho = \rho_L - \rho_G$, $\Delta \phi^* = \phi_L^* - \phi_G^*$, and $\bar{\phi}^* = (\phi_L^* - \phi_G^*)/2$. In Inamuro et al. (2004), the cut-off values ϕ_L^* and ϕ_G^* are selected to ensure better numerical stability. The dynamic viscosity η in the interface is calculated through

$$\eta = \frac{\rho - \rho_G}{\rho_L - \rho_G}(\eta_L - \eta_G) + \eta_G, \tag{6.14}$$

where η_G and η_L are the dynamic viscosities of the gas and liquid phases, respectively.

6.1.2 Comment on the presentation

To make the physical units of the above presentation consistent with our standard notation in this book we suggest writing Eq. (6.7) in the form that follows. It should be kept in mind that usually the distribution function in the LBM has units of density. Because g_i^{eq} does not have density units, it is assumed that g_i is a non-dimensional variable. It is noted that the coefficients E_i, F_i, and H_i are non-dimensional. Here g_i^{eq} is written as

$$g_i^{eq} = E_i \left[1 + \frac{e_{i\alpha} u_\alpha}{c_s^2} + \frac{e_{i\alpha} u_\alpha e_{i\beta} u_\beta}{2c_s^4} - \frac{u_\alpha u_\alpha}{2c_s^2} + \frac{3}{2} \left(\tau_g - 0.5 \right) \frac{\Delta x}{c^3} (\partial_\alpha u_\beta + \partial_\beta u_\alpha) e_{i\alpha} e_{i\beta} \right]$$
$$+ E_i \frac{k_g}{\rho c^2} G_{\alpha\beta}(\rho) e_{i\alpha} e_{i\beta} - \frac{2}{3} F_i \frac{k_g}{\rho} \left(\frac{\partial \rho}{\partial x_\alpha} \right)^2. \tag{6.15}$$

To be consistent in dimension Eq. (6.3) should be written as

$$S_i = 3E_i e_{i\alpha} \frac{\Delta x}{\rho c^3} \{\partial_\beta [\eta(\partial_\alpha u_\beta + \partial_\beta u_\alpha)]\}. \tag{6.16}$$

The other equations in Section 6.1.1 can be rewritten based on dimensional analysis.

6.1.3   Chapman–Enskog expansion analysis

Following the Chapman–Enskog expansion routine we have

$$O(\varepsilon^0) : (g_i^{(0)} - g_i^{eq})/\tau = 0 \tag{6.17}$$

$$O(\varepsilon^1) : (\partial_{t_1} + e_{i\alpha}\partial_\alpha)g_i^{(0)} + \frac{1}{\tau}g_i^{(1)} = S_i \tag{6.18}$$

$$O(\varepsilon^2) : \partial_{t_2}g_i^{(0)} + \left(1 - \frac{1}{2\tau}\right)(\partial_{t_1} + e_{i\alpha}\partial_\alpha)g_i^{(1)} + \frac{1}{2}(\partial_{t1} + e_{i\alpha}\partial_\alpha)S_i + \frac{1}{\tau}g_i^{(2)} = 0 \tag{6.19}$$

From the definition of E_i we have

$$\sum_i E_i = 1, \quad \sum_i E_i e_{i\alpha} = 0, \quad \sum_i E_i e_{i\alpha}e_{i\beta} = c_s^2 \delta_{\alpha\beta} \tag{6.20}$$

The zeroth, first, and second moments of g_i^{eq} are

$$\sum_i g_i^{eq} = 1 - \frac{3}{2}u^2 + \frac{9}{2}c_s^2 u^2 + \frac{3}{2}(\tau_g - 0.5)\Delta x c_s^2 (2\partial_\alpha u_\alpha) + \frac{k_g}{\rho}c_s^2 \left[\frac{9}{2}(\partial_\alpha\rho)^2 - \frac{3}{2}(\partial_\alpha\rho)^2\right]$$

$$= 1 + (\tau_g - 0.5)\Delta x(\partial_\alpha u_\alpha) + \frac{k_g}{\rho}(\partial\rho)^2, \tag{6.21}$$

$$\sum_i g_i^{eq} e_{i\gamma} = u_\gamma, \tag{6.22}$$

and

$$\sum_i g_i^{eq} e_{i\gamma}e_{i\delta} = c_s^2 \delta_{\gamma\delta} - \frac{1}{2}u_\alpha^2 \delta_{\gamma\delta} + \frac{9}{2} \cdot \frac{1}{9}(u_\alpha^2 \delta_{\gamma\delta} + 2u_\gamma u_\delta)$$

$$+ \frac{3}{2}(\tau_g - 0.5)\Delta x \frac{1}{9}[2\partial_\alpha u_\alpha \delta_{\gamma\delta} + 2(\partial_\gamma u_\delta + \partial_\delta u_\gamma)]$$

$$+ \underbrace{\frac{k_g}{\rho}\left\{\partial_\gamma\phi\partial_\delta\phi - \frac{1}{3}(\partial_\lambda\phi)^2 \delta_{\gamma\delta}\right\} - \frac{2}{3}\frac{k_g}{\rho}(\partial_\alpha\rho)^2 \delta_{\gamma\delta}}$$

$$= c_s^2 \delta_{\gamma\delta} + u_\gamma u_\delta + \frac{1}{3}(\tau - 0.5)\Delta x \left[\partial_\alpha u_\alpha \delta_{\gamma\delta} + (\partial_\gamma u_\delta + \partial_\delta u_\gamma)\right]$$

$$+ \frac{k_g}{\rho}[\partial_\gamma\rho\partial_\delta\rho - (\partial_\alpha\rho)^2 \delta_{\gamma\delta}]. \tag{6.23}$$

We will illustrate how the underbraced term emerges in the following two equations. In the above derivation it should be noted that

$$\sum_i E_i e_{i\alpha} e_{i\beta} e_{i\gamma} e_{i\delta} = c_s^4 (\delta_{\alpha\beta}\delta_{\gamma\delta} + \delta_{\alpha\gamma}\delta_{\beta\delta} + \delta_{\alpha\delta}\delta_{\beta\gamma}). \tag{6.24}$$

From the definition of $G_{\alpha\beta}$ (Eq. (6.8)) and the above formula we have

$$
\begin{aligned}
&\sum_i E_i \frac{k_g}{\rho} G_{\alpha\beta}(\rho) e_{i\alpha} e_{i\beta} e_{i\delta} e_{i\gamma} \\
&= \sum_i E_i \frac{k_g}{\rho} \left[\frac{9}{2}\partial_\alpha\phi\partial_\beta\phi - \frac{3}{2}\partial_\gamma\phi\partial_\gamma\phi\delta_{\alpha\beta} \right] e_{i\alpha} e_{i\beta} e_{i\delta} e_{i\gamma} \\
&= c_s^4 \frac{k_g}{\rho}(\delta_{\alpha\beta}\delta_{\gamma\delta} + \delta_{\alpha\gamma}\delta_{\beta\delta} + \delta_{\alpha\delta}\delta_{\beta\gamma}) \left[\frac{9}{2}\partial_\alpha\phi\partial_\beta\phi - \frac{3}{2}\partial_\gamma\phi\partial_\gamma\phi\delta_{\alpha\beta} \right] \\
&= c_s^4 \frac{k_g}{\rho}\delta_{\alpha\beta}\delta_{\gamma\delta} \left[\frac{9}{2}\partial_\alpha\phi\partial_\beta\phi - \frac{3}{2}(\partial_\lambda\phi)^2\delta_{\alpha\beta} \right] \\
&\quad + c_s^4 \frac{k_g}{\rho} \left\{ 9\partial_\gamma\phi\partial_\delta\phi - (\delta_{\alpha\gamma}\delta_{\beta\delta} + \delta_{\alpha\delta}\delta_{\beta\gamma})\frac{3}{2}(\partial_\lambda\phi)^2\delta_{\alpha\beta} \right\} \\
&= c_s^4 \frac{k_g}{\rho} \left[\frac{9}{2}(\partial_\alpha\phi)^2\delta_{\gamma\delta} - \frac{9}{2}(\partial_\lambda\phi)^2\delta_{\gamma\delta} \right] + c_s^4 \frac{k_g}{\rho}\{9\partial_\gamma\phi\partial_\delta\phi - 3(\partial_\lambda\phi)^2\delta_{\gamma\delta}\} \\
&= \frac{k_g}{\rho} \left\{ \partial_\gamma\phi\partial_\delta\phi - \frac{1}{3}(\partial_\lambda\phi)^2\delta_{\gamma\delta} \right\}.
\end{aligned}
\tag{6.25}
$$

In the above derivation we should note the fact that for the 3D cases $\delta_{\alpha\beta}\delta_{\alpha\beta} = 3$ and $\delta_{\alpha\beta}\delta_{\alpha\gamma} = \delta_{\beta\gamma}$ (refer to Chapter 1 for more details).

For the third moment of g_i^{eq} we have

$$\sum_i e_{i\alpha} e_{i\beta} e_{i\gamma} g_i^{eq} = \sum_i E_i 3 e_{i\alpha} e_{i\beta} e_{i\gamma} e_{i\delta} u_\delta = c_s^2 (u_\beta\delta_{\alpha\gamma} + u_\gamma\delta_{\alpha\beta} + u_\alpha\delta_{\beta\gamma}). \tag{6.26}$$

The term $\sum(e_{i\alpha} e_{i\beta} g_i^{(1)})$ is necessary to derive the momentum equation. From Eq. (6.18) we have $g_i^{(1)} = -\tau_g(\partial_{t1} + e_{i\alpha}\partial_\alpha)g_i^{(0)} + \tau_g S_i$. Hence,

$$
\begin{aligned}
\sum e_{i\alpha} e_{i\beta} g_i^{(1)} &= -\tau_g \sum_i e_{i\alpha} e_{i\beta}(\partial_t + e_{i\gamma}\partial_\gamma)g_i^{(0)} + \tau_g \sum_i e_{i\alpha} e_{i\beta} S_i \\
&= -\tau_g \left[\frac{1}{3}(\partial_\alpha u_\beta + \partial_\beta u_\alpha + \partial_\gamma u_\gamma\delta_{\alpha\beta}) \right].
\end{aligned}
\tag{6.27}
$$

In the above derivation, $\partial_t \sum_i e_{i\alpha} e_{i\beta} g_i^{(0)}$ is assumed to be a high-order term. This temporal derivative can be substituted by the spatial derivatives. Refer to Section 5.2 or Amir et al. (2015) for more details, which are complicated and tedious.

For the moments of the source term S_i we have

$$\sum_i S_i = 0, \quad \sum_i S_i e_{i\gamma} = \frac{1}{\rho}\partial_\beta[\eta(\partial_\gamma u_\beta + \partial_\beta u_\gamma)], \quad \sum_i S_i e_{i\beta} e_{i\gamma} = 0. \tag{6.28}$$

With the above equations we can derive the momentum equation through the Chapman–Enskog expansion (Eqs (6.18) and (6.19)). Multipling Eq. (6.17) by $e_{i\beta}$ and summing on i (notice Eq. (6.23)) we have

$$\partial_{t1} u_\beta + \partial_\alpha \left\{ c_s^2 \delta_{\alpha\beta} + u_\beta u_\alpha + \frac{1}{3}(\tau - 0.5)\Delta x [\partial_\gamma u_\gamma \delta_{\alpha\beta} + (\partial_\alpha u_\beta + \partial_\beta u_\alpha)] \right\}$$

$$+ \partial_\alpha \left\{ \frac{k_g}{\rho} \left[\partial_\alpha \rho \partial_\beta \rho - (\partial_\gamma \rho)^2 \delta_{\alpha\beta} \right] \right\} = \frac{1}{\rho} \partial_\alpha [\eta(\partial_\alpha u_\beta + \partial_\beta u_\alpha)]. \qquad (6.29)$$

It is noted that in the above derivation $\sum_i e_{i\beta} g_i^{(1)} = 0$ is used. It is easy to derive: from Eqs (6.5) and (6.22) we know $\sum_i e_{i\beta} (g_i^{eq} + \varepsilon g_i^{(1)} + \varepsilon^2 g_i^{(2)}) = u_\beta = \sum_i e_{i\beta} g_i^{eq}$. Hence, $\sum_i e_{i\beta} g_i^{(1)} = 0$ and $\sum_i e_{i\beta} g_i^{(2)} = 0$.

Let us proceed to $O(\varepsilon^2)$. Multiplying Eq. (6.19) by $e_{i\beta}$ and summing on i (notice Eq. (6.27)) we have

$$\partial_{t2} u_\beta + \left(1 - \frac{1}{2\tau_g} \right) \partial_\alpha \left(\sum_i e_{i\alpha} e_{i\beta} g_i^{(1)} \right) + \frac{1}{2} \partial_{t1} \left(\sum_i e_{i\beta} S_i \right) = 0. \qquad (6.30)$$

Noticing Eq. (6.27) we have

$$\partial_{t2} u_\beta - \frac{1}{3}(\tau_g - 0.5)\partial_\alpha (\partial_\alpha u_\beta + \partial_\beta u_\alpha + \partial_\gamma u_\gamma \delta_{\alpha\beta}) + \frac{1}{2} \partial_{t1} \left(\sum_i e_{i\beta} S_i \right) = 0. \qquad (6.31)$$

Combining Eqs (6.29) and (6.31), i.e., Eq. (6.29) + $\varepsilon \times$Eq. (6.31), where $\varepsilon = \Delta x = \Delta t$, and omitting terms $\frac{1}{2} \partial_{t1} (\sum e_{i\beta} S_i)$ we have

$$\partial_t u_\beta + \partial_\alpha \{ c_s^2 \delta_{\alpha\beta} + u_\beta u_\alpha \}$$

$$= \frac{1}{\rho} \partial_\alpha [\eta(\partial_\alpha u_\beta + \partial_\beta u_\alpha)] - \partial_\alpha \left\{ \frac{k_g}{\rho} \left[\partial_\alpha \rho \partial_\beta \rho - (\partial_\gamma \rho)^2 \delta_{\alpha\beta} \right] \right\}. \qquad (6.32)$$

This is almost identical to the non-dimensional form found in Eq. (27) in Inamuro et al. (2004).

As we know, the viscous tensor can be obtained through the summation of the second moments of the non-equilibrium distribution function in the LBM. However, in the above derivation we found that this advantage of the LBM is lost because the viscous stress tensor is obtained through the finite difference method instead of through summation of the second moments of the non-equilibrium distribution function. The viscous tensor obtained through second moments of the non-equilibrium distribution function is canceled by the last term inside the square bracket in Eq. (6.7); the final recovered viscous term comes from the source term of g_i in the LBE.

This scheme eliminates the advantage of the LBM. In this scheme τ_g is not related to viscosity as in the common LBM. The dynamic viscosity is determined by the order parameter ϕ (Eq. (6.14)) and the dynamic viscosities η_L and η_G, which can be set independently.

6.1.4   Cahn–Hilliard equation (equation for order parameter)

The Cahn–Hilliard (CH) equation is usually used to track the interface explicitly (Jacqmin 1999). One may ask that why there are no CH equations in Chapters 1 to 5. Actually in Chapters 1, 2, 3, and 5 the thermodynamic pressure or the non-ideal EOS is involved in the recovered N–S equations. The models can track interfaces implicitly. Hence, the pressure term in the N–S equation partially takes the role of the CH equation. Even in Chapter 4 (RK model) we believe that the complicated second collision term and the recoloring step may lead to a non-ideal EOS. On the other hand, if there are no thermodynamic effects (non-ideal EOS) in the pressure term of the recovered N–S equations, an extra CH equation is required. In all of the following chapters the CH equation is used.

The CH equation (evolution equation of the interface) is recovered from the evolution equation of f_i (Eq. (6.1)). Through the Chapman–Enskog expansion we obtained similar equations at different scales as Eqs (6.17)–(6.19). Summing Eq. (6.18) on i (g_i is replaced with f_i and $S_i = 0$) we have

$$\partial_{t1}\phi + \partial_\alpha(\phi u_\alpha) = 0. \tag{6.33}$$

In the above derivation the following formulae are used:

$$\sum_i f_i^{eq} = \phi, \quad \sum_i f_i^{(1)} = 0, \quad \text{and} \sum_i f_i^{eq} e_{i\beta} = \phi u_\beta. \tag{6.34}$$

Summing Eq. (6.19) on i (g_i is replaced with f_i and $S_i = 0$) we have

$$\partial_{t2}(\sum_i f_i^{(0)}) + \left(1 - \frac{1}{2\tau_f}\right)\partial_\alpha(\sum_i e_{i\alpha} f_i^{(1)}) = 0. \tag{6.35}$$

We have to get $\sum_i e_{i\alpha} f_i^{(1)}$. From Eq. (6.18) we have

$$f_i^{(1)} = -\tau_f \partial_{t1}(f_i^{(0)}) - \tau_f \partial_\beta(e_{i\beta} f_i^{(0)}). \tag{6.36}$$

Substituting Eq. (6.36) into Eq. (6.35) we can obtain

$$\partial_{t2}\phi - (\tau_f - 0.5)\partial_\alpha(\partial_{t1}\phi u_\alpha + \partial_\beta P_{\alpha\beta}) = 0, \tag{6.37}$$

where $P_{\alpha\beta}$ is the pressure tensor derived in the following:

$$P_{\alpha\beta} = \sum_i f_i^{eq} e_{i\alpha} e_{i\beta} = \left[p_0 - k_f \phi \partial_\lambda^2 \phi - \frac{k_f}{6}(\partial_\lambda \phi)^2\right]\delta_{\alpha\beta} + k_f \underbrace{\left\{\partial_\alpha \phi \partial_\beta \phi - \frac{1}{3}(\partial_\lambda \phi)^2 \delta_{\alpha\beta}\right\}}$$

$$= \left[p_0 - k_f \phi \partial_\lambda^2 \phi - \frac{k_f}{2}(\partial_\lambda \phi)^2\right]\delta_{\alpha\beta} + k_f \partial_\alpha \phi \partial_\beta \phi. \tag{6.38}$$

The derivation of the underbraced term is identical to that in Eq. (6.25). The pressure tensor $P_{\alpha\beta}$ is exactly that defined in Eq. (19) in Inamuro et al. (2004).

Because the incompressibility condition is satisfied in this method, i.e., $\partial_\alpha u_\alpha = 0$, we have

$$\partial_\alpha \partial_{t1} \phi u_\alpha = \partial_{t1}(u_\alpha \partial_\alpha \phi). \tag{6.39}$$

Since the term $\partial_{t1}(u_\alpha \partial_\alpha \phi)$ is higher than $O(u^2)$ this term can be omitted.

Combining Eqs (6.33) and (6.37) and omitting the above term, finally we have the CH equation, which appears as Eq. (25) in Inamuro et al. (2004):

$$\partial_t \phi + \partial_\alpha(\phi u_\alpha) = \partial_t \phi + u_\alpha \partial_\alpha(\phi) = \lambda \partial_\alpha \partial_\beta P_{\alpha\beta}, \tag{6.40}$$

where

$$\lambda = (\tau_f - 0.5)\Delta x \tag{6.41}$$

is the mobility in the CH equation. Here we note $\phi \partial_\alpha u_\alpha = 0$ for incompressible flows.

6.1.5 Poisson equation

Because $\mathbf{u}^*$ is not divergence free ($\nabla \cdot \mathbf{u}^* \neq 0$), to get the correct $\mathbf{u}$ the Poisson equation has to be solved (Inamuro et al. 2004). It can be solved by the finite difference scheme. However, to avoid having to apply a hybrid strategy, the Poisson equation is still solved by the LBM. The third distribution function h_i is used to solve the Poisson equation. The evolution of h_i is (Inamuro et al. 2004)

$$h_i^{n+1}(\mathbf{x} + \mathbf{e}_i \Delta t) = h_i^n(\mathbf{x}) - \frac{1}{\tau_h}[h_i^n(\mathbf{x} - E_i p^n(\mathbf{x}))] - \frac{1}{3}E_i \frac{\partial u_\alpha^*}{\partial x_\alpha}\Delta x, \tag{6.42}$$

where n is the number of iterations and the relaxation time is

$$\tau_h = \frac{1}{\rho} + \frac{1}{2}. \tag{6.43}$$

The pressure is obtained by

$$p = \sum_{i=0}^{14} h_i. \tag{6.44}$$

Through the Chapman–Engskog expansion we can see that the recovered macroscopic equation is the Poisson equation. It can be derived in the following way. In this derivation Eqs (6.17)–(6.19) are used but g_i is replaced by h_i and $S_i = \frac{-E_i}{3}\partial_\alpha u_\alpha \Delta x$. Summing Eq. (6.18) over i we get

$$\partial_{t1} p = -\frac{1}{3}\partial_\alpha u_\alpha \Delta x. \tag{6.45}$$

Note that in the derivation $\sum_i e_{i\alpha} h_i^{(0)} = 0$, $\sum_i h_i^{(1)} = 0$, and $\sum_i S_i = -\frac{1}{3}\partial_\alpha u_\alpha \Delta x$ are used. In the simulations $\partial_{t1} p = 0$ is ensured because in each time step sub-iterations of h_i are performed to obtain a converged pressure field at the time step. Hence, the incompressible flow condition

$$\partial_\alpha u_\alpha = 0 \tag{6.46}$$

is ensured. This is the macroscopic equation recovered from the Lattice Boltzmann equation for h_i, i.e., Eq. (6.42).

6.2 Droplet collision

In our simulation two liquid droplets having the same diameter D are placed $2D$ apart in a gas phase, and they collide with the relative velocity V. The densities of the liquid and gas are set to be $\rho_L = 50$ and $\rho_G = 1$. The Weber number is defined as $We = \frac{\rho_L D V^2}{\sigma}$ and the Reynolds number $Re = \frac{\rho_L D V}{\eta_L}$. Periodic boundary conditions are used on all sides of the domain. The computational domain is $128 \times 96 \times 96$.

The parameters in the EOS Eq. (6.10) are chosen as $a = 1, b = 6.7$, and $T = 3.5 \times 10^{-2}$. Because of this choice the corresponding $\phi_{max} = 9.714 \times 10^{-2}$ and $\phi_{min} = 1.134 \times 10^{-2}$. The values of ϕ_{max} and ϕ_{min} can be determined through the Maxwell construction. Here ϕ in the EOS (Eq. (6.10)) is analogous to ρ and $\frac{1}{\phi}$ is analogous to molar volume v. The p–v diagram is shown in Figure 6.1. In the diagram the points l and g represent the liquid and gas phases and we can see that when the Maxwell construction is satisfied at equilibrium, approximately $p_0 = 0.0003$. In the diagram $\frac{1}{\phi_l} = 10.29$ and $\frac{1}{\phi_g} = 89$ at equilibrium and we obtained the corresponding $\phi_{max} = \phi_l$ and $\phi_{min} = \phi_g$ mentioned above.

The corresponding cut-off values of the order parameter are $\phi_L^* = 9.2 \times 10^{-2}$ and $\phi_G^* = 1.5 \times 10^{-2}$. The other parameters are fixed at $k_f = 0.5(\Delta x)^2, \tau_f = \tau_g = 1$, $C = 0$ (C is defined in Eq.(6.47)), $\varepsilon = 10^{-5}, D = 32$, and $k_g = 0.0005$ (the corresponding surface tension $\sigma = 0.4024$).

First a case with $We = 19.5$ and $Re = 14,000$ is simulated. In this case the viscosities of the droplet and the gas are $\eta_L = 8 \times 10^{-3}$ and $\eta_G = 1.6 \times 10^{-4}$. The

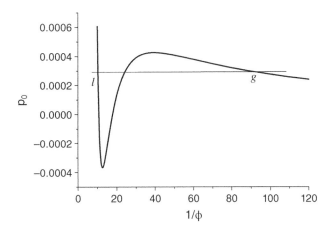

Figure 6.1 p–v diagram of the EOS (Eq. (6.10)). Here ϕ and $\frac{1}{\phi}$ are analogous to density ρ and molar volume v, respectively.

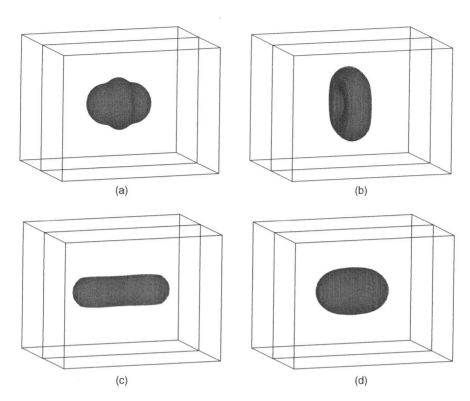

Figure 6.2 Two-droplet collision (coalescence mode): $V = 0.07$ lu/ts, $Re = 14,000$, $We = 19.5$. (a) Dimensionless time $t^* = 1.75$, (b) $t^* = 3.06$, (c) $t^* = 5.69$, and (d) $t^* = 17.5$.

relative velocity $V = 0.07$. That means that, in the head-on collision, one droplet moves with $\frac{V}{2}$ and the other moves with $-\frac{V}{2}$ in the horizontal direction. The result is shown in Figure 6.2. In the figure time is normalized by $t^* = \frac{tV}{D}$. It is seen that at $t^* = 1.75$ the two droplets coalesce. Because of the momentum of the two droplets, the coalesced droplets assume a toroidal shape ($t^* = 3.06$). Thus, due to kinetic energy, the droplet stretched ($t^* = 5.69$). Later the coalesced droplet enters the oscillation mode and finally reaches an equilibrium state (a perfect sphere). Figure 6.2(d) shows a snapshot of the process. This observation is consistent with the expectation that the kinetic energy of the droplet will ultimately be damped by the viscosity of the fluids. Here, although Re is much higher than that simulated in Inamuro et al. (2004) (case $Re = 2000$ and $We = 20.2$ in Figure 2 of Inamuro et al. 2004), the movement mode is identical (coalescence). For the simulated We and Re this observation is consistent with experimental results in Ashgriz and Poo (1990).

A case with $We = 39.7$ and $Re = 2000$ is also simulated. The parameters are almost identical to those above except that in this case the viscosities of the droplet and the gas are $\eta_L = 8 \times 10^{-2}$ and $\eta_G = 1.6 \times 10^{-3}$. The relative velocity

$V = 0.10$ lu/ts. However, the droplets still coalesce and the expected reflexive separation is not observed (Ashgriz and Poo 1990).

Through a detailed check we found that an important detail may contribute to this "unsuccessful" simulation. There is a comment in Inamuro et al. (2004) that a large mobility may damp flows (smaller mobility is preferred). In Zheng et al. (2006) using smaller mobility is also suggested. Here, smaller mobility as given by Eq. (6.41) is possible if τ_f is close to $\frac{1}{2}$. On the other hand, when τ_f is close to $\frac{1}{2}$ simulations are unstable due to numerical instability. Inamuro et al. (2004) suggested an alternate way to decrease the mobility: adding the term

$$S_i' = E_i C \partial_\beta P_{\gamma\beta} e_{i\gamma} \Delta x \tag{6.47}$$

to the function f^{eq} where C is a constant (Inamuro et al. 2004). In this way, the mobility λ becomes

$$\lambda = \left(\tau_f - \frac{1}{2} - \frac{1}{3}C \right) \Delta x. \tag{6.48}$$

This point is easy to understand because when the term S_i' is added into f^{eq}, in Eq. (6.34), we have

$$\sum_i S_i' e_{i\alpha} = c_s^2 C \partial_\beta P_{\alpha\beta}, \quad \text{and} \quad \sum_i f^{eq} e_{i\alpha} = \phi u_\alpha + c_s^2 C \partial_\beta P_{\alpha\beta}. \tag{6.49}$$

Meanwhile $\sum_i S_i' = 0$ and $\sum_i S_i' e_{i\beta} e_{i\delta} = 0$. Hence, Eq. (6.33) becomes

$$\partial_{t1}\phi + \partial_\alpha(\phi u_\alpha) = -c_s^2 C \partial_\alpha \partial_\beta P_{\alpha\beta}. \tag{6.50}$$

After it is combined with Eq. (6.37), we obtain the new expression for λ, i.e., Eq. (6.48).

Inamuro et al. (2004) mentioned that "By choosing a proper value of C, we can decrease the mobility even with $\tau_f = 1$". C is mentioned only once and never mentioned in the context of their simulations. Here we guess that our possible "unsuccessful" simulation originates from the $C = 0$ in our simulation. We also tried to perform simulations with possible reasonable C for the case $Re = 2000$ and $We = 39.7$. However, $C = 1.5, 1.0$, and 0.7 all result in simulations that blow up.

Overall, Inamuro's method seems to be using the LBM only as an iteration tool and does not take significant advantage of the LBM at all. The reason is that first, the actual viscous stress tensor is obtained through a finite difference scheme and the natural viscous stress tensor that comes from the second moments of the non-equilibrium distribution function is canceled. Hence, the viscosity of the simulated fluid is independent of τ_g. This is very different from the common LBMs. Second, in this scheme the Poisson equation has to be solved. That advantage of the LBM is totally lost. Because of the computational effort required to solve the Poisson equation, this scheme is far from an efficient one. Using LBM-style code to do iteration is not efficient compared to common iteration schemes. Overall, we do not recommend that readers apply this method to simulate 3D multiphase flow problems because of the heavy computational load.

6.3 **Appendix**

```
NOTES:
! 1. This programme is a 3D Inamuro model code using D3Q15 model to
!    simulate two droplets collision.
! 2. The whole project is composed of head.inc, main.for,
!    initial.for, macro.for,
!    streamcollision.for, utility.for, and output.for.
! 3. head.inc is a header file, where some common blocks are defined.
! 4. Before running this project, please create a new sub-folder
!    'out' under the
!    working directory.
! 5. The computational domain is symmetric about $x=1$. That means
!    only half the whole computational domain is used. For more details
!    please refer to the JCP paper (Inamuro, 2004).
! 6. There is an input file 'params.in' and the first line in the
!    file is
!    "48  96  128 !lx, ly, lz", instead of "c==============..."


FILE  head.inc
C=======================================================
      integer lx,ly,lz
      PARAMETER(lx=48, ly=96, lz=128)
      common /exyz/ ex(0:14), ey(0:14), ez(0:14)
      integer ex,ey, ez

      common /uvwc/ uc(0:14), vc(0:14), wc(0:14), t_k(0:14)
      real*8 uc,vc,wc,t_k

      common/AA/ tau, Ei(0:14), Hi(0:14), Fi(0:14)
      real*8 tau, Ei, Hi, Fi

      common/b/ error,vel, eps, DD,t_max, Nwri, time, period
      real*8 error,vel, eps, DD
      integer Nwri, time, t_max, period

      common/para/ k_f, k_g,  tauf, taug, TT, a,b,rho_L, rho_G
      real*8 k_f, k_g,  tauf, taug , TT, a,b, rho_L, rho_G
      common/para2/ mu_L, mu_G, fei_Ls, fei_Gs, fei_max,fei_min
      real*8 mu_L, mu_G, fei_Ls, fei_Gs,  fei_max, fei_min
      common/par3/ gam_l, gam_g
      real*8 gam_l, gam_g
      common/rb/ ff(0:14,lx,ly,lz),rho(lx,ly,0:lz+1)
      real*8 ff, rho
      common/rb3/ u_x(lx,ly,0:lz+1)
      real*8  u_x
      common/rb2/  u_y(lx,ly,0:lz+1), u_z(lx,ly,0:lz+1),
     &p(lx,ly,0:lz+1)
      real*8  u_y, u_z , p

      common/fai/ gg(0:14,lx,ly,lz),fei(lx,ly,0:lz+1)
      real*8 gg, fei
      common/obs/ obst(lx,ly,lz)
      integer obst
      common/ho/ hh(0:14,lx,ly,lz)
      real*8 hh
```

```
      common/fo2/ fei_x(lx,ly,lz),fei_y(lx,ly,lz), fei_z(lx,ly,lz)
      real*8 fei_x, fei_y, fei_z
      common/fo2/ rho_x(lx,ly,lz),rho_y(lx,ly,lz), rho_z(lx,ly,lz)
      real*8 rho_x, rho_y, rho_z

      common/fo2/ upx(lx,ly,lz),upy(lx,ly,lz), upz(lx,ly,lz)
      real*8 upx, upy, upz
      common/fo2/ vpx(lx,ly,lz),vpy(lx,ly,lz), vpz(lx,ly,lz)
      real*8 vpx, vpy, vpz
      common/fo2/ wpx(lx,ly,lz),wpy(lx,ly,lz), wpz(lx,ly,lz)
      real*8 wpx, wpy, wpz

      common/fo2/ upx2(lx,ly,lz),upy2(lx,ly,lz), upz2(lx,ly,lz)
      real*8 upx2, upy2, upz2
      common/fo2/ vpx2(lx,ly,lz),vpy2(lx,ly,lz), vpz2(lx,ly,lz)
      real*8 vpx2, vpy2, vpz2
      common/fo2/ wpx2(lx,ly,lz),wpy2(lx,ly,lz), wpz2(lx,ly,lz)
      real*8 wpx2, wpy2, wpz2


FILE main.for
C========================================================
! Inamuro's model,  Inamuro, T., Ogata, T., Tajima, S., &
! Konishi, N. (2004).
! A Lattice Boltzmann method for incompressible two-phase flows
! with large density differences. JCP, 2004
      program Inamuro_Model_D3Q15
! The definition of array range should be consistent!
C----------------------------------------
!   Author: Haibo Huang, huanghb@ustc.edu.cn
C----------------------------------------
      implicit none
      include "head.inc"
      integer k, m, n
      real*8 pp(lx,ly,lz), err1, err2
      integer*4 now(3)

      integer x,y,z, i
! Velocities' x,y,z components of the D3Q15 velocity model (real*8)
      data uc/0.d0, 1.d0, 0.d0, 0.d0, -1.d0, 0.d0, 0.d0, 1.d0, -1.d0,
     & 1.d0, 1.d0, -1.d0,  1.d0,-1.d0, -1.d0  /, !ex
     &  vc/0.d0, 0.d0, 1.d0, 0.d0, 0 .d0,-1.d0, 0.d0, 1.d0,  1.d0,
     & -1.d0, 1.d0, -1.d0, -1.d0, 1.d0, -1.d0  /, !ey
     &  wc/0.d0, 0.d0, 0.d0, 1.d0, 0 .d0, 0.d0,-1.d0, 1.d0,  1.d0,
     & 1.d0,-1.d0, -1.d0, -1.d0,-1.d0,  1.d0  / !ez

! Velocities' x,y,z components of the D3Q15 velocity model in
! Integer format

!         0  1  2  3  4   5  6  7  8  9 10  11 12  13  14
     & ex/ 0, 1, 0, 0, -1, 0, 0, 1, -1, 1, 1, -1,  1,-1, -1/,
     & ey/ 0, 0, 1, 0, 0 ,-1, 0, 1,  1,-1, 1, -1, -1, 1, -1/,
     & ez/ 0, 0, 0, 1, 0 , 0,-1, 1,  1, 1,-1, -1, -1,-1,  1/

      open(40,file='out/p_err.dat')
```

```fortran
      write (6,*) '@@@__Inamuro_D3Q15_starting_..._@@@'
! Output domain size
      write (6,*) '@@@__lattice_size_lx_=_',lx
      write (6,*) '@@@__lattice_size_ly_=_',ly
      write (6,*) '@@@__lattice_size_lz_=_',lz
C-------------------------
! Initialisation
C-------------------------
      call read_parametrs()
      call read_obstacles()

      call init_density()
C-------------------------
! Begin iterations
C-------------------------
      do 100 time = 0, t_max

        if ( mod(time, Nwri) .eq. 0) then
        write(*,*) time
        call itime(now)
        write(*,"(i2.2,_':',_i2.2,_':',_i2.2)") now
          call write_results( time)
        end if
! Collision step
      call collision( )

! Streaming step (Propagation)
      call stream(ff )
      call stream(gg )
! Implement the symmetric boundary condition
      call slip_bounceback(ff)
      call slip_bounceback(gg)

! Get macro variables
      call getuv()

! Pressure correction (Solving Poisson equation)
C----------------------
      do 66  i = 1, 1000
! Solving the Poisson equation using LBM iteration (collision and
! stream)
      call correction()
      call stream(hh )
      call slip_bounceback(hh )
      call getp()
! To check whether the pressure field is converged
      if(mod(i,100) .eq. 0) then
      err1 = 0.d0
      err2 = 0.d0
      do z = 1, lz
        do y = 1, ly
          do x = 1, lx
            err1 = err1+ dabs( p(x,y,z)- pp(x,y,z))
            err2 = err2+ dabs( p(x,y,z) )
            pp(x,y,z) = p(x,y,z)
            enddo
```

```
          enddo
        enddo
        eps = err1/err2
c   write(40,'(i8,f13.8)')
c      &                 time, eps
! When a criterion is reached (pressure correction is finished),
! go to label 99
        if(eps .lt. 0.001d0) goto 99
        endif
c-----------------------
    66 continue
! To get correct velocity field, which satisfy the incompressibility
! condition
    99      call correct_uvw()
! Update distribution function 'hh'
        do z = 1, lz
          do y = 1, ly
          do x = 1, lx
            do i = 0, 14
              hh(i,x,y,z) = p(x,y,z) * Ei(i)
            enddo
          enddo
         enddo
        enddo

    100 continue
c-------------------------------------------------------
        write (6,*) '------___end___---------'
        close(40)
        end



FILE initial.for
c======================================================
        subroutine read_parametrs()
        implicit none
        include "head.inc"
        integer x1, y1, z1, n

        Nwri= 100
! Weighting coefficients (Eq.~(6.9))
        Ei(0) = 2.d0/9.d0
        Hi(0) = 1.d0
        Fi(0) = -7.d0/3.d0

        do 1  n=1,6
        Ei(n) = 1.d0/9.d0
        Hi(n) = 0.d0
        Fi(n) = 3.d0* Ei(n)
      1 continue

        do 2  n=7,14
        Ei(n) = 1.d0/72.d0
        Hi(n) = 0.d0
        Fi(n) = 3.d0* Ei(n)
```

```
   2 continue
! Read main parameters that control the droplet collision
      open(1,file='params.in')
! Read the domain size specified in the params.in, and
! check whether the size is consistent with that in head.inc
      read(1,*)  x1, y1, z1
      if(lx .ne. x1 .or. ly .ne. y1 .or. lz .ne. z1) then
      write(*,*) "error!-------"
      stop
      endif

! Periodic boundary condition ?
      read(1,*)  period
! Maximum time steps
      read(1,*)  t_max
! Dump result every 'Nwri' time step.
      read(1,*)  Nwri
! Specify the densities of the liquid and gas
      read(1,*)  rho_L, rho_G
! Relaxation time for LBE of f_i and g_i
      read(1,*)  tauf, taug
! Specify the dynamic viscosities of teh liquid and gas
      read(1,*)  mu_L, mu_G
! k_f control the width of the interface
! k_g determine the surface tension strength
      read(1,*)  k_f, k_g
      read(1,*)
! Parameters in the equation of state (EOS)
      read(1,*)  TT  !
      read(1,*)  a  !
      read(1,*)  b  !
      read(1,*)
! Coexisting \phi obtained from the EOS
      read(1,*)  fei_max, fei_min
! Truncation of the above two values (purely due numerical stability)
      read(1,*)  fei_Ls, fei_Gs
! The diameter of the droplets
      read(1,*)  DD

      close(1)

      gam_l = (tau-0.5d0)/3.d0
      gam_g = (tau-0.5d0)/3.d0
c------------------------------------
! Dump the parameters to keep record
      open(2,file='./out/params.dat')
      write(2,'("lx=",_i5,"_ly=",_i5,"_lz=",_i5)')  lx, ly, lz
      write(2,'("t_max=",_i10)')  t_max    !
      write(2,'("Nwri=",_i9)')  Nwri    !
      write(2,'("Inamuro_multiphase_model")')
      write(2,'("rho_h=",_f12.5,"_rho_l=",_f12.5)')  rho_L, rho_G
      write(2,'("tau_f=",_f12.5,"_tau_g=",_f12.5)')  tauf, taug
      write(2,'("mu_L=",_f12.5,"_mu_G=",_f12.5)')
     &    mu_L, mu_G
      write(2,'("Dyn_vis_ratio=mu_L/mu_G=",_f12.5)')
     &    (mu_L)/mu_G
```

```
      write(2,'("k_f=",_e12.7)')  k_f
      write(2,'("k_g=",_f12.7)')  k_g
      write(2,'("TT=",_e15.7)')   TT
      write(2,'("a=",_f12.5)')  a  !
      write(2,'("b=",_f12.5)')  b  !
      write(2,'("fei_max_=_",e12.5)')  fei_max
      write(2,'("fei_min_=_",e12.5)')  fei_min
      write(2,'("fei_Ls_=_",e12.5)')  fei_Ls
      write(2,'("fei_Gs_=_",e12.5)')  fei_Gs
      write(2,'("DD=",_f12.5)')  DD  !
      close(2)

      end
c-------------------------------------------------
      subroutine read_obstacles()
      implicit none
      include "head.inc"
      integer  x,y,z
! Initial the whole computational domain: obst=0 denotes fluid node.
      do 10 z = 1, lz
        do 20 y = 1, ly
          do 40 x = 1, lx
            obst(x,y,z) =  0
  40      continue
  20    continue
  10 continue
      end
c-----------------------------------------------------
      subroutine init_density()
      implicit none
      include "head.inc"
      integer i,j,x,y,z,k,k1, n
      real*8  u_squ, xx, u_n(0:14),ff1(0:14)
      real*8 ran

      do 10 z = 0, lz+1
        do 10 y = 1, ly
          do 10 x = 1, lx
! Initialize the flow field
      u_x(x,y,z) = 0.d0
      u_y(x,y,z) = 0.d0
      u_z(x,y,z) = 0.d0
      p(x,y,z) = 0.d0
      fei(x,y,z) = fei_min
      rho(x,y,z) = rho_G
  10 continue

c---------------------------------------
      do 40 z = 1, lz
        do 30 y = 1, ly
          do 20 x = 1, lx
! Initialize the first droplet moving with velocity -0.035 lu/ts
      if(  sqrt(real(x-1)**2+real(y-ly/2)**2
     &   +real(z-lz/2-DD)**2 ) .lt. DD/2.) then !
          fei(x,y,z) = fei_max
```

```
          rho(x,y,z) = rho_L
      u_z(x,y,z) = -0.035d0
      endif
! Initialize the second droplet moving with velocity 0.035 lu/ts
         if(  sqrt(real(x-1)**2+real(y-ly/2)**2
   &     +real(z-lz/2+DD)**2 ) .lt. DD/2.) then   !
            fei(x,y,z) = fei_max
            rho(x,y,z) = rho_L
      u_z(x,y,z) = 0.035d0
      endif
C-------------------

   20 continue
   30 continue
   40 continue

      return
      end



FILE macro.for
C======================================================

! Get density (Eq.~(6.13)) and velocities (refer to Eq.~(6.4), (6.5))
      subroutine getuv()
      implicit none
      include "head.inc"
      integer x,y,z, i
      real*8 fei_ave

      fei_ave = (fei_Ls + fei_Gs)/2.d0

      do 25 z = 1, lz
         do 15 y = 1, ly
          do 5 x = 1, lx

       fei(x,y,z)=
   &     ff(0,x,y,z)+ff(1,x,y,z)+ff(2,x,y,z)+ff(3,x,y,z)
   &     +ff(4,x,y,z)+ff(5,x,y,z)+ff(6,x,y,z)
   &     +ff(7,x,y,z)+ff(8,x,y,z)+ff(9,x,y,z)+ff(10,x,y,z)
   &     +ff(11,x,y,z)+ff(12,x,y,z)+ff(13,x,y,z)+ff(14,x,y,z)

! Refer to Eq.~(6.13)
          if(fei(x,y,z) .lt. fei_Gs) then
       rho(x,y,z) = rho_G
            else if(fei(x,y,z) .gt. fei_Ls) then
       rho(x,y,z) = rho_L
         else
         rho(x,y,z) = (rho_L-rho_G)/2.d0 *(dsin(
   &       (fei(x,y,z)- fei_ave) /(fei_Ls- fei_Gs)*3.1415926 ) +1.d0)
   &          + rho_G
         endif
```

```
      u_x(x,y,z)= 0.d0
      u_y(x,y,z)= 0.d0
      u_z(x,y,z)= 0.d0

      do i = 1, 14
      u_x(x,y,z)= u_x(x,y,z) + uc(i)*gg(i,x,y,z)
      u_y(x,y,z)= u_y(x,y,z) + vc(i)*gg(i,x,y,z)
      u_z(x,y,z)= u_z(x,y,z) + wc(i)*gg(i,x,y,z)
      enddo

  5   continue
 15   continue
 25   continue
      end

c-------------------------------------------------------
      subroutine getp()
      implicit none
      include "head.inc"
      integer x,y,z, i
      real*8  pcr

! Get pressure, see Eq.~(6.44)
      do 25 z = 1, lz
        do 15 y = 1, ly
          do  5 x = 1, lx
      p(x,y,z)=
    &     hh(0,x,y,z)+hh(1,x,y,z)+hh(2,x,y,z)+hh(3,x,y,z)
    &    +hh(4,x,y,z)+hh(5,x,y,z)+hh(6,x,y,z)
    &    +hh(7,x,y,z)+hh(8,x,y,z)+hh(9,x,y,z)+hh(10,x,y,z)
    &    +hh(11,x,y,z)+hh(12,x,y,z)+hh(13,x,y,z)+hh(14,x,y,z)
  5     continue
 15   continue
 25 continue

      end
c----------------------------------------
      subroutine correction()
      implicit none
      include "head.inc"
      integer  l, x_e,x_w,y_n,y_s,z_n,z_s,zmin, zmax
      integer  x,y,z,k
      real*8  u_n(0:14),hequ(0:14),u_squ
      real*8    usqu
      real*8    tauh

      if(period .eq. 1) then
      zmin = 1
      zmax = lz
      endif

      do 5 z = zmin, zmax  ! change
      do 5 y = 1, ly
      do 5 x = 1, lx
```

```
! Relaxation time in the LBE for 'hh', see Eq.~(6.43)
      tauh= 1.d0/rho(x,y,z) + 0.5d0

      do 13 k = 0, 14
! Equilibrium distribution function for 'hh'
        hequ(k) = Ei(k)*p(x,y,z)
! Collision step for 'hh', see Eq.~(6.42)
          hh(k,x,y,z) = hh(k,x,y,z) -1.d0/tauh*( hh(k,x,y,z) -hequ(k) )
     &      - 1.d0/3.d0 *Ei(k)*( upx(x,y,z) + vpy(x,y,z)+ wpz(x,y,z) )
   13     continue

   5  continue

      return
      end


C----------------------------

      subroutine correct_uvw()
      implicit none
      include "head.inc"
      integer x,y,z, i,x_e,x_w,y_n,y_s,z_n,z_s
      do 25 z = 1, lz
         do 15 y = 1, ly
          do 5 x = 1, lx
          z_n = mod(z,lz) + 1   ! Periodic BC in z direction
          y_n = y + 1
          x_e = x + 1
          z_s = lz - mod(lz + 1 - z, lz)   ! Periodic BC in z
!         direction
          y_s = y-1
          x_w = x-1
      if(y_n .gt. ly) y_n=1    ! Periodic BC
      if(y_s .lt. 1)  y_s=ly   ! Periodic BC
      if(x_e .gt. lx) x_e=lx-1 ! Symmetric BC
      if(x_w .lt. 1)  x_w=2    ! Symmetric BC
! Velocities are corrected.
      u_x(x,y,z)=u_x(x,y,z) - ( p(x_e,y,z) -p(x_w,y,z) )/2.d0/rho(x,y,z)
      u_y(x,y,z)=u_y(x,y,z) - ( p(x,y_n,z) -p(x,y_s,z) )/2.d0/rho(x,y,z)
      u_z(x,y,z)=u_z(x,y,z) - ( p(x,y,z_n) -p(x,y,z_s) )/2.d0/rho(x,y,z)

   5     continue
   15  continue
   25 continue

      end


FILE streamcollision.for
C====================================================
! Streaming step
      subroutine stream(f)
      implicit none
      include "head.inc"
      integer  x,y,z,x_e,x_w,y_n,y_s,z_n,z_s, k
      real*8  f(0:14,lx,ly,lz), fp(0:14,lx,ly,lz)
```

```
c----------------
      do 12 z = 1, lz
        do 11 y = 1, ly
          do 10 x = 1, lx

            z_n = mod(z,lz) + 1
            y_n = mod(y,ly) + 1
            x_e = mod(x,lx) + 1
            z_s = lz - mod(lz + 1 - z, lz)
            y_s = ly - mod(ly + 1 - y, ly)
            x_w = lx - mod(lx + 1 - x, lx)

            fp(0 ,x  ,y  ,z  ) = f(0,x,y,z)
            fp(1 ,x_e,y  ,z  ) = f(1,x,y,z)
            fp(2 ,x  ,y_n,z  ) = f(2,x,y,z)
            fp(3 ,x  ,y  ,z_n) = f(3,x,y,z)
            fp(4 ,x_w,y  ,z  ) = f(4,x,y,z)
            fp(5 ,x  ,y_s,z  ) = f(5,x,y,z)
            fp(6 ,x  ,y  ,z_s) = f(6,x,y,z)
            fp(7 ,x_e,y_n,z_n) = f(7,x,y,z)
            fp(8 ,x_w,y_n,z_n) = f(8,x,y,z)
            fp(9 ,x_e,y_s,z_n) = f(9,x,y,z)
            fp(10,x_e,y_n,z_s) = f(10,x,y,z)
            fp(11,x_w,y_s,z_s) = f(11,x,y,z)
            fp(12,x_e,y_s,z_s) = f(12,x,y,z)
            fp(13,x_w,y_n,z_s) = f(13,x,y,z)
            fp(14,x_w,y_s,z_n) = f(14,x,y,z)
 10      continue
 11    continue
 12 continue

c
      do 22 z = 1, lz
        do 21 y = 1, ly
          do 20 x = 1, lx
          do k =1, 14
          f(k,x,y,z) = fp(k,x,y,z)
      enddo
 20      continue
 21    continue
 22 continue

      return
      end

c----------------------------------------------------
! Collision step for distribution functions 'ff' and 'gg'
      subroutine collision()
      implicit none
      include "head.inc"
      integer l, x_e,x_w,y_n,y_s,z_n,z_s, zmin, zmax
      integer x,y,z,k,ip,jp, i
      real*8 u_n(0:14),gequ(0:14),fequ2(0:14),u_squ,invtau,fequ(0:14)
      real*8    fei2sum , sum_fei, usqu, sum_rho
      real*8    p0, fei_ei(0:14), Gfei(0:14), tmp1, mu
      real*8    mueiapb2ua(0:14), paub_pbua(0:14), Grho(0:14)
```

```
      &   , rho_ei(0:14)
      real*8 ulap(lx,ly,lz), vlap(lx,ly,lz), wlap(lx,ly,lz)
      real*8 feilap(lx,ly,lz), diverg(lx,ly,0:lz+1), divx(lx,ly,lz)
      real*8 divy(lx,ly,lz), divz(lx,ly,lz)

      tmp1 = (mu_L-mu_G)/(rho_L- rho_G)

! Get the first and second derivatives for macrovariables
      call firstord(u_x, upx, upy, upz)
      call firstord(u_y, vpx, vpy, vpz)
      call firstord(u_z, wpx, wpy, wpz)
      call firstord(rho, rho_x, rho_y, rho_z)
      call firstord(fei, fei_x, fei_y, fei_z)
      call secondord(u_x, ulap)
      call secondord(u_y, vlap)
      call secondord(u_z, wlap)
      call secondord(fei, feilap)

      if(period .eq. 1) then
      zmin = 1
      zmax = lz
      endif

      do 5 z = zmin, zmax
        do 15 y = 1, ly
        do 25 x = 1, lx
! Get dynamic viscosity $\eta$ (refer to Eq.~(6.14))
      mu = (rho(x,y,z)- rho_G) *tmp1 + mu_G

c------------------------
      sum_fei = fei_x(x,y,z)*fei_x(x,y,z) +fei_y(x,y,z)*fei_y(x,y,z) +
     &          +fei_z(x,y,z)*fei_z(x,y,z)

      sum_rho = rho_x(x,y,z)*rho_x(x,y,z) +rho_y(x,y,z)*rho_y(x,y,z) +
     &          +rho_z(x,y,z)*rho_z(x,y,z)

!   fei2sum= fei_x2(x,y,z)+ fei_y2(x,y,z) + fei_z2(x,y,z)
      fei2sum= feilap(x,y,z)

      usqu= u_x(x,y,z)*u_x(x,y,z)+ u_y(x,y,z)*u_y(x,y,z) +
     &      u_z(x,y,z)*u_z(x,y,z)

! Equation of state
      call EOS(fei(x,y,z), p0)

      do 6  i = 0, 14
      u_n(i) = uc(i)*u_x(x,y,z) + vc(i)*u_y(x,y,z) + wc(i)*u_z(x,y,z)

      fei_ei(i) = uc(i)*fei_x(x,y,z)
     &            + vc(i)*fei_y(x,y,z)
     &            + wc(i)*fei_z(x,y,z)

      rho_ei(i) = uc(i)*rho_x(x,y,z)
     &            + vc(i)*rho_y(x,y,z)
     &            + wc(i)*rho_z(x,y,z)
! The last term in Eq.~(6.6)
      Gfei(i)=    (
```

```
     &         9.d0/2.d0* (fei_ei(i)* fei_ei(i))
     &        -3.d0/2.d0* sum_fei *( uc(i)*uc(i) + vc(i)*vc(i) + wc(i)*wc(i))
     &         )
! See Eq.~(6.8) and (6.7)
     Grho(i)=   (
     &         9.d0/2.d0* (rho_ei(i)* rho_ei(i))
     &        -3.d0/2.d0* sum_rho
     &        *( uc(i)*uc(i) + vc(i)*vc(i) + wc(i)*wc(i) )
     &         )

! Refer to Eq.~(6.6)
     fequ(i) = Hi(i)*fei(x,y,z) + Fi(i)*(p0- k_f*fei(x,y,z)* fei2sum
     &  -k_f/6.d0*sum_fei )
     &   + 3.d0 * Ei(i)*fei(x,y,z) * u_n(i) + Ei(i)*k_f
     &   * Gfei(i)
C--------------------------------------------------
! Get the equilibrium distribution function, refer to Eq.~(6.7)
     gequ(i) = Ei(i)*( 1.d0 + 3.d0* u_n(i) -1.5d0 *usqu +
     &         4.5* u_n(i)*u_n(i) + 1.5d0 *(taug-0.5)*
     & 2.d0*(
     & upx(x,y,z)*uc(i)*uc(i) +upy(x,y,z)*uc(i)*vc(i)
     & +upz(x,y,z)*uc(i)*wc(i) +
     & vpx(x,y,z)*vc(i)*uc(i) +vpy(x,y,z)*vc(i)*vc(i)
     & +vpz(x,y,z)*vc(i)*wc(i) +
     & wpx(x,y,z)*wc(i)*uc(i) +wpy(x,y,z)*wc(i)*vc(i)
     & +wpz(x,y,z)*wc(i)*wc(i)
     &)
     & )
     &         + Ei(i)*k_g/rho(x,y,z)* Grho(i)
     &         -2.d0/3.d0 *Fi(i) *k_g/rho(x,y,z) *sum_rho
   6    continue
C--------------------
! Initialization
     if(time .eq. 0) then
     do k = 0, 14
     ff(k,x,y,z) = fequ(k)
     gg(k,x,y,z) = gequ(k)
     enddo
     endif
C----------------------
     if(time .ne. 0) then

     do 13 k = 0, 14
! Collision step for the distribution function 'ff'
     ff(k,x,y,z) = ff(k,x,y,z) -1.d0/tauf*( ff(k,x,y,z) - fequ(k) )

! Collision step for the distribution function 'gg'
     gg(k,x,y,z) = gg(k,x,y,z) -1.d0/taug*( gg(k,x,y,z) - gequ(k) ) +
     &         3.d0 * Ei(k)/rho(x,y,z)*
     &         ( mu*(
     &         uc(k)* (ulap(x,y,z) )
     &         +vc(k)* (vlap(x,y,z) )
     &         +wc(k)* (wlap(x,y,z) )
     &                          )
     &         )
  13 continue

     endif
```

```
C---------------------
   25 continue
   15 continue
    5 continue

      return
      end


FILE utility.for
C=======================================================
! Get the first-order derivative for a macrovariable 'va'
! Refer to Eq.~(6.11)
      subroutine firstord(va, px, py, pz)
      implicit none
      include 'head.inc'
      real*8 va(lx,ly,0:lz+1), px(lx,ly,lz), py(lx,ly,lz), pz(lx,ly,lz)
      integer zp, yp , xp, x, y, z, k

      do 25 z = 1, lz
         do 15 y = 1, ly
          do  5 x = 1, lx

          px(x,y,z)=0.d0
          py(x,y,z)=0.d0
          pz(x,y,z)=0.d0

         do k = 1, 14

! Symmetric boundary condition in x
          xp = x+ex(k)
          if(xp .gt. lx) xp = lx-1
          if(xp .lt. 1) xp = 2
! Periodic boundary conditions in y and z direction
          yp = y+ey(k)
          if(yp .gt. ly) yp = 1
          if(yp .lt. 1) yp = ly
          zp = z+ez(k)
          if(zp .gt. lz) zp = 1
          if(zp .lt. 1) zp = lz

          px(x,y,z)=px(x,y,z)  + va(xp,yp,zp)* uc(k)
          py(x,y,z)=py(x,y,z)  + va(xp,yp,zp)* vc(k)
          pz(x,y,z)=pz(x,y,z)  + va(xp,yp,zp)* wc(k)
         enddo

          px(x,y,z)  = px(x,y,z)/10.d0
          py(x,y,z)  = py(x,y,z)/10.d0
          pz(x,y,z)  = pz(x,y,z)/10.d0
    5       continue
   15     continue
   25 continue
      end
```

```
C---------------------------------------------------
! Get the second-order derivative (Laplacian) for a macrovariable 'va'
! Refer to Eq.~(6.12)
      subroutine secondord(va, p2)
      implicit none
      include 'head.inc'
      real*8 va(lx,ly,0:lz+1), p2(lx,ly,lz)
      integer zp, yp , xp, x, y, z, k

      do 25 z = 1, lz
         do 15 y = 1, ly
          do 5  x = 1, lx

           p2(x,y,z)=0.d0

           do k = 1, 14
! Symmetric boundary condition in x direction
             xp = x+ex(k)
             if(xp .gt. lx) xp = lx-1
             if(xp .lt. 1) xp = 2
! Periodic boundary conditions in y and z direction
             yp = y+ey(k)
             if(yp .gt. ly) yp = 1
             if(yp .lt. 1) yp = ly
             zp = z+ez(k)
             if(zp .gt. lz) zp = 1
             if(zp .lt. 1) zp = lz
             p2(x,y,z)=p2(x,y,z) + va(xp,yp,zp)
           enddo

             p2(x,y,z) = (p2(x,y,z)- 14.d0*va(x,y,z) )/5.d0
    5        continue
   15    continue
   25 continue

      end


C-----------------------------------------------------------
! Equation of state, Eq.~(6.10)
      subroutine EOS(rh, tmp)
      include 'head.inc'
      real*8 rh, tmp
        tmp= rh*TT/(1.d0 -rh*b) -a *rh *rh
      return
      end

C-----------------------------------------
! Specular reflection (to fulfill the slip wall
! boundary condition or symmetric boundary condition)
C-----------------------------------------
      subroutine  slip_bounceback(f )
      implicit none
      include "head.inc"
      integer x, y, z
      real*8  f(0:14,lx,ly,lz)
```

```
      x = 1

      do 10 z = 1, lz
       do 11 y = 1, ly
           f(1,x,y,z) = f(4,x,y,z)
           f(7,x,y,z) = f(8,x,y,z)
           f(9,x,y,z) = f(14,x,y,z)
           f(10,x,y,z) = f(13,x,y,z)
           f(12,x,y,z) = f(11,x,y,z)
   11  continue
   10  continue
c---------------------
      x = lx

      do 20 z = 1, lz
       do 21 y = 1, ly
           f(4,x,y,z)   = f(1,x,y,z)
           f(8,x,y,z)   = f(7,x,y,z)
           f(14,x,y,z)  = f(9,x,y,z)
           f(13,x,y,z)  = f(10,x,y,z)
           f(11,x,y,z)  = f(12,x,y,z)
   21  continue
   20  continue

      end


FILE output.for
c=====================================================
! Write a binary Tecplot file,
c This is a good example about how to write the binary output file
c for TECPLOT. Binary file is the smallest file but contains
! all information,
c which can save much disk than the ASCII file.
! To write a qualified binary file, we must follow the rules
! illustrated here.
c Alternatively, you can also write as a ASCII file as illustrated
! in Chapter 2.

      subroutine  write_results( n)
      implicit none
      include "head.inc"
      integer  x,y,z,i,j,k, n
      real*8   dx,dy,temp,dist, press, tmp1, tmp2

      character filename*16,  B2*9
      REAL*4 ZONEMARKER, EOHMARKER

      character*40 Title,var
      character*40 Va1,Va2,Va3, V3,V4, V5, V6, V7, V8
      character*40 Zonename1

      write(B2,'(i9.9)') n
      open(41,file='out/3D'//B2//'.plt', form="BINARY")
```

```
c----------------------------------------------

      ZONEMARKER=            299.0
      EOHMARKER =            357.0
c----------------------------------------
      write(41) "#!TDV101"
      write(41) 1

      Title="Inamuro"
      call dumpstring(Title)

! Number of variables in this data file (here 9 variables)
      write(41) 9
c-- Variable names.
      Va1='X'
      call dumpstring(Va1)
      Va2='Y'
      call dumpstring(Va2)
      Va3='Z'
      call dumpstring(Va3)
      V3='u'
      call dumpstring(V3)
      V4='v'
      call dumpstring(V4)
      V8='w'
      call dumpstring(V8)
      V5='rho'
      call dumpstring(V5)
      V6='fei'
      call dumpstring(V6)
      V7='press'
      call dumpstring(V7)
c-----Zones-------------------------
c--------Zone marker. Value = 299.0

      write(41) ZONEMARKER

c--------Zone name.

      Zonename1="ZONE 001"
      call dumpstring(Zonename1)

c---------Zone Color
      write(41) -1
c---------ZoneType
      write(41) 0
c---------DataPacking 0=Block, 1=Point
      write(41) 1
c---------Specify Var Location. 0 = Don't specify, all data
c---------is located at the nodes. 1 = Specify
      write(41) 0
c---------Number of user defined face neighbor connections
! (value >= 0)
      write(41) 0
c---------IMax,JMax,KMax
      write(41) lx
```

```
        write(41) ly
        write(41) lz
c----------1=Auxiliary name/value pair to follow
c----------0=No more Auxiliar name/value pairs.
        write(41) 0
        write(41) EOHMARKER
c----zone ------------------------------------------------------------
        write(41) Zonemarker
C--------variable data format, 1=Float, 2=Double, 3=LongInt,
c--------4=ShortInt, 5=Byte, 6=Bit
        write(41) 3
        write(41) 3
        write(41) 3
        write(41) 1
        write(41) 1
        write(41) 1
        write(41) 1
        write(41) 1
        write(41) 1
C--------Has variable sharing 0 = no, 1 = yes.
        write(41) 0
C----------Zone number to share connectivity list with (-1 = no
! sharing).
        write(41) -1

      do k=1,lz
      do j=1,ly
      do i=1,lx
      if(obst(i,j,k).ne. 0) then
      press = 0.d0
      tmp1 = -1.d0
      tmp2 = -1.d0
      else

      press= p(i,j,k)  !rho(i,j,k)/3.d0
      tmp1 = rho(i,j,k)
      tmp2 = fei(i,j,k)
      endif

      write(41) i
      write(41) j
      write(41) k
      write(41) real(u_x(i,j,k))
      write(41) real(u_y(i,j,k))
      write(41) real(u_z(i,j,k))
      write(41) real(tmp1)
      write(41) real(tmp2)
      write(41) real(press)
        end do
      end do
      end do

      close(41)
      end
```

```
C-------------------------------
      subroutine dumpstring(instring)
      character(40) instring
      integer len

      len=LEN_TRIM(instring)
      do ii=1,len
      I=ICHAR(instring(ii:ii))
      write(41) I
      end do
      write(41) 0
      return
      end

FILE params.in
C=================================================
48       96      128    ! lx, ly, lz
1                       ! period
8000                    ! tmax
200                     ! Nwri
50.d0      1.d0            ! rho_L    rho_G
1.d0       1.d0            ! tauf    taug
0.008d0    0.00016d0       ! mu_L    mu_G
0.5d0       0.0005d0       ! k_f    k_g
-------------------------
0.035d0      ! TT    EOS
1.d0         ! a     EOS
6.7d0        ! b
-------------------------
0.09714d0    0.01134d0  ! fei_max    fei_min
0.092d0      0.015d0         ! fei_Ls  fei_Gs
32.d0                        ! DD
```

CHAPTER 7

He–Chen–Zhang multiphase Lattice Boltzmann model

7.1 Introduction

He et al. (1999) proposed a single-component multiphase Lattice Boltzmann model for simulation of multiphase flow in the incompressible limit. It is referred to as the He–Cheng–Zhang or HCZ model. In the model the interfacial dynamics are modeled by incorporating molecular interactions (He et al. 1999). Similar to the model in the last chapter, a pressure distribution function and a distribution function for an index function are introduced into the model. The Lattice Boltzmann equations for the pressure and index function are able to recover the N–S equation and the CH interface-tracking equation, respectively. Applying the HCZ model and adding an energy equation, thermal two-phase flows (e.g., Rayleigh–Benard convection) can be simulated successfully (Chang and Alexander 2006). This model has also been applied to simulate droplet spreading (Frank and Perré 2012), 3D droplet oscillation (Premnath and Abraham 2007), Rayleigh–Taylor mixing (Clark 2003), droplet impact on dry walls (Mukherjee and Abraham 2007b), etc.

In this chapter, after the method is introduced, example simulations of phase separation, the Rayleigh–Taylor instability, and contact line movement are provided.

7.2 HCZ model

Here we briefly introduce the model of He et al. (1999). In this model, two distribution functions $\bar{g}_i$ and $\bar{f}_i$ are introduced to recover the incompressible N–S equation and a macro interface-tracking equation (a CH-like equation), respectively. The distribution functions satisfy the following Lattice Boltzmann

Multiphase Lattice Boltzmann Methods: Theory and Application, First Edition.
Haibo Huang, Michael C. Sukop and Xi-Yun Lu.

equations (He et al. 1999):

$$\bar{g}_i(\mathbf{x} + \mathbf{e}_i \Delta t, t + \Delta t) = \bar{g}_i(\mathbf{x}, t) - \frac{1}{\tau_1}(\bar{g}_i(\mathbf{x}, t) - \bar{g}_i^{eq}(\mathbf{x}, t)) + S_i(\mathbf{x}, t)\Delta t, \tag{7.1}$$

$$\bar{f}_i(\mathbf{x} + \mathbf{e}_i \Delta t, t + \Delta t) = \bar{f}_i(\mathbf{x}, t) - \frac{1}{\tau_2}(\bar{f}_i(\mathbf{x}, t) - \bar{f}_i^{eq}(\mathbf{x}, t)) + S_i'(\mathbf{x}, t)\Delta t, \tag{7.2}$$

where $f_i(\mathbf{x}, t)$ is the density distribution function in the ith velocity direction at position $\mathbf{x}$ at time step t. τ_1 is a relaxation time that is related to the kinematic viscosity as $v = c_s^2(\tau_1 - 0.5)\Delta t$. τ_2 is related to the mobility in the CH equation (see Chapter 6). Usually in our simulations

$$\tau_2 = \tau_1 \tag{7.3}$$

is adopted. $S_i(\mathbf{x}, t)$ and $S_i'(\mathbf{x}, t)$ are the source terms in Eqs (7.1) and (7.2), respectively. The equilibrium distribution functions $\bar{g}_i^{eq}(\mathbf{x}, t)$ and $\bar{f}_i^{eq}(\mathbf{x}, t)$ can be calculated as (He et al. 1999)

$$\bar{g}_i^{eq}(\mathbf{x}, t) = w_i \left[p + c_s^2 \rho \left(\frac{e_{i\alpha} u_\alpha}{c_s^2} + \frac{e_{i\alpha} u_\alpha e_{i\beta} u_\beta}{2c_s^4} - \frac{u_\alpha u_\alpha}{2c_s^2} \right) \right] \tag{7.4}$$

and

$$\bar{f}_i^{eq}(\mathbf{x}, t) = w_i \phi \left[1 + \frac{e_{i\alpha} u_\alpha}{c_s^2} + \frac{e_{i\alpha} u_\alpha e_{i\beta} u_\beta}{2c_s^4} - \frac{u_\alpha u_\alpha}{2c_s^2} \right], \tag{7.5}$$

respectively, where p and ρ are the hydrodynamics pressure and density of the fluid, respectively. ϕ is the index function.

In Eqs (7.1) and (7.2) the $\mathbf{e}_i$ s are the discrete velocities. For the 2D case the D2Q9 model (refer to Chapter 1) is used here. The weighting coefficients w_i are also described in Chapter 1.

The macroscopic variables are given by (He et al. 1999)

$$\phi = \sum_i \bar{f}_i \tag{7.6}$$

and

$$p = \sum_i \bar{g}_i + \frac{\Delta t}{2} u_\beta E_\beta, \tag{7.7}$$

where

$$E_\beta = -\frac{\partial \psi(\rho)}{\partial \beta} \tag{7.8}$$

and

$$\rho u_\alpha c_s^2 = \sum_i e_{i\alpha} \bar{g}_i + \frac{\Delta t}{2} c_s^2 F_\alpha. \tag{7.9}$$

ψ is a function of ρ or ϕ. $\psi(\rho)$ and $\psi(\phi)$ are related to the hydrodynamic pressures p and thermodynamic pressure p_{th} by (Chao et al. 2011)

$$\psi(\rho) = p - c_s^2 \rho \quad \text{and} \tag{7.10}$$

$$\psi(\phi) = p_{th} - c_s^2 \phi, \tag{7.11}$$

where p_{th} is calculated from the Carnahan–Starling EOS (Chao et al. 2011; He et al. 1999), i.e.,

$$p_{th} = \phi c_s^2 \frac{1 + b\phi/4 + (b\phi/4)^2 - (b\phi/4)^3}{(1 - b\phi/4)^3} - a\phi^2. \tag{7.12}$$

It is worth mentioning that in He et al. (1999) we are unable to distinguish the thermodynamic and hydrodynamic pressures. This may confuse readers because on the one hand we can obtain the pressure from Eq. (7.7) and on the other hand the pressure can be calculated from the above EOS once ρ is known. The pressures obtained in these two ways may not be the same. Later Zhang et al. (2000) and Chao et al. (2011) clarified this issue. Both the hydrodynamic and thermodynamic pressure are described clearly and we follow their descriptions.

Identical to the choice in He et al. (1999), in our simulations a and b are set to be $12RT$ and 4, respectively. From the Maxwell construction we get the coexisting index function values $\phi_l = 0.251$ and $\phi_g = 0.024$, and they represent ϕ values for liquid and gas states, respectively.

Once the index function $\phi(\mathbf{x}, t)$ is known, we can easily obtain the density (ρ) and kinematic viscosity (ν) of the fluids and the relaxation factor τ_1, according to the following formulations:

$$\rho(\phi) = \rho_g + \frac{\phi - \phi_g}{\phi_l - \phi_g}(\rho_l - \rho_g), \tag{7.13}$$

$$\nu(\phi) = \nu_g + \frac{\phi - \phi_g}{\phi_l - \phi_g}(\nu_l - \nu_g), \tag{7.14}$$

and

$$\tau_1(\phi) = \tau_g + \frac{\phi - \phi_g}{\phi_l - \phi_g}(\tau_l - \tau_g), \tag{7.15}$$

where ρ_l and ρ_g are the densities of liquid and gas, respectively, $\nu_l = c_s^2(\tau_l - 0.5)\Delta t$ and $\nu_g = c_s^2(\tau_g - 0.5)\Delta t$ denote the kinematic viscosities of liquid and gas, respectively, and τ_l and τ_g are the relaxation times of liquid and gas, respectively. ϕ_l and ϕ_g are the maximum and minimum values of the index function, which can be obtained theoretically through Eq. (7.12) and the Maxwell construction or numerically (Zhang et al. 2000). The maximum density ratio (approximately 10) is achieved by choosing $\rho_l = 0.251$ and $\rho_g = 0.024$ in the simulations (Zhang et al. 2000).

The source terms appearing in Eqs (7.1) and (7.2) are (He et al. 1999)

$$S_i = \left(1 - \frac{1}{2\tau_1}\right)(e_{i\alpha} - u_\alpha)F_\alpha\Gamma_i(\mathbf{u}) + \left(1 - \frac{1}{2\tau_1}\right)(e_{i\alpha} - u_\alpha)E_\alpha[\Gamma_i(\mathbf{u}) - \Gamma_i(0)], \tag{7.16}$$

where

$$\Gamma_i(\mathbf{u}) = \bar{f}_i^{eq}/\phi, \tag{7.17}$$

$$F_\alpha = \kappa \rho \partial_\alpha (\partial_\delta^2 \rho), \tag{7.18}$$

and

$$S_i' = \left(1 - \frac{1}{2\tau_2}\right) \frac{(e_{i\alpha} - u_\alpha)F_\alpha'}{c_s^2 \rho} \bar{f}_i^{eq}, \tag{7.19}$$

where $F_\alpha' = -\partial_\alpha \psi(\phi)$.

7.3 Chapman–Enskog analysis

7.3.1 N–S equations

Applying the Taylor expansion in Eq. (7.1) and using the Chapman–Enskog expansion $\partial_t = \partial_{t_1} + \varepsilon \partial_{t_2} + \dots$ and $\bar{g}_i = \bar{g}_i^{(0)} + \varepsilon \bar{g}_i^{(1)} + \varepsilon^2 \bar{g}_i^{(2)}$, where $\varepsilon = \Delta t$, we have

$$
\begin{aligned}
&\varepsilon(\partial_{t_1} + \varepsilon\partial_{t_2} + e_{i\alpha}\partial_\alpha)\left(\bar{g}_i^{(0)} + \varepsilon \bar{g}_i^{(1)} + \varepsilon^2 \bar{g}_i^{(2)}\right) \\
&+ \frac{\varepsilon^2}{2}(\partial_{t_1} + \varepsilon\partial_{t_2} + e_{i\alpha}\partial_\alpha)^2 \left(\bar{g}_i^{(0)} + \varepsilon \bar{g}_i^{(1)}\right) \\
&= -\frac{1}{\tau_1}\left(\bar{g}_i^{(0)} + \varepsilon \bar{g}_i^{(1)} + \varepsilon^2 \bar{g}_i^{(2)} - \bar{g}_i^{(eq)}\right) + S_i \Delta t
\end{aligned}
\tag{7.20}
$$

Retaining terms to $O(\varepsilon^2)$ in scales $O(1)$, $O(\varepsilon)$, and $O(\varepsilon^2)$, Eq. (7.20) yields

$$O(\varepsilon^0): (\bar{g}_i^{(0)} - \bar{g}_i^{eq})/\tau_1 = 0, \tag{7.21}$$

$$O(\varepsilon^1): (\partial_{t_1} + e_{i\alpha}\partial_\alpha)\bar{g}_i^{(0)} + \frac{1}{\tau_1}\bar{g}_i^{(1)} - S_i = 0, \tag{7.22}$$

$$O(\varepsilon^2): \partial_{t_2}\bar{g}_i^{(0)} + \left(1 - \frac{1}{2\tau_1}\right)(\partial_{t_1} + e_{i\alpha}\partial_\alpha)\bar{g}_i^{(1)} + \frac{1}{2}(\partial_{t_1} + e_{i\alpha}\partial_\alpha)S_i \tag{7.23}$$

$$+ \frac{1}{\tau_1}\bar{g}_i^{(2)} = 0.$$

We note that Eq. (7.7) yields $\sum \bar{g}_i = \sum \bar{g}_i^{(0)} + \Delta t \sum \bar{g}_i^{(1)} + \Delta t^2 \sum \bar{g}_i^{(2)} + \dots = p - \frac{\Delta t}{2}u_\beta E_\beta$.

From Eqs (7.21) and (7.4) we have

$$\sum \bar{g}_i^{(0)} = \sum \bar{g}_i^{eq} = p. \tag{7.24}$$

Hence, from the above two equations we have

$$\sum \bar{g}_i^{(1)} = -\frac{1}{2}u_\beta E_\beta \quad \text{and} \quad \sum \bar{g}_i^{(2)} = 0. \tag{7.25}$$

We note that in Mukherjee and Abraham (2007c) $\sum \bar{g}_i^{(n)}$ is always assumed to be zero for $n \geq 1$. However, here we see that this is not true.

Similarly, from Eq. (7.9) we can obtain $\sum e_{i\alpha}\bar{g}_i = \sum e_{i\alpha}\bar{g}_i^{(0)} + \Delta t \sum e_{i\alpha}\bar{g}_i^{(1)} = c_s^2(\rho u_\alpha - \frac{\Delta t}{2}F_\alpha)$. Hence, the first moments of $\bar{g}_i^{(0)}$ and $\bar{g}_i^{(1)}$ are

$$\sum e_{i\alpha}\bar{g}_i^{(0)} = c_s^2 \rho u_\alpha \quad \text{and} \quad \sum e_{i\alpha}\bar{g}_i^{(1)} = -\frac{1}{2}F_\alpha c_s^2. \qquad (7.26)$$

The zeroth and first moments of the source term S_i take the form

$$\sum_i S_i = \left(1 - \frac{1}{2\tau_1}\right)u_\alpha E_\alpha \quad \text{and} \quad \sum_i e_{i\beta}S_i = c_s^2\left(1 - \frac{1}{2\tau_1}\right)F_\beta. \qquad (7.27)$$

To be prepared for further derivation of the expansion, we also write down the second moments of $\bar{g}_i^{(0)}$ and S_i. From Eq. (7.4) we have

$$\sum g_i^{(0)} e_{i\alpha} e_{i\beta} = c_s^2(p\delta_{\alpha\beta} + \rho u_\alpha u_\beta). \qquad (7.28)$$

Using Eq. (7.16) and omitting higher order terms of $O(u^3)$ we have

$$\sum e_{i\alpha}e_{i\beta}S_i = c_s^2\left(1 - \frac{1}{2\tau_1}\right)[E_\gamma(u_\alpha\delta_{\beta\gamma} + u_\beta\delta_{\alpha\gamma} + u_\gamma\delta_{\beta\alpha}) + F_\beta u_\alpha + F_\alpha u_\beta]. \qquad (7.29)$$

Summing both sides of Eq. (7.22) over i and using Eqs (7.24), (7.25), and (7.27) gives

$$\partial_{t_1}(p) + c_s^2\partial_\alpha(\rho u_\alpha) - u_\alpha E_\alpha = (\partial_{t_1} + u_\alpha\partial_\alpha)p + c_s^2\rho\partial_\alpha u_\alpha = 0. \qquad (7.30)$$

Then we proceed to $O(\varepsilon^2)$. Using Eqs (7.24), (7.25), (7.26), and (7.27), and summing both sides of Eq. (7.23) over i gives

$$\partial_{t_2}p = 0. \qquad (7.31)$$

Combining Eqs(7.30) and (7.31) we obtain

$$(\partial_t + u_\alpha\partial_\alpha)p + c_s^2\rho\partial_\alpha u_\alpha = 0. \qquad (7.32)$$

Here we can see that only when $(\partial_t + u_\alpha\partial_\alpha)p = 0$ can the incompressibility condition $\partial_\alpha u_\alpha = 0$ be satisfied.

Multiplying Eq. (7.22) by $e_{i\beta}$, summing over i, and using Eqs (7.27) and (7.28) gives

$$O(\varepsilon): \quad \partial_{t_1}(\rho u_\beta) + \partial_\alpha(p\delta_{\alpha\beta} + \rho u_\alpha u_\beta) - F_\beta = 0. \qquad (7.33)$$

Multiplying Eq. (7.23) by $e_{i\beta}$ and summing over i gives

$$O(\varepsilon^2): c_s^2\partial_{t_2}(\rho u_\beta) + \left(1 - \frac{1}{2\tau_1}\right)\left[\partial_{t_1}\left(\sum e_{i\beta}\bar{g}_i^{(1)}\right) + \partial_\alpha\left(\sum e_{i\alpha}e_{i\beta}\bar{g}_i^{(1)}\right)\right]$$
$$+ \frac{1}{2}\left[\partial_{t_1}\left(\sum e_{i\beta}S_i\right) + \partial_\alpha\sum(e_{i\beta}e_{i\alpha}S_i)\right] = 0. \qquad (7.34)$$

From Eqs (7.26) and (7.27) we know that

$$\left(1 - \frac{1}{2\tau_1}\right)\partial_{t_1}\left(\sum e_{i\beta}\bar{g}_i^{(1)}\right) = c_s^2\left(1 - \frac{1}{2\tau_1}\right)\partial_{t_1}\left(-\frac{1}{2}F_\alpha\right) = -\frac{1}{2}\partial_{t_1}\left(\sum e_{i\beta}S_i\right).$$
(7.35)

Hence, Eq. (7.34) can be simplified to

$$c_s^2\partial_{t_2}(\rho u_\beta) + \left(1 - \frac{1}{2\tau_1}\right)\partial_\alpha \sum \left(e_{i\alpha}e_{i\beta}\bar{g}_i^{(1)}\right) + \frac{1}{2}\partial_\alpha \sum (e_{i\beta}e_{i\alpha}S_i) = 0.$$
(7.36)

Now our task is to get $\sum(e_{i\alpha}e_{i\beta}\bar{g}_i^{(1)})$. We note from Eq. (7.22) that $\bar{g}_i^{(1)} = -\tau_1(\partial_{t_1} + e_{i\alpha}\partial_\alpha)\bar{g}_i^{(0)} + \tau_1 S_i$. Substituting this equation into Eq. (7.36) further simplifies it to

$$c_s^2\partial_{t_2}(\rho u_\beta) - (\tau_1 - 0.5)\partial_\alpha \sum \left(e_{i\alpha}e_{i\beta}(\partial_{t_1} + e_{i\gamma}\partial_\gamma)\bar{g}_i^{(0)}\right) + \tau_1\partial_\alpha \sum (e_{i\beta}e_{i\alpha}S_i) = 0. \quad (7.37)$$

Eq. (7.37) includes the third moments of $\bar{g}_i^{(0)}$. From Eq. (7.4) we have

$$\partial_\gamma \sum \left(e_{i\alpha}e_{i\beta}e_{i\gamma}\bar{g}_i^{(0)}\right) = c_s^4\partial_\gamma[\rho(u_\alpha\delta_{\beta\gamma} + u_\beta\delta_{\alpha\gamma} + u_\gamma\delta_{\beta\alpha})].$$
(7.38)

Substituting Eqs (7.28), (7.29), and (7.38) into Eq. (7.37) gives

$$\partial_{t_2}(\rho u_\beta)$$
$$-(\tau_1 - 0.5)\left\{\partial_\alpha\partial_{t_1}\left(p\delta_{\alpha\beta}\right) + \partial_\alpha\partial_{t_1}(\rho u_\alpha u_\beta) + c_s^2\partial_\alpha\partial_\gamma[\rho(u_\alpha\delta_{\beta\gamma} + u_\beta\delta_{\alpha\gamma} + u_\gamma\delta_{\beta\alpha})]\right\}$$
$$+(\tau_1 - 0.5)\left\{\partial_\alpha\left[F_\beta u_\alpha + F_\alpha u_\beta\right] + \partial_\alpha[E_\gamma(u_\alpha\delta_{\beta\gamma} + u_\beta\delta_{\alpha\gamma} + u_\gamma\delta_{\beta\alpha})]\right\} = 0. \quad (7.39)$$

On the other hand, from Eq. (7.30) we know

$$\partial_{t_1}(p\delta_{\alpha\beta}) = \left(-u_\gamma\partial_\gamma p - c_s^2\rho\partial_\gamma u_\gamma\right)\delta_{\alpha\beta}.$$
(7.40)

Substituting this equation and $E_\gamma = -\partial_\gamma(p - c_s^2\rho)$ into Eq. (7.39) we have

$$\partial_{t_2}(\rho u_\beta)$$
$$-(\tau_1 - 0.5)\left[\underbrace{\partial_\alpha\left(u_\alpha\partial_\beta p + u_\beta\partial_\alpha p\right)} + \underline{\partial_\alpha\partial_{t_1}(\rho u_\alpha u_\beta)} + c_s^2\partial_\alpha(\rho\partial_\beta u_\alpha + \rho\partial_\alpha u_\beta)\right]$$
$$+(\tau_1 - 0.5)\left[\underbrace{\partial_\alpha\left(F_\beta u_\alpha + F_\alpha u_\beta\right)}\right] = 0.$$
(7.41)

In the above equation terms such as $\partial_\alpha(u_\beta\partial_\alpha p)$ or $\partial_\alpha(u_\alpha\partial_\beta p)$ will break Galilean invariance if the pressure gradient is large (Holdych et al. 1998). Fortunately, in Appendix A of this chapter we will see that the underbraced terms can be canceled with the underlined term $\partial_\alpha\partial_{t1}(p\delta_{\alpha\beta})$. Finally, all these terms that may break Galilean invariance will disappear.

Eq. (7.39) yields

$$\partial_{t_2}(\rho u_\beta) - c_s^2(\tau_1 - 0.5)\partial_\alpha\{\rho[\partial_\beta u_\alpha + \partial_\alpha u_\beta]\} = 0.$$
(7.42)

From Eqs (7.33) and (7.42), using $\partial_t(\rho u_\beta) = \partial_{t_1}(\rho u_\beta) + \Delta t \partial_{t_2}(\rho u_\beta)$ and $v = c_s^2(\tau_1 - 0.5)\Delta t$, we have the N–S equations:

$$\partial_t(\rho u_\beta) + \partial_\alpha(\rho u_\alpha u_\beta) = -\partial_\beta p + v\partial_\alpha\{\rho[\partial_\beta u_\alpha + \partial_\alpha u_\beta]\} + F_\beta. \tag{7.43}$$

Hence, the choice of $E_\gamma = -\partial_\gamma(p - c_s^2\rho)$ does ensure Galilean invariance of the model. However, if E_γ takes the form $E_\gamma = \partial_\gamma\rho$ in the other models (e.g., Lee and Lin 2005), although terms such as $\partial_\alpha(u_\alpha\partial_\beta\rho)$ will be canceled, terms related to the derivatives of pressure are unable to be canceled and the model may not be Galilean invariant.

7.3.2   CH equation

To obtain the equation that is used to track interfaces we start from the equations analogous to Eqs (7.21)–(7.23) for the distribution function $\bar{f}_i$. From Eqs (7.6) and (7.21) we know that $\sum_i \bar{f}_i^{(0)} = \phi$ and $\sum_i \bar{f}_i^{(n)} = 0$, for $n \geq 1$. Summing both sides of Eq. (7.22) over i yields

$$\partial_{t_1}\phi + \partial_\alpha(\phi u_\alpha) = 0. \tag{7.44}$$

Summing Eq. (7.23) over i yields

$$\partial_{t_2}\phi + \left(1 - \frac{1}{2\tau_2}\right)\partial_\alpha\left(\sum_i e_{i\alpha}\bar{f}_i^{(1)}\right) + \frac{1}{2}\partial_\alpha\left(\sum_i e_{i\alpha}S_i'\right) = 0. \tag{7.45}$$

In the simulations, since $\bar{f}_i e_{i\alpha}$ is not evaluated, the term of $\sum_i e_{i\alpha}\bar{f}_i^{(1)}$ cannot be calculated. If we assume this term can be omitted and substitute $\sum_i e_{i\alpha}S_i' = \left(1 - \frac{1}{2\tau_2}\right)F_\alpha'$ into Eq. (7.45), we have

$$\partial_{t_2}\phi + \frac{1}{2}\left(1 - \frac{1}{2\tau_2}\right)\partial_\alpha F_\alpha' = 0. \tag{7.46}$$

Substituting $F_\alpha' = -\partial_\alpha(p - c_s^2\phi)$ into the above equation we have

$$\partial_{t_2}\phi - \frac{1}{2}\left(1 - \frac{1}{2\tau_2}\right)\partial_\alpha\left(\partial_\alpha\left(p - c_s^2\phi\right)\right) = 0. \tag{7.47}$$

Combining Eqs (7.44) and (7.47) together gives

$$\partial_t\phi + \partial_\alpha(\phi u_\alpha) = \frac{1}{2}\left(1 - \frac{1}{2\tau_2}\right)\partial_\alpha^2\left(p - c_s^2\phi\right). \tag{7.48}$$

This is the macroscopic interface-tracking equation, which is a CH-like equation in the HCZ model (Zhang et al. 2000).

7.4   Surface tension and phase separation

Before starting simulations, we must know that ϕ_l and ϕ_g are determined by Eq. (7.12). ϕ_l and ϕ_g represent the coexisting "densities". They can be obtained theoretically (Maxwell construction) or numerically (Zhang et al. 2000). If the constants in the EOS are chosen as $a = 12c_s^2 = 12RT$ and $b = 4$, then the coexisting index function values are $\phi_l = 0.251$ and $\phi_g = 0.024$ (Zhang et al. 2000). These parameters are fixed in the following simulations.

In Zhang et al. (2000) the authors suggest that the surface tension σ should be calculated through ϕ rather than ρ, i.e.,

$$F_\alpha = \kappa \phi \partial_\alpha \left(\partial_\delta^2 \phi \right), \tag{7.49}$$

where k can be used to adjust the strength of surface tension. σ increases with κ.

We simulated several cases of a droplet immersed in gas and the Laplace law was used to calculate the surface tension numerically, i.e., after measuring the pressure inside and outside the droplet, Δp, and the equilibrium radius $R, \sigma = \Delta p R$. Then we obtained the surface tension as a linear function of κ. Using Eq. (7.49) and the constants mentioned in Eq. (7.12), we obtained approximately $\sigma = 0.0111\kappa$. The equilibrium state of one simulation is illustrated in Figure 7.1. In the simulation $\rho_l = 0.251, \rho_g = 0.024$, and $\kappa = 5 \times 10^{-4}$. The measured radius of the droplet is $R = 20.25$ lu and the pressure difference between the inside and outside of the droplet is $\Delta p = 2.8 \times 10^{-7}$ mu/luts^2. Hence, the calculated $\sigma = \Delta p R = 5.67 \times 10^{-6}$. It obeys the formula $\sigma = 0.0111\kappa$ well.

Since ρ_l and ρ_g can be set somewhat arbitrarily, if Eq. (7.18) is used to calculate the surface tension σ, then the relationship between the σ and κ is not fixed as mentioned above, but it is also a function of a density. Hence, Eq. (7.18) is not as convenient as Eq. (7.49). In the following discussion Eq. (7.49) is used to calculate the surface tension (Zhang et al. 2000).

Figure 7.1 Equilibrium state of a droplet immersed in gas. In the simulation $\rho_l = 0.251, \rho_g = 0.024$, and $\kappa = 5 \times 10^{-4}$. The maximum velocity magnitude measured is 1.616×10^{-6} lu/ts.

7.5   Layered two-phase flow in a channel

Layered two-phase flow in a channel has analytical solutions. Hence, it is a good benchmark test for different two-phase models. The flow set-up and analytical solution are identical to those in previous chapters.

In our simulations the non-wetting phase flows in the central region $0 < |y| < a$, while the wetting phase flows in the region $a < |y| < b$. The computational domain is 10×100. Because the periodic boundary condition is used on the left and right boundaries, the number of mesh nodes used in the x direction can be much smaller. In the present simulations $a = 20$ lu. The error between the numerical and analytical solutions is defined as

$$E(t) = \frac{\sum_j |u(j,t) - u_0(j)|}{\sum_j |u_0(j)|}, \qquad (7.50)$$

where the summation is over the lattice nodes j in the slice $x = 5$ and u_0 is the analytical solution. The convergence criterion is $|\frac{E(t) - E(t-10^4 \Delta t)}{E(t - 10^4 \Delta t)}| < 10^{-4}$.

Figure 7.2 shows the velocity profile across the middle vertical section of the channel for different kinematic viscosity ratios $M = \frac{v_n}{v_w} = \frac{v_l}{v_g}$ (density ratio $\frac{\rho_l}{\rho_g} = 3$ is fixed), where v_n and v_w are the kinematic viscosities of non-wetting and wetting fluids, respectively. In the figure velocity profiles in (a) and (b) are obtained through applying a body force $G = 1.0 \times 10^{-8}$ mu lu/ts^2 to both liquid and gas. From Figure 7.2 we can see that the numerical solutions agree well with the analytical ones except for small oscillations near the interfaces. Through inspection

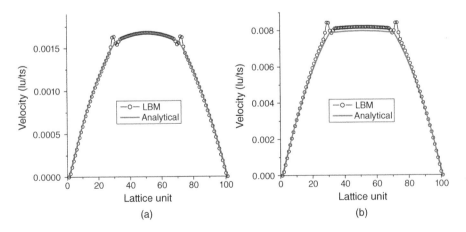

Figure 7.2   Layered two-phase flow simulation with $\kappa = 0.005$. Fluid in the center of the channel is liquid with $\rho_l = 0.12$. There is gas flow near the channel walls with $\rho_g = 0.04$. (a) $\tau_l = \tau_g = 1.0$ and (b) $\tau_l = 1.0, \tau_g = 0.6$.

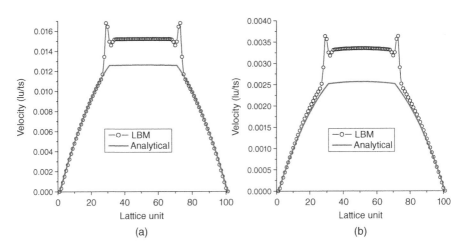

Figure 7.3   Layered two-phase flow simulation with $\kappa = 0.001$. Fluid in the center of the channel is liquid with $\rho_l = 0.251$. There is gas flow near the channel walls with $\rho_g = 0.024$. (a) $\tau_l = 1.0, \tau_g = 0.6$, and (b) $\tau_l = \tau_g = 1.0$.

of the pressure field, we found that there are small oscillations of pressure near the interfaces.

Figure 7.3 shows the velocity profiles for cases of higher density ratio ($\rho_l = 0.251, \rho_g = 0.024$). The density ratio $\frac{\rho_l}{\rho_g}$ is fixed in the two simulations (a) and (b). In Case (a), the kinematic viscosity ratios $M = \frac{v_l}{v_g} = 5$ (the ratio of dynamic viscosities is about 50), while in Case (b), $M = 1$ and the ratio of dynamic viscosities is about 10. The velocity profiles are all obtained by applying a body force $G = 1.0 \times 10^{-8}$ mu lu/ts^2 to both fluids. From Figure 7.3 we can see that the velocity in the gas region agrees well with the analytical solutions. However, for the interface and liquid regions there are significant discrepancies between the numerical and theoretical solutions. Hence, this model appears not to work very well for high-density ratio or high-viscosity ratio cases.

7.6 Rayleigh–Taylor instability

In this section the Rayleigh–Taylor instability is simulated. Rayleigh–Taylor instability means that when denser fluid is put above a less dense fluid, gravity and a small amplitude perturbation in the interface may cause the dense fluid to fall down while lower density fluid will rise. Of course, if the surface tension force is large enough or the perturbation is too small, the perturbation of the interface would be damped out to stabilize as a flat interface (an equilibrium state).

First the case of $Re = 256$ is simulated. Our simulations are conducted on a computational domain $W \times L = 128 \times 512$. Periodic boundary conditions are

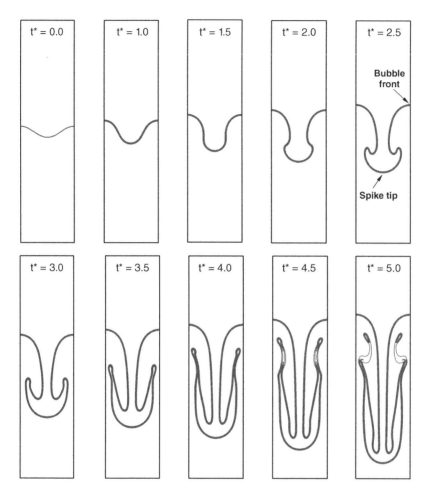

Figure 7.4 Interface evolution for Rayleigh–Taylor instability simulation with $Re = 256, \kappa = 0.01, U = \sqrt{Wg} = 0.04$ lu/ts, $\rho_l = 0.12$, and $\rho_g = 0.04$. The non-dimensional time, which is normalized by $\sqrt{W/g}$, is labeled in each figure. Fifteen equal-interval contours at $\rho = 0.045, 0.05, 0.055 \dots 0.115$ are drawn in the figures.

applied to the left and right boundaries. For the upper and lower boundaries, the wall boundary condition is applied. Hence, the simple bounce-back scheme is applied to f_i and g_i on these boundaries.

A 10% initial perturbation in the fluid interface, which is located in the middle of the domain, is set up (Figure 7.4). The function describing the location of the perturbed interface is

$$y = \frac{L}{2} + 0.1W \cos\left(\frac{2\pi x}{W}\right), \tag{7.51}$$

where x, y, L, and W are all in lattice units. Initially, there is a zero velocity field.

In the simulations we set $\rho_l = 0.12$ mu/lu^3 and $\rho_g = 0.04$ mu/lu^3, and the Atwood number ($At = \frac{\rho_l - \rho_g}{\rho_l + \rho_g}$) is 0.5.

On the one hand the gravity should be chosen to be as small as possible to make the characteristic velocity $U = \sqrt{Wg}$ small because the maximum velocity in the flow field increases with U. To eliminate compressibility effects smaller U is preferred. On the other hand, setting smaller U leads a lower relaxation time (close to 0.5), which also may induce numerical instability in the HCZ LBM. Here $U = 0.04$ lu/ts is adopted in our simulation. The corresponding gravitational acceleration is $g = -1.25 \times 10^{-5}$ lu/ts^2. The negative sign means the gravity force is in the $-y$ direction. The Reynolds number in the simulation is defined as $Re = \frac{UW}{\nu} = 256$ (He et al. 1999). Hence $\nu = 0.02$ lu^2/ts and $\tau = 0.56$.

In our code the gravity force $\mathcal{G}$ (in the y direction) is added to each fluid node in the computational domain with

$$\mathcal{G}(\mathbf{x}) = g\rho. \tag{7.52}$$

In the code it is also necessary to calculate the surface tension $F_\alpha = \kappa \rho \partial_\alpha (\partial_\delta^2 \rho)$ (He et al. 1999). $\partial_\delta^2 \rho$ is usually calculated through

$$
\begin{aligned}
\partial_\delta^2 \rho|_{(i,j)} = {} & \frac{4}{6} \left(\rho_{(i+1,j)} + \rho_{(i-1,j)} + \rho_{(i,j+1)} + \rho_{(i,j-1)} \right) \\
& + \frac{1}{6} \left(\rho_{(i+i,j+1)} + \rho_{(i-1,j-1)} + \rho_{(i+1,j-1)} + \rho_{(i-1,j+1)} \right) \\
& - \frac{20}{6} \rho_{(i,j)},
\end{aligned}
\tag{7.53}
$$

or equivalently, but more conveniently,

$$\partial_\delta^2 \rho|_{(i,j)} = \frac{2}{c_s^2 \Delta x^2} \sum_a w_a [\rho(\mathbf{x} + \mathbf{e}_a \Delta t) - \rho(\mathbf{x})]. \tag{7.54}$$

Here we can see that to calculate $\partial_\delta^2 \rho$ information from the surrounding eight nodes is required. For the lattice nodes on the left and right boundaries, the calculation is not a problem because the periodic boundary condition is applied. For the nodes just above the lower boundary ($j = 2$) or the nodes just below the upper boundary ($j = ly - 1$) the calculation is also straightforward. In this simulation the $\partial_\delta^2 \rho$ do not need to be calculated at $j = 1$ or $j = ly$ because for nodes on $j = 1$ and $j = ly$ the collision (which requires $\partial_\delta^2 \rho$) is not implemented at all. It is noted that the ρ in nodes $j = 1$ and $j = ly$ should be specified for convenience in the above calculation.

In our simulation the first-order derivatives $\partial_x \rho$ and $\partial_y \rho$ can be calculated through

$$\partial_\delta \rho|_{(i,j)} = \frac{1}{c_s^2 \Delta x} \sum_a w_a e_{a\delta} \rho(\mathbf{x} + \mathbf{e}_a \Delta t), \tag{7.55}$$

or the following equivalent expression: suppose for example that δ refers to the x coordinate, then we have

$$\partial_x \rho|_{(i,j)} = \frac{1}{3} \left(\rho_{(i+1,j)} - \rho_{(i-1,j)} \right) + \frac{1}{12} \left(\rho_{(i+1,j+1)} - \rho_{(i-1,j-1)} \right)$$

$$+ \frac{1}{12} \left(\rho_{(i+1,j-1)} - \rho_{(i-1,j+1)} \right). \tag{7.56}$$

The result of case $Re = 256$ is shown in Figure 7.4. At this lower Re, before the non-dimensional time (which is normalized $\sqrt{W/g}$) $t^* = 1.5$, the perturbation grows gradually but the fluid interface remains symmetrical up-and-down (He et al. 1999). The dense fluid falls as a spike and the less dense fluid rises. Compared to Figure 6 in He et al. (1999), the evolution of the interface is indistinguishable from their result computed on a finer mesh with domain size 256×1024.

A case with $Re = 2048$ is also simulated. In this case the computational domain is $W \times L = 256 \times 1024$. The boundary condition set-up and the initial perturbation in the fluid interface (Eq. (7.51)) are identical to those in the $Re = 256$ case.

In the simulations we set $\rho_l = 0.12$ and $\rho_g = 0.04$, and the Atwood number $\left(\frac{\rho_l - \rho_g}{\rho_l + \rho_g} \right)$ is 0.5. Here it is worth mentioning that the choice of ρ_l and ρ_g also affects the numerical stability. For example, if $\rho_l = 0.15$ and $\rho_g = 0.05$, the characteristic velocity U must be set to a larger value (e.g., even if $U = 0.1$ lu/ts, corresponding to $\tau = 0.5375$, the simulation may be unstable). However, as stated before, the larger U is, the stronger the compressibility effect is. The numerical stability seems to be sensitive to the choice of ρ_l and ρ_g, but that was not mentioned in He et al. (1999).

In this case the characteristic velocity $U = \sqrt{Wg} = 0.04$ lu/ts, which is small to eliminate compressibility effects. Hence, the gravity force acceleration in our simulation is $g = -6.25 \times 10^{-6} lu/ts^2$. The corresponding kinematic viscosity is $v = 0.005$ and the relaxation parameter $\tau = 0.515$ for both liquid and gas. In the simulation, the gravity force is applied to all fluid nodes (including liquid and gas phases). Further, $\kappa = 0.01$ and the surface tension is $\sigma = 1.12 \times 10^{-4} \kappa = 1.12 \times 10^{-6}$. The Weber number is about

$$\frac{U\eta}{\sigma} = \frac{0.04 \times 0.12 \times 0.005}{1.12 \times 10^{-6}} \approx 21. \tag{7.57}$$

Our simulated result is shown in Figure 7.5. It is seen that after $t^* = 1.5$, the main spike has formed and when it falls down two sides of the spike will roll up ($t^* = 3.0$). After that the flow field becomes more complex. After $t^* = 4.5$, at the tail of the roll-ups, some droplets formed due to pinch-off (He et al. 1999; Zhang et al. 2000). The evolution of the interface fundamentally agrees with Figure 3 of He et al. (1999) and Figure 6 of Zhang et al. (2000). The difference is that the unstable roll-ups in Figure 3 of He et al. (1999) are suppressed here by the surface tension (Eq. (7.49)) that was set to zero in He et al. (1999), and consequently complex structures in the flow become fewer and smoother (Zhang et al. 2000).

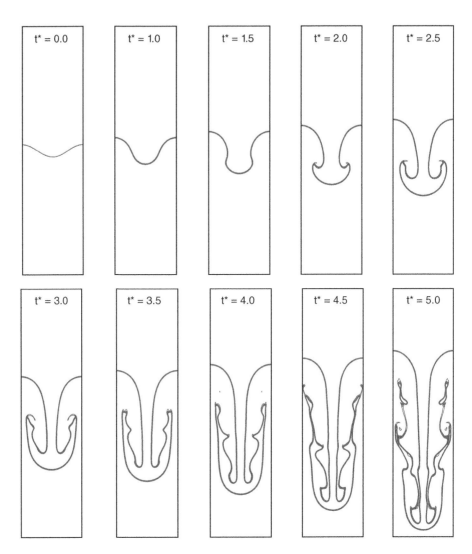

Figure 7.5 Interface evolution for Rayleigh–Taylor instability simulation with $Re = 2048, \kappa = 0.01, U = \sqrt{Wg} = 0.04$ lu/ts, $\rho_l = 0.12$, and $\rho_g = 0.04$. The non-dimensional time, which is normalized by $\sqrt{W/g}$, is labeled in the figures. Fifteen equal-interval contours $\rho = 0.045, 0.05, 0.055 \ldots 0.115$ are drawn in each figure.

The positions and velocities of the spike tip and bubble front (see Figure 7.4) as a function of time are shown in Figure 7.6. Basically, the curves are consistent with those in Figure 7 of He et al. (1999). There are small amplitude oscillations in the velocity curves and a possible reason is compressibility of the fluid system. Another possible reason is that in He et al. (1999) the $\nabla\psi$ is calculated through a third-order differencing scheme but here our differencing scheme is second-order in space. From Figure 7.6(b) it is seen that although Re is very

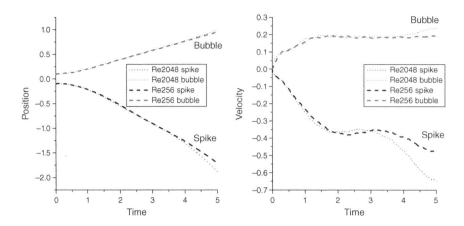

Figure 7.6   (a) The normalized positions and (b) velocities of the spike tip and bubble front as a function of time (both the above simulated cases $Re = 256$ and $Re = 2048$ are shown, $\kappa = 0.01, At = 0.5$). Time and length are normalized by $\sqrt{W/g}$ and W, respectively.

different, the normalized bubble rise velocity is almost a constant (about 0.2) in both cases. For the spike, between the normalized times $t^* \in (2, 3.5)$, the velocity is almost a constant, but after that the spike's velocity increases quickly.

In Zhang et al. (2000) the surface tension effect on the Rayleigh–Taylor instability is also discussed. For very small surface tension the positions and velocities of the bubble and spike fronts versus time are almost not affected at all. When the surface tension is larger ($\kappa > 0.2$), positions and velocities will be slightly different from cases with smaller surface tension (Zhang et al. 2000). Similar behavior is observed in our simulations.

7.7   Contact angle

Yiotis et al. (2007) proposed a scheme to achieve desired contact angles using the HCZ model. They adopted the assumption of Rowlinson and Widom (1982) that the solid is made up of rigid molecules of given density. Different contact angles can be achieved by assigning an "effective" ρ for the solid nodes in the range between ρ_g and ρ_l (Yiotis et al. 2007). The "effective" ρ is used because in their simulations the original surface tension calculation (He et al. 1999) is used, i.e., Eq. (7.18). Here, because the revised surface tension is adopted in our code, i.e., Eq. (7.49) (Zhang et al. 2000), we should specify ϕ_w rather than ρ_w. This is the minor difference between our scheme and theirs.

In the following we will illustrate how to specify a wall's wetting property in the HCZ Lattice Boltzmann simulations in detail. In our simulations the computational domain is 200×60. Periodic boundary conditions are applied on

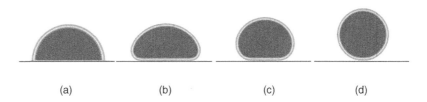

Figure 7.7 Evolution of a droplet contacting with a wall (because $\phi_w = 0.041$, the expected equilibrium contact angle is about 180°). (a) $t = 0$, initially a semi-circle droplet is initialized, (b) $t = 2000$ ts, (c) $t = 10,000$ ts, and (d) $t = 100,000$ ts. In the simulation $\phi_{max} = 0.251, \phi_{min} = 0.024, \rho_l = 0.12$, and $\rho_g = 0.04$.

Figure 7.8 Equilibrium contact angles (initial condition is a semi-circular droplet as shown in Figure 7.7(a)). (a) $\phi_w = 0.20$, measured $\theta = 43°$, (b) $\phi_w = 0.14$, measured $\theta = 92°$, (c) $\phi_w = 0.10$, measured $\theta = 125°$, and (d) $\phi_w = 0.06$, measured $\theta = 156°$. In the simulation $\phi_{max} = 0.251, \phi_{min} = 0.024, \rho_l = 0.12$, and $\rho_g = 0.04$.

the left and right boundaries. The upper and lower boundaries are walls. The bounce-back boundary condition is applied instead of the collision step on all lattice nodes that are solid walls.

A semi-circle with radius of 25 lu is initialized in all of our contact angle simulations. Here, an "effective" ϕ for the solid nodes in the range between ϕ_{min} and ϕ_{max} is specified to account for attractive (adhesive) forces between fluid particles and "molecules" of the solid surface. The ϕ_w takes effect through Eq. (7.49) and $F'_\alpha = -\partial_\alpha \psi(\phi)$ in Eq. (7.19).

First a case with $\phi_w = 0.041$ is simulated in which the contact angle is expected to be approximately 180°. In this simulation $\kappa = 0.01, \rho_l = 0.12$, and $\rho_g = 0.04$. Figure 7.7 shows the evolution of the droplet. Initially the droplet is a semi-circle and gradually it rises up with less surface contacting with the wall. Finally the droplet reaches a equilibrium state with a contact angle of approximately 180°.

More cases with different ϕ_w are simulated. The other settings are identical to the above case. Figure 7.8 shows several typical results of the simulations. By specifying different ϕ_ws, different equilibrium contact angles are obtained.

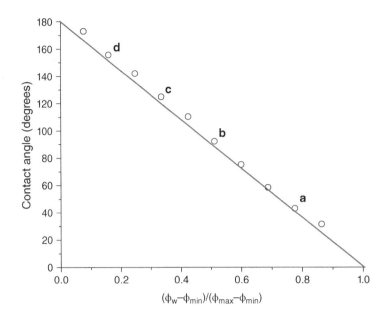

Figure 7.9 The equilibrium contact angle as a function of $\frac{\phi_w - \phi_{min}}{\phi_{max} - \phi_{min}}$. The circles show results measured from LBM simulations, including the simulations in Figure 7.8, labeled as a, b, c, and d. The solid line connects points $(0, 180°)$ and $(1, 0)$ in the coordinate system. In the simulation $\phi_{max} = 0.251$, $\phi_{min} = 0.024$, $\rho_l = 0.12$, and $\rho_g = 0.04$.

The equilibrium contact angle as a function of $\frac{\phi_w - \phi_{min}}{\phi_{max} - \phi_{min}}$ is shown in Figure 7.9. In the figure, a solid line connecting points $(0, 180°)$ and $(1, 0)$ is drawn to guide the eye. It appears that the contact angle is almost a linear function of $\frac{\phi_w - \phi_{min}}{\phi_{max} - \phi_{min}}$. It is worth mentioning that, compared to Figure 4 of Yiotis et al. (2007), here the wetting angle exhibits a much better linear dependence on the adjustable parameter.

We note that the contact angle is only a function of ϕ_w ($\phi_w \in [\phi_{min}, \phi_{max}]$). Through many simulations with other κ and density ratios, we confirmed that the contact angle obtained is independent of κ and density ratio, i.e., of ρ_l and ρ_g. In the other words, Figure 7.9 is uniformly valid for all κ, ρ_l, and ρ_g.

We would like to mention what has to be accommodated in the implementation of this model to solve contact angle problems. The simulation code is basically identical to that in Appendix B for the Rayleigh–Taylor instability in Section 7.12. The differences are shown in detail in the following.

The code segment for the initialization of the ϕ field in initial.for should be:

```
      do 21 y = 1, ly
      do 31 x = 1, lx
        rh(x,y) = 0.5d0*(psi_min +psi_max) + 0.5d0*(psi_max-psi_min)
    &    *tanh( (25.d0-sqrt((y-1.d0)*(y-1.d0))
```

```
&   +(x-lx/2)*(x-lx/2)) )*2.d0/3.d0)

    if(obst(x,y) .eq. 1) rh(x,y)= rh_w   ! 0.041
31 continue
21 continue
```

Here $rh(x, y)$ represents the ϕ field. We can see that not only ϕ in the fluid nodes but also ϕ in the solid nodes (nodes of $obst(x, y) = 1$) is specified. After the ϕ field is specified, the density field is fully determined through Eq. (7.13).

The streaming and collision steps are identical to those in Appendix B. It is noted there is no regular collision step for the solid nodes and in the "collision" subroutine the distribution functions in the solid nodes only implement the bounce-back rule.

The ϕ_w in solid nodes will normally be kept at its initial specified value. This means that the calculation of ϕ is only applicable to the fluid nodes. However, the calculation of $\psi(\phi)$ and $\psi(\rho)$ should be applied to all nodes, including solid nodes. Hence, with $\psi(\phi)$ and $\psi(\rho)$ on the $y = 1$ (bottom) and $y = ly$ (top) layers, the first derivatives $\partial_y \psi(\phi)$ and $\partial_y \psi(\rho)$ can be calculated correctly for lattice nodes on $y = 2$ and $y = ly - 1$. When performing contact angle simulations, the Fortran code listed in Appendix B can be used directly with specification of the $\phi = \phi_w$ using the code fragment given above.

7.8 Capillary rise

Here we use the HCZ model to simulate capillary rise in a simple capillary. As in Chapter 2, the dimensionless Bond number Bo is used to relate capillary and gravitational forces (Sukop and Thorne 2006). We simulate the same physical problem simulated in Chapter 2, which has $Bo = r/h = 1/7.36$. Defining an analogous LBM system begins with capillary tube size. Suppose the radius of tube in our simulation is 12 lu, then the maximum equilibrium capillary rise we expect in our simulation is $r = 12$ lu $\times 7.36 = 88.5$ lu.

As shown in Figure 7.10, we chose a domain size of 200×300 with a wall on the bottom. Because bounce-back is easier to implement than the open boundary on the top, another wall is placed on the upper boundary. It does not affect the equilibrium rise height. Periodic boundary conditions are applied on the left and right boundaries.

The length of the tube is about 270 lu. The tube is put 15 lu above the bottom wall. In the analogous system

$$g = \frac{\sigma}{(\rho_l - \rho_g)hr}. \tag{7.58}$$

In our LBM simulation $\kappa = 0.1$, the surface tension is $\sigma = 1.11 \times 10^{-3}$ (see Section 7.4), and the coexisting densities are set to be $\rho_l = 0.12$ and $\rho_g = 0.04$.

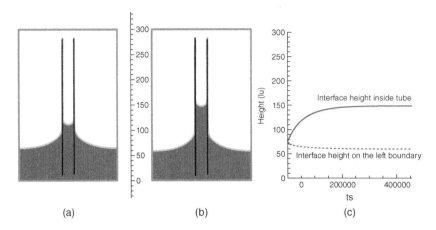

(a)   (b)   (c)

Figure 7.10   Capillary rise. (a) Capillary rise at $t = 40,000$ ts, (b) equilibrium rise height at $t = 440,000$ ts, and (c) the height of liquid inside and outside (on the left boundary) the capillary tube as a function of time. The coexisting densities are $\rho_l = 0.12$, $\rho_g = 0.04$, and $\phi_w = 0.251$.

Hence, according to Eq. (7.58), the gravity acceleration in our simulation should be $g = 1.318 \times 10^{-5}$ lu/ts² for our model problem. Since the liquid is completely wetting, the "density" of the solid nodes is set to be $\phi_w = 0.251$.

We begin with the liquid filling the region below $y = 80$ lu both inside and outside the capillary tube. The results are shown in Figure 7.10. Figure 7.10(a) shows the capillary rise at $t = 40,000$ ts, when the liquid is in the process of rising. Figure 7.10(c) shows the interface heights inside and outside (on the left boundary) the capillary as a function of time. We can see that the liquid column inside the tube rises quickly at beginning but later slows down and finally reaches an equilibrium height. The equilibrium height difference between the inside and outside of the tube is 89 lu, which is consistent with the expected value of 88.5 lu computed analytically above. For the capillary rise problem, the HCZ model appears to be more accurate than the SC model (c.f., Chapter 2).

To track the interface position, the following functional subroutine can be inserted into the Fortran code. In this subroutine the interface position represented by $\rho = \frac{\rho_l + \rho_g}{2}$ is obtained through an interpolation scheme.

```
subroutine track_interface(lx,ly,rho,n)
implicit none
include "head.inc"
integer lx,ly, n, x,y
real*8   rho(lx,ly), rho_d, pos_d,rho_u, pos_u, pos, pos2

open(45,file='./out/posi.dat', access='append')
x=lx/2
do y=1, ly
  if(rho(x,y) .lt. (rho_h+rho_l)/2.d0 ) then
    rho_d=rho(x,y)
```

```
      pos_d=y
     endif
  enddo

  do y=ly, 1, -1
     if(rho(x,y) .gt. (rho_h+rho_l)/2.d0 ) then
       rho_u=rho(x,y)
       pos_u=y
     endif
  enddo

  pos=pos_d+ (pos_u-pos_d)/(rho_u-rho_d)*((rho_h+rho_l)/2.-rho_d)

  x=1
  do y=1, ly
     if(rho(x,y) .lt. (rho_h+rho_l)/2.d0 ) then
       rho_d=rho(x,y)
       pos_d=y
     endif
  enddo

  do y=ly, 1, -1
  if(rho(x,y) .gt. (rho_h+rho_l)/2.d0 ) then
    rho_u=rho(x,y)
    pos_u=y
  endif
  enddo
  pos2=pos_d+ (pos_u-pos_d)/(rho_u-rho_d)*((rho_h+rho_l)/2.-rho_d)
  write(45,'(i8,_2f15.4)') n, pos, pos2

  end
```

7.9 Geometric scheme to specify the contact angle and its hysteresis

Besides the above scheme of Yiotis et al. (2007), a simple geometric scheme (Wang et al. 2013) is also able to specify the wetting properties of walls in the simulations. The geometric scheme makes it easier to simulate contact angle hysteresis compared to the above scheme (Wang et al. 2013). The geometric formulation is (Ding and Spelt 2007),

$$\phi_{i,1} = \phi_{i,3} + tan\left(\frac{\pi}{2} - \theta\right)|\phi_{i+1,2} - \phi_{i-1,2}|, \qquad (7.59)$$

where the first and second subscripts denote the coordinates along and normal to the solid boundary, respectively. θ is the desired contact angle. In order to calculate the terms of $\nabla\psi(\phi)$ and $\nabla\psi(\rho)$, a layer of ghost cells adjacent to the solid boundary is necessary. The values of $\phi_{i,0}$ that are defined on the ghost cells are given by $\phi_{i,0} = \phi_{i,1}$.

For 3D simulations the extension of Eq. (7.59) is straightforward. A schematic of the 3D situation is shown in Figure 7.11. In the figure the x–y plane is supposed

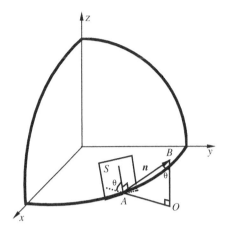

Figure 7.11 Schematic of the contact angle in 3D situations. Reprinted from Wang et al. (2013), copyright (2013), with permission from APS.

to be a wall and A is a point on the contact line. The gradient of ϕ at point A is $-\mathbf{n} = (\frac{\partial \phi}{\partial x}, \frac{\partial \phi}{\partial y}, \frac{\partial \phi}{\partial z})$ for cases of a droplet inside gas. A tangent plane S on the droplet surface, which passes through point A is also illustrated. The plane is perpendicular to the vector **n**. As illustrated in Figure 7.11, the contact angle between the wall and the interface, i.e., θ, satisfies $\tan(\frac{\pi}{2} - \theta) = \frac{\overline{OB}}{\overline{OA}} = -\frac{\partial \phi}{\partial z}/\sqrt{(\frac{\partial \phi}{\partial x})^2 + (\frac{\partial \phi}{\partial y})^2}$, where the overline means the length of the line. This equation would leads to the following discrete form,

$$\phi_{i,j,1} = \phi_{i,j,3} + tan\left(\frac{\pi}{2} - \theta\right)\xi, \tag{7.60}$$

where

$$\xi = \sqrt{\left(\phi_{i+1,j,2} - \phi_{i-1,j,2}\right)^2 + \left(\phi_{i,j+1,2} - \phi_{i,j-1,2}\right)^2}. \tag{7.61}$$

In this way we can achieve any wettability between fluids and solids through inputting a specified contact angle ($0° \leq \theta \leq 180°$) into Eqs (7.59) and (7.60) for 2D and 3D cases, respectively. It is noted that Eqs (7.59) and (7.60) are applied only at the contact point and the contact line, respectively. The simulation of multiphase flow is in the incompressible limit.

On the other hand, in many natural and industrial systems the solid walls are usually rough and chemically inhomogeneous (de Gennes 1985), so we further extend the above geometric formulation to non-ideal surfaces. In other words, contact angle hysteresis is taken into consideration. Because of hysteresis, the contact line remains pinned when the local contact angle θ is within a hysteresis window:

$$\theta_R \leq \theta \leq \theta_A, \tag{7.62}$$

where θ_R and θ_A denote the receding contact angle and advancing contact angle, respectively. If θ is not inside the hysteresis window, the contact line is allowed to move (Ding and Spelt 2008). When θ is greater than θ_A, the contact line moves forward; when θ is less than θ_R, the contact line moves backward. To realize this effect, at each time step of computation we should first obtain the local apparent contact angle at the contact points. Then comparisons of θ with θ_R and θ_A are required. If $\theta \leq \theta_R$, θ in Eq. (7.59) should be replaced by θ_R; if $\theta \geq \theta_A$, θ in Eq. (7.59) should be replaced by θ_A. Otherwise θ in Eq. (7.59) remains unchanged (Ding and Spelt 2008). Here we can see that the hysteresis effect is prescribed by two parameters: advancing angle and receding angle, which depend on the properties of the fluids and solids (de Gennes 1985). This simple implementation can simulate the contact angle hysteresis effect independently of whether the hysteresis is caused by roughness or chemical inhomogeneities.

First, droplet contacts with an ideal wall for six different angles are simulated. In our simulations, the top and bottom boundaries are solid walls and simple bounce-back is applied. Periodic boundary conditions are applied to the left and right boundaries. In these simulations a semicircular droplet was initialized in a computational domain 201×101. The angle θ was specified in Eq. (7.59). After an evolution process, the droplet reaches an equilibrium state. Figure 7.12 shows the six different equilibrium contact angles achieved. The specified θ is shown below each part of the figure. We find that the final contact angles obtained are consistent with the specified ones. In the six cases the maximum discrepancy between the measured and the specified contact angles is less than $1°$. This

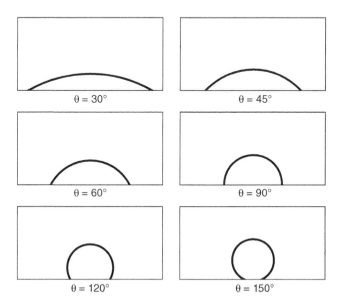

Figure 7.12 Different contact angles are achieved by putting different specified θ into Eq. (7.59). Reprinted from (Wang et al. (2013), copyright (2013), with permission from APS.

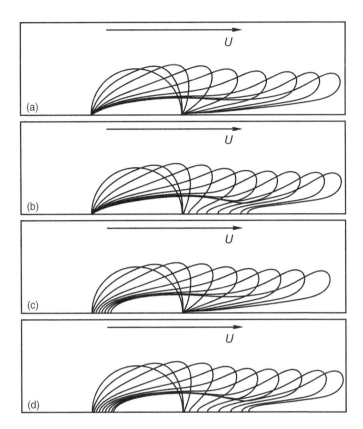

Figure 7.13 Four typical motion modes of contact points obtained by specifying the contact angle hysteresis effect in the HCZ model. (a) $\theta_R = 0°, \theta_A = 180°$; (b) $\theta_R = 0°, \theta_A = 110°$; (c) $\theta_R = 70°, \theta_A = 180°$; (d) $\theta_R = 70°, \theta_A = 110°$. Reprinted from Wang et al. (2013), copyright (2013), with permission from APS.

demonstrates that the scheme is accurate for the specification of wettability on an ideal wall.

For the validation of 3D cases, refer to Wang et al. (2013).

7.9.1 Examples of droplet slipping in shear flows

In this flow an initially semicircular droplet deforms and/or slips in a shear flow driven by an upper plane moving with a constant velocity U.

Qualitative simulation results showing the effect of contact angle hysteresis are given in Figure 7.13. There are four typical motion modes of the contact points due to four specified hysteresis windows. In Figure 7.13(a) both the upstream and downstream contact angles are always inside the effectively infinite hysteresis window $(0°, 180°)$, so the two corresponding contact points are pinned on the wall at all times. In Figure 7.13(b) the upstream contact angle varies from $90°$ to acute angles, so it is always within the specified window range $(0°, 110°)$, consequently the upstream contact point cannot move. For the downstream contact point, at early stages it is pinned as the contact angle is less than $110°$, and later it moves due to the contact angle being greater than $110°$. In Figure 7.13(c) the downstream contact angle varies from $90°$ to obtuse angles which are always in the range $(70°, 180°)$, so the downstream contact point remains immobile. For the upstream contact point, at early stages it is pinned due to the contact angle being greater than $70°$, and later it moves due to the contact angle becoming less than $70°$. Figure 7.13(d) shows that at early stages the droplet deforms and remains pinned, while subsequently the two contact angles are not in the specified hysteresis window and the whole droplet slips.

The effect of contact angle hysteresis is also tested in 3D cases. In Wang et al. (2013) initially a 3D hemispherical droplet is at rest on a solid plane in a Couette flow, where the upper wall moves rightward and the bottom wall remains stationary. In the simulations four typical motion modes for the contact lines equivalent to those in the 2D study (i.e., Figure 7.13) are also observed. One of the four modes is shown in Figure 7.14. In the figure the shapes of contact lines at different times are also shown in the right-hand column. Figure 7.14 is obtained through setting the hysteresis window $(0°, 110°)$. The motion mode is similar to that of Figure 7.13(b), in which the upstream contact line remains stationary while the downstream section moves. In this motion mode the contact area increases.

More results from 3D simulations of the typical contact angle hysteresis phenomena can be found in Wang et al. (2013).

7.10   Oscillation of an initially ellipsoidal droplet

First, 3D simulations were performed to get the coefficient between the surface tension and κ. The 3D HCZ code is illustrated in Appendix C in this chapter. In the simulations the computational domain is $120 \times 120 \times 120$. Initially a sphere with radius $R = 30$ lu is filled with fluid $\rho = \rho_l$ ($\phi = \phi_l$) and the other region is filled with gas. Periodic boundary conditions are applied. In the simulations, $\rho_l = 0.12, \rho_g = 0.04$, and $\tau_g = \tau_f = 0.7$.

The surface tension can be calculated from the pressure difference between the inside and the outside of the drop, i.e., $\sigma = \frac{1}{2}(p_{in} - p_{out})R$. From the calculated surface tension as a function of κ in three simulations, we have

$$\sigma = 0.0115\kappa. \tag{7.63}$$

Then oscillation of an initially ellipsoidal droplet was simulated. An initially ellipsoidal drop suspended in gas will oscillate and due to surface tension and viscous dissipation the drop will eventually reach a spherical drop equilibrium

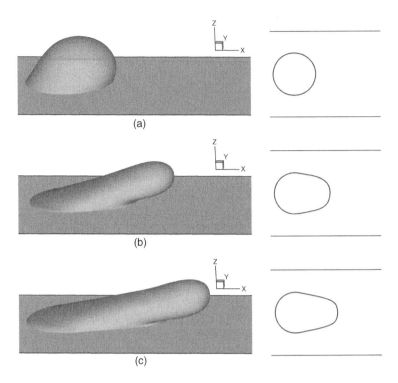

Figure 7.14 Movement of a 3D hemispherical droplet, which is driven by a shear flow at (a) $t = 2000$ ts, (b) $t = 6500$ ts, and (c) $t = 8000$ ts. The hysteresis window is $\theta_R = 0°, \theta_A = 110°$. The shapes of contact lines at different times are shown in the right-hand column. Reprinted from Wang et al. (2013), copyright (2013), with permission from APS.

state. There are many theoretical studies on this topic, e.g., Miller and Scriven (1968) and Lamb (1932).

The oscillation frequency of the ellipsoidal drop can be predicted by the theory of Miller and Scriven (1968), which includes viscous dissipation effects in the boundary layer at the interface (Premnath and Abraham 2005a). Here the frequency predicted from theory will be compared with the LBM result to validate the LBM code. The frequency for the nth mode of oscillation for a drop is (Miller and Scriven 1968)

$$\omega_n = \omega_n^* - \frac{1}{2}\alpha\omega_n^{*\frac{1}{2}} + \frac{1}{4}\alpha^2, \tag{7.64}$$

where ω_n is the angular response frequency and ω_n^* is Lamb's natural resonance frequency (Lamb 1932):

$$(\omega_n^*)^2 = \frac{n(n+1)(n-1)(n+2)}{R_d^3[n\rho_g + (n+1)\rho_l]}\sigma. \tag{7.65}$$

Table 7.1 Theoretical and simulated oscillation period T of ellipsoidal drops ($R = 30$ lu).

Case	ρ_l	ρ_g	κ	τ	T (Eq. (7.64))	T (LBM)	Error of T
I	0.12	0.03	0.02	0.53	8872	9440	6.4%
II	0.12	0.04	0.01	0.555	13560	14650	8.0%

R_d is the equilibrium radius of the drop. The parameter α is given by (Miller and Scriven 1968)

$$\alpha = \frac{(2n+1)^2 \sqrt{\eta_l \eta_g \rho_l \rho_g}}{\sqrt{2} R_d [n\rho_g + (n+1)\rho_l] \left[\sqrt{\eta_l \rho_l} + \sqrt{\eta_g \rho_g}\right]}, \tag{7.66}$$

where η_l and η_g are the dynamic viscosities of the liquid and gas, respectively. The time period calculated from the second mode of oscillation in Eq. (7.64) is used as the theoretical prediction (Premnath and Abraham 2005a).

In the simulation an ellipsoidal drop instead of a spherical drop is used for the initial condition. The mathematical description of the ellipsoid is

$$\frac{(x-x_c)^2}{R^2} + \frac{(y-y_c)^2}{R^2} + \frac{(z-z_c)^2}{(0.85R)^2} = 1, \tag{7.67}$$

where (x_c, y_c, z_c) is the center of the ellipsoid and the computational domain. It is noted that R_d can be calculated through mass conservation, i.e., $\frac{4}{3}\pi R_d^3 = \frac{4}{3}\pi R^3 \times 0.85$. In the simulations $R = 30$ lu. The ellipsoid is initially placed in the center of the 100^3 lu^3 computational domain. Here we carried out simulations of two cases. The parameters are listed in Table 8.1. Cases I and II are all simulated using the BGK-based code given in Appendix C.

The periods of oscillation for the four cases are listed in Table 7.1. The period has units of ts. The theoretical period is obtained from Eq. (7.64), i.e., $T = \frac{2\pi}{\omega_n}$, where $n = 2$. From the table we can see that oscillation periods measured from the LBM are consistent with the theoretical predictions (with maximum error less than 8%).

The radii of the spheroid as a function of time are also shown in Figure 7.15. In each frame the dashed and solid lines represent the radius measured from the center of the droplet in horizontal and vertical directions, respectively. Obviously, when the two curves (the dashed and solid lines) cross, the drop is spherical. The oscillatory period can be easily measured from these plots. From the figure we can see that due to viscous dissipation, the amplitude of the oscillation decreases with time.

In Figure 7.15, if we connect the crossings of the two curves, the resulting line is almost horizontal, which means that the mass of the drop is almost conserved.

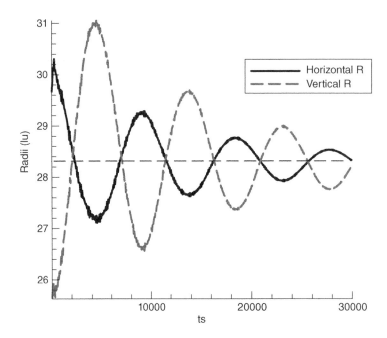

Figure 7.15 Radii of the oscillatory spheroid as a function of time for Case I. The dashed and solid lines represent the radius measured from the center of the droplet in the z and x directions, respectively. Quantities are in lattice units.

7.11 Appendix A

Here we prove that the underlined term $\partial_{t_1}(\rho u_\alpha u_\beta)$ and the underbraced terms in Eq. (7.41) can be canceled.

From Eq. (7.33) we know that

$$u_\alpha \partial_{t_1}(\rho u_\beta) = -u_\alpha \partial_\alpha (p\delta_{\alpha\beta} + \rho u_\alpha u_\beta) + u_\alpha F_\beta \tag{7.68}$$

and

$$u_\beta \partial_{t_1}(\rho u_\alpha) = -u_\beta \partial_\beta (p\delta_{\alpha\beta} + \rho u_\alpha u_\beta) + u_\beta F_\alpha. \tag{7.69}$$

Substituting the above two equations into

$$\partial_{t_1}(\rho u_\alpha u_\beta) = u_\alpha \partial_{t_1}(\rho u_\beta) + u_\beta \partial_{t_1}(\rho u_\alpha) - u_\alpha u_\beta \partial_{t_1}\rho \tag{7.70}$$

we have

$$\partial_{t_1}(\rho u_\alpha u_\beta) = -u_\alpha \partial_\alpha (p\delta_{\alpha\beta} + \rho u_\alpha u_\beta) - u_\beta \partial_\beta (p\delta_{\alpha\beta} + \rho u_\alpha u_\beta)$$
$$+ (u_\beta F_\alpha + u_\alpha F_\beta) - u_\alpha u_\beta \partial_{t_1}\rho. \tag{7.71}$$

We note that $\partial_{t_1}\rho = -\partial_\alpha(\rho u_\alpha)$. Hence, the last term in the above equation is of order $O(u^3)$, which means it can be omitted. Omitting all terms of order $O(u^3)$ we have

$$\partial_\alpha\partial_{t_1}(\rho u_\alpha u_\beta) = \partial_\alpha(-u_\alpha\partial_\beta p - u_\beta\partial_\alpha p) + \partial_\alpha(F_\beta u_\alpha + F_\alpha u_\beta) + O(u^3). \qquad (7.72)$$

Hence, the underlined term $\partial_\alpha\partial_{t_1}(\rho u_\alpha u_\beta)$ and the underbraced terms in Eq. (7.41) can be canceled.

7.12 Appendix B: 2D code

```
c 1. This program is aimed to simulate the Rayleigh-Taylor
! instability phenomena.
c 2. The fortran files: head.inc,  main.for, IO.for, initial.for,
! fgequ.for collisionStr.for
c   should be put in a identical folder and compile together.
c 3. A new subfolder .\out should be created before run the code
! because
c   all output files are supposed to write under ".\out".
c 4. The file 'params.in' should put with *.for and you can change the
! parameters.
c   'params.in' begins with parameters and "c===========..." is not
! the first line.
c 5. The subroutine "track_interface" is listed in Section 7.8.
! Please copy
c   it to any of *.for files.
c 6. If an error "forrt1: severe <170>: Program Exception -- stack
! overflow" appear,
c   please modify the following option and compile again. In
! "Project->Setting->Link->
c   Output->Reserve", please set the reserve value to be at least
! 200,000,000.

params.in:
c=================================================================
0.251       ! psi_max
0.024       ! psi_min
0.251       ! psi_max0 <psi_max
0.024       ! psi_min0 >psi_min
0.12        ! rho_h
0.04        ! rho_l
-------------------
256.        ! Re
0.04        ! V1 characteristic vel.
20000       ! t_max
1600        ! Nwri
0.01        ! Kappa

head.inc
c Here some common blocks are defined.
c The domain size is xA \times yA
!----------------------------------
```

```fortran
      integer xA,yA
      parameter(xA=128,yA=512)
      real*8 Rref(3), h_ref
      PARAMETER(h_ref=128.d0)

      common/AA/ Nwri,   BD, RAYLEIGH, opp(8)
      integer Nwri,    BD, RAYLEIGH, opp

      common/b/ error,vel,tau_g,tau_f,xv(0:8),yv(0:8),t_k(0:8)
      real*8 error,vel,tau_g, tau_f,xv,yv,t_k

      common/app/ limi,lam,gam,t_0,t_1,t_2,c_squ,visc,pai, con2
      real*8 limi,lam,gam,t_0,t_1,t_2,c_squ,visc,pai, con2

      common /shp/  rho_h, rho_l,psi_max,psi_min, const, Tn
      real*8 rho_h, rho_l,psi_max,psi_min, const, Tn

      common/psi/ psi_min0, psi_max0
      real*8  psi_min0, psi_max0
      common /velo/ Kappa, gforce,  We, R, Eo
      real*8  Kappa, gforce,  We, R , Eo

      common /nu/  dfai_x(xA,yA),dfai_y(xA,yA)
      real*8 dfai_x,dfai_y
      common /prho/ dprho_x(xA,yA), dprho_y(xA,yA)
      real*8 dprho_x,dprho_y
```

main.for
```fortran
C==================================================================
      program d2q9lbm
      implicit none
      include "head.inc"
      integer t_max,time,k,i, j, x, y
      integer obstA(xA,yA)
c      a 9-speed lattice is used here, other geometries are possible
c             6    2    5
c              \ | /
c             3 - 0 - 1
c              / | \
c             7    4    8
c
c    the lattice nodes are numbered as follows:
      real*8   deltaA(8,xA,yA),UA(xA,yA),VA(xA,yA),rhoA(xA,yA)
      real*8   noteqA(0:8,xA,yA), ffA(0:8,xA,yA), gpA(0:8,xA,yA)
      real*8    rhA(xA,yA), up(xA,yA), vp(xA,yA)
      real*8   Force1(xA,yA)
      real*8   Force2(xA,yA),pA(xA,yA)
      character BT*10
      integer*4 now(3)
C-----------------------------
c  Author: Haibo Huang   email: huanghb@ustc.edu.cn
C-----------------------------
      data xv/0.d0,1.d0,0.d0, -1.d0, 0.d0, 1.d0, -1.d0, -1.d0, 1.d0/,
     &     yv/0.d0,0.d0,1.d0, 0.d0, -1.d0, 1.d0, 1.d0, -1.d0, -1.d0/
```

```
      data opp/3,4,1,2,7,8,5,6/  ! opposite direction of each
! velocity
! Opposite direction of each velocity in the D2Q9 model, opp(1)=3,
! opp(2)=4, ...
! This array is serve for the bounce back implementation.
C-----------------------
      BD = 0

      pai = datan(1.d0)*4.d0
      c_squ = 1.d0 / 3.d0
      t_0 =  4.d0 / 9.d0
      t_1 =  1.d0 / 9.d0
      t_2 =  1.d0 / 36.d0

      t_k(0) = t_0
      do 1 k =1,4
        t_k(k) = t_1
    1 continue
      do 2 k =5,8
        t_k(k) = t_2
    2 continue
! t_k(k) are the constant coefficient $\omega_i$ in $f^{eq}$,
!  for D2Q9,
! when k=0, $\omega_i=4/9$, when i=1,..4, $\omega_i=1/9$, ......

      con2 =  2.d0 * c_squ * c_squ
! Another constant will be used in calculation of $f^{eq}$.

      write (6,*) '@@@_He-Chen-Zhang_multiphase_LBM_..._____@@@'

! Initialization
      call read_parametrs(yA,t_max)
      call read_obst(xA,yA,obstA)

      call init_density(xA,yA,obstA,UA,VA,rhA,
     &  rhoA,pA,ffA,gpA,Force1,Force2)
      call write_results(xA,yA,obstA,rhoA,UA,VA,pA,0)
!     Nwri = 0.5*int(Tn)  ! every 0.5 non-dimensional time

      call comp_rey(xA,yA,time,BT)

      open(39,file='./out/residue.dat')
      open(38,file='./out/mass.dat')

C--------------------------------
c    Begin iterations
C--------------------------------
      do 100 time = 1, t_max

      if(mod(time,Nwri) .eq. 0) then
       write(*,*) time
       call itime(now)
       write(*,"(i2.2,':',_i2.2,_':',_i2.2)") now
       call write_results(xA,yA,obstA,rhoA,UA,VA,pA,time)
      endif
        if ( mod(time,20) .eq. 0)
```

```
      &   call track_interface(xA,yA,rhoA,time)
! this subroutine has demonstrate in the section "Capillary rise".
! Insert the subroutine to this fortran file is ok to compile.
c--------------
      call stream(xA,yA,obstA,ffA,gpA)
      call getuv(xA,yA,obstA,UA,VA,rhA, rhoA,ffA, pA,Force1, Force2)
      call geten(xA,yA,obstA,UA,VA,rhoA,gpA,pA, Force1,Force2)
      call collision(xA,yA,obstA,UA,VA,rhA,rhoA,pA,ffA,gpA,
      &   Force1,Force2)

  100 continue

      close(39)
      end
c--------------------------------------------
      subroutine index_function(lx,ly,rh,rho)
      implicit none
      include "head.inc"
      integer lx,ly,x,y
      real*8 rh(lx,ly), rho(lx,ly)
      do 1 x = 1,lx
      do 1 y = 1,ly
        rho(x,y) = rho_l + ( rh(x,y)- psi_min0 )*const
! From the index function, we can get the
! density of the fluid at (x,y).
    1 continue
      end
c--------------------------------------------


initial.for
c================================================================

      subroutine read_parametrs(ly,t_max)
      implicit none
      include "head.inc"
      real*8  Re1, V1
      integer  t_max,ly

! Default values
      Kappa = 0.0001d0
      t_max  = 15000
      tau_f = 0.515d0
      tau_g = 0.515d0
      gforce = 0.d0   !-0.00008d0
      psi_max = 0.251260d0
      psi_min = 0.02428d0
      psi_max0 = 0.230d0
      psi_min0 = 0.027d0
      rho_h = 0.251260d0
      rho_l = 0.02428d0

      open(1,file='./params.in')
        read(1,*) psi_max
        read(1,*) psi_min
        read(1,*) psi_max0
```

```
      read(1,*) psi_min0
      read(1,*) rho_h
      read(1,*) rho_l
      read(1,*)           ! a blank line
      read(1,*) Re1
      read(1,*) V1
      read(1,*) t_max
      read(1,*) Nwri
      read(1,*) Kappa
   close(1)

   const=  (rho_h - rho_l)/(psi_max0 - psi_min0)

c    V1 = 0.10d0
     gforce = -V1*V1/h_ref      !
     tau_f= 1.d0/(Re1/V1/real(xA)*c_squ) +0.5d0   !
     tau_g = tau_f     !
     Tn = sqrt(-xA/gforce)! characteristic time

     RAYLEIGH =1
     BD =1
     write(*,'("tau_=_",_f13.7,_4X,_"gforce=",_f13.7)') tau_f,gforce
     write(*,'("V1=",_1f13.7,_10X,_"Re1=",1f13.7)') V1,Re1
     write(*, '("time_is_normalized_by:",_f15.4_)') sqrt(-xA/gforce)

     end
c--------------------------------------------
     subroutine read_obst(lx,ly,obst)
     implicit none
     include "head.inc"
     integer  x,y,lx,ly,obst(lx,ly)
      do 10 y = 1, ly
        do 40 x = 1, lx
          obst(x,y) = 0
 40     continue
 10   continue

     if (BD .eq. 1) then
     do 20 y = 1, ly, ly-1
       do 20 x = 1, lx
         obst(x,y) = 1
 20  continue
     endif

     return
     end
c--------------------------------
     subroutine init_density(lx,ly,obst,u_x,u_y,rh,rho,p,ff,gp,
    &  Force1,Force2)
     implicit none
     include "head.inc"
     integer lx,ly,i,j,x,y,k,n,obst(lx,ly)
     real*8  u_squ,u_n(0:8),fequi(0:8),u_x(lx,ly),u_y(lx,ly),
    & rho(lx,ly),ff(0:8,lx,ly), gp(0:8,lx,ly), p(lx,ly), gequi(0:8),
    & Force1(lx,ly), Force2(lx,ly),rh(lx,ly), rhoh,rhol, xx
```

```
      do 10 y = 1, ly
        do 10 x = 1, lx
          u_x(x,y) = 0.d0
          u_y(x,y) = 0.d0
          rh(x,y) = psi_min   !psi_max
          Force1(x,y) = 0.d0
          Force2(x,y) = 0.d0
   10 continue

C--------------------------------
      if(RAYLEIGH .eq. 1) then
      do 21 x = 1, lx
      do 31 y = 1, ly
      if(y.gt. real(ly/2)+h_ref* 0.1d0 *dcos(real(x)/h_ref* 2.d0* pai))
     &   then
             rh(x,y) = psi_max !psi_min ! !
      endif

   31 continue
   21 continue
      endif

      call index_function(lx,ly, rh,rho)

      do y = 1, ly
      do x = 1, lx
        p(x,y) = rho(x,y)*rho(x,y) *c_squ*( 4.d0 - 2.d0*rho(x,y) )
     *             /(1.d0- rho(x,y))/(1.d0- rho(x,y))/(1.d0- rho(x,y))
     *             - 12.d0* c_squ* rho(x,y)* rho(x,y) + rho(x,y)*c_squ
      enddo
      enddo
C--------------------------------
      do 80 y = 1, ly
      do 80 x = 1, lx

        call f_eq( rh(x,y),u_x(x,y),u_y(x,y),fequi)
        call g_eq(p(x,y),rho(x,y),fequi,gequi)
        do 60 k = 0,8
          ff(k,x,y) = fequi(k)*rh(x,y)
          gp(k,x,y) = gequi(k)
   60 continue

   80 continue
      end


fgequ.for
C================================================================

      subroutine g_eq(p,rho,fequ,gequ)
      implicit none
      include "head.inc"
      real*8 u_x,u_y,gequ(0:8),fequ(0:8),u_n(0:8),u_squ,rho,p, tmp1
      integer k
      tmp1 = rho*c_squ
```

```
      do k= 0, 8
      gequ(k) = (fequ(k)-t_k(k) )*tmp1 +t_k(k)*p
      enddo
      end
c---------------------------------------
      subroutine f_eq(rho,u_x,u_y,fequ)
      implicit none
      include "head.inc"
      real*8 u_x,u_y,fequ(0:8),u_n(0:8),rho,u_squ, tmp1
      integer k
          u_squ = u_x * u_x + u_y * u_y

          tmp1 = u_squ *1.5d0
        fequ(0) = t_k(0)* ( -tmp1 + 1.d0)
      do 10 k = 1, 8
          u_n(k) =  xv(k)* u_x + yv(k)* u_y
          fequ(k) = t_k(k)*  ( u_n(k) / c_squ
     &                + u_n(k) * u_n(k) / (con2)
     &                - tmp1 + 1.d0)
   10 continue
      end
c---------------------------------------


collisionStr.for
c==================================================================

      subroutine stream(lx,ly,obst,fp,gp)
      implicit none
      integer  k,lx,ly,obst(lx,ly)
      real*8 fp(0:8,lx,ly), gp(0:8,lx,ly), fx(lx), fy(ly)
      real*8 gx(lx), gy(ly)
      integer x,y,x_e,x_w,y_n,y_s,l,m,n,xi,yi
      real*8 f_hlp(0:8,lx,ly)

      do 10 y = 1, ly
         do 10 x = 1, lx
         y_n = mod(y,ly) + 1
         x_e = mod(x,lx) + 1
         y_s = ly - mod(ly + 1 - y, ly)
         x_w = lx - mod(lx + 1 - x, lx)

         f_hlp(1,x_e,y  ) = fp(1,x,y)
         f_hlp(2,x  ,y_n) = fp(2,x,y)
         f_hlp(3,x_w,y  ) = fp(3,x,y)
         f_hlp(4,x  ,y_s) = fp(4,x,y)
         f_hlp(5,x_e,y_n) = fp(5,x,y)
         f_hlp(6,x_w,y_n) = fp(6,x,y)
         f_hlp(7,x_w,y_s) = fp(7,x,y)
         f_hlp(8,x_e,y_s) = fp(8,x,y)

   10 continue

      do 20 y = 1, ly
         do 20 x = 1, lx
```

```
          do 20 k = 1, 8
            fp(k,x,y) = f_hlp(k,x,y)
   20 continue


c--------------
      do 30 y = 1, ly
         do 30 x = 1, lx
         y_n = mod(y,ly) + 1
         x_e = mod(x,lx) + 1
         y_s = ly - mod(ly + 1 - y, ly)
         x_w = lx - mod(lx + 1 - x, lx)

         f_hlp(1,x_e,y  ) = gp(1,x,y)
         f_hlp(2,x  ,y_n) = gp(2,x,y)
         f_hlp(3,x_w,y  ) = gp(3,x,y)
         f_hlp(4,x  ,y_s) = gp(4,x,y)
         f_hlp(5,x_e,y_n) = gp(5,x,y)
         f_hlp(6,x_w,y_n) = gp(6,x,y)
         f_hlp(7,x_w,y_s) = gp(7,x,y)
         f_hlp(8,x_e,y_s) = gp(8,x,y)

   30 continue

      do 40 y = 1, ly
         do 40 x = 1, lx
           do 40 k = 1, 8
             gp(k,x,y) = f_hlp(k,x,y)
   40 continue

      return
      end


c------------------------------------------------------------------
      subroutine getuv(lx,ly,obst,u_x,u_y,rh, rho,fp, p, Force1,
     & Force2)
      implicit none
      include "head.inc"
      integer x,y,lx,ly,obst(lx,ly),ip,jp,l,k,xp,yp, xn,yn
      real*8  temp(lx,ly),
     &  u_x(lx,ly),u_y(lx,ly),rho(lx,ly),rh(lx,ly),
     &  fp(0:8,lx,ly), Force1(lx,ly), fai(lx,ly),
     &  Force2(lx,ly), p(lx,ly), prho(lx,ly)

      do 10 y = 1, ly
      do 10 x = 1, lx
      if(obst(x,y) .eq. 0) then
          rh(x,y) = fp(0,x,y) +fp(1,x,y) +fp(2,x,y) +fp(3,x,y)
     &           +fp(4,x,y) +fp(5,x,y) +fp(6,x,y) +fp(7,x,y) +fp(8,x,y)
      endif

   10 continue

      call index_function(lx,ly, rh,rho)
c---------------------------------------------
      do 11 y = 1, ly
       do 11 x = 1, lx
```

```
          if(obst(x,y) .eq. 0) then
           fai(x,y) =
     &     rh(x,y)*rh(x,y) *c_squ*( 4.d0 - 2.d0*rh(x,y) )
     %         /(  (1.d0- rh(x,y))*(1.d0- rh(x,y))*(1.d0- rh(x,y))  )
     %     - 12.d0* c_squ* rh(x,y)* rh(x,y)

           prho(x,y) = p(x,y) - rho(x,y) *c_squ
          endif
   11 continue

ccccc===================================
      do 15 y = 1, ly
       do 15 x = 1, lx

       xp = x+1
       yp = y+1
       xn = x-1
       yn = y-1
       if (xp.gt.lx )  xp = 1
       if (xn.lt.1 )   xn = lx
       if (yp.gt.ly )  yp = 1
       if (yn.lt.1 )   yn = ly

       temp(x,y) =
     *  ( ( rh(xp,y) + rh(xn,y) + rh(x,yp)+ rh(x,yn) )*4.d0/6.d0
     *     +( rh(xp,yp)+ rh(xn,yn)
     *     +  rh(xp,yn)+ rh(xn,yp) )/6.d0
     *     - 20.d0* rh(x,y)/6.d0
     *     )
   15 continue

      do 14 x = 1, lx
       temp(x,1) = temp(x,2)
       temp(x,ly) = temp(x,ly-1)
   14 continue

c-----------------------------------------
      do 20 y = 1, ly
      do 20 x = 1, lx
       xp = x+1
       yp = y+1
       xn = x-1
       yn = y-1
       if (xp.gt.lx )  xp = 1
       if (xn.lt.1 )   xn = lx
       if (yp.gt.ly )  yp = 1
       if (yn.lt.1 )   yn = ly

       Force1(x,y) = Kappa *rh(x,y) *
     *  ( (temp(xp,y)- temp(xn,y) )/3.d0
     *     +(temp(xp,yp)- temp(xn,yn) )/12.d0
     *     +(temp(xp,yn)- temp(xn,yp) )/12.d0
     *     )

       Force2(x,y) = Kappa *rh(x,y) *
     &  ( (temp(x,yp)- temp(x,yn) )/3.d0
```

```
     &       +(temp(xp,yp)- temp(xn,yn) )/12.d0
     &       +(temp(xn,yp)- temp(xp,yn) )/12.d0   )
     &       + rho(x,y)* gforce
c---------! the gravity force !!!!
   20 continue

      do 30 y = 1, ly
       do 30 x = 1, lx

        xp = x+1
        yp = y+1
        xn = x-1
        yn = y-1
        if (xp.gt.lx )   xp = 1
        if (xn.lt.1 )    xn = lx
        if (yp.gt.ly )   yp = 1
        if (yn.lt.1 )    yn = ly

        dfai_x(x,y) =
     *  ( (fai(xp,y)- fai(xn,y) )/3.d0
     *     +(fai(xp,yp)- fai(xn,yn) )/12.d0
     *     +(fai(xp,yn)- fai(xn,yp) )/12.d0
     *     )
        dfai_y(x,y) =
     *  ( (fai(x,yp)- fai(x,yn) )/3.d0
     *     +(fai(xp,yp)- fai(xn,yn) )/12.d0
     *     +(fai(xn,yp)- fai(xp,yn) )/12.d0
     *     )

        dprho_x(x,y) =
     *  ( (prho(xp,y)- prho(xn,y) )/3.d0
     *     +(prho(xp,yp)- prho(xn,yn) )/12.d0
     *     +(prho(xp,yn)- prho(xn,yp) )/12.d0
     *     )
        dprho_y(x,y) =
     *  ( (prho(x,yp)- prho(x,yn) )/3.d0
     *     +(prho(xp,yp)- prho(xn,yn) )/12.d0
     *     +(prho(xn,yp)- prho(xp,yn) )/12.d0
     *     )

   30 continue

! Derivatives in the nodes most near to the bottom and upper
! boundary, e.g.,
! dfai_x(x,2), dfai_y(x,2), ... are extrapolated from inner fluid
! nodes.
      if(BD .eq. 1) then
       do 13 x = 1, lx

   dfai_x(x,2)  = dfai_x(x,3)
   dfai_y(x,2)  = dfai_y(x,3)
   dprho_x(x,2) = dprho_x(x,3)
   dprho_y(x,2) = dprho_y(x,3)

   dfai_x(x,ly-1) = dfai_x(x,ly-2)
   dfai_y(x,ly-1) = dfai_y(x,ly-2)
```

```
      dprho_x(x,ly-1) = dprho_x(x,ly-2)
      dprho_y(x,ly-1) = dprho_y(x,ly-2)

  13    continue
        endif


c------------------------------------------------------------------
c It is noted that the calculations of  dfai_x, dfai_y, dprho_x,
! dprho_y
c for the lattice node in bottom and upper boundaries (walls) may be
c not necessary because in the collision step,  dfai_x, dfai_y,
! dprho_x,
c dprho_y on the wall nodes are not used at all. ONLY bounce back is
c necessary to implement for the wall nodes.

        end
c------------------------------------------------------------------
        subroutine geten(lx,ly,obst,u_x,u_y,rho,gp, p, Force1,Force2)
        implicit none
        include "head.inc"
        integer x,y,lx,ly,obst(lx,ly),ip,jp,k,l
        real*8  u_x(lx,ly),u_y(lx,ly),rho(lx,ly),
      & gp(0:8,lx,ly), p(lx,ly),Force1(lx,ly),
      & Force2(lx,ly),tmp

        do 10 y = 1, ly
          do 10 x = 1, lx
          if(obst(x,y) .eq. 0) then

            u_x(x,y)= (gp(1,x,y) + gp(5,x,y) + gp(8,x,y)
      &             -(gp(3,x,y) + gp(6,x,y) + gp(7,x,y)) )
      &             + 0.5d0 * c_squ *( Force1(x,y) +0.d0 )

            u_y(x,y)= (gp(2,x,y) + gp(5,x,y) + gp(6,x,y)
      &             -(gp(4,x,y) + gp(7,x,y) + gp(8,x,y)) )
      &             + 0.5d0 * c_squ *( Force2(x,y)  )

             u_x(x,y) = u_x(x,y)/rho(x,y)/c_squ
             u_y(x,y) = u_y(x,y)/rho(x,y)/c_squ

             p(x,y)=gp(0,x,y) +gp(1,x,y) +gp(2,x,y)
      &          +gp(3,x,y) +gp(4,x,y) +gp(5,x,y)
      &          +gp(6,x,y) +gp(7,x,y) +gp(8,x,y)
      %      + 0.5d0 * ( u_x(x,y)* (-dprho_x(x,y))
      %              + u_y(x,y)* (-dprho_y(x,y)) )
          endif
  10    continue

        end
c------------------------------------------------------------------
        subroutine collision(lx,ly,obst,u_x,u_y,rh,rho,p,fp,gp,
      & Force1,Force2)
        implicit none
        include "head.inc"
        integer  l,lx,ly,obst(lx,ly),x,y,k,ip,jp
        real*8   u_x(lx,ly),u_y(lx,ly)
```

```
     &, fp(0:8,lx,ly),rho(lx,ly),Force1(lx,ly),Force2(lx,ly),
     &  gp(0:8,lx,ly), p(lx,ly),rh(lx,ly),
     &  gneq(0:8),fneq(0:8), gequ(0:8), temp(8),temp2(8)
        real*8  u_n(0:8),fequ(0:8),u_squ, div
c----------------------------------
        do 5 y = 1, ly
        do 5 x = 1, lx

        if(obst(x,y) .eq. 0) then

        call f_eq( rh(x,y), u_x(x,y), u_y(x,y), fequ)
        call g_eq( p(x,y), rho(x,y), fequ, gequ)

          do 60 k = 0,8

            fneq(k)= fp(k,x,y) - rh(x,y)*fequ(k)

              fp(k,x,y) = fp(k,x,y) -1.d0/tau_f * fneq(k)
     &          + (tau_f-0.5d0)/tau_f*
     &          ( (xv(k)-u_x(x,y))* (-dfai_x(x,y))
     &           +(yv(k)-u_y(x,y))* (-dfai_y(x,y)) )*fequ(k)/c_squ
c--------------------------------------------------------------
        gneq(k)= gp(k,x,y) - gequ(k)

        gp(k,x,y) = gp(k,x,y) -1.d0/tau_g *gneq(k)
     &          + (tau_g-0.5d0)/tau_g *
     *          (
     &            ( (xv(k)-u_x(x,y))*  Force1(x,y)
     &            +(yv(k)-u_y(x,y))*  Force2(x,y)
     &            ) *fequ(k)
     &            + (xv(k)-u_x(x,y))*( fequ(k)-t_k(k) )*(-dprho_x(x,y))
     &            + (yv(k)-u_y(x,y))*( fequ(k)-t_k(k) )*(-dprho_y(x,y))

     *          )
   60 continue

        endif

c-----------------
        if( BD .eq. 1 .and. obst(x,y) .eq. 1) then
        do k = 1, 8
        temp(k)    = fp(k,x,y)
        temp2(k)   = gp(k,x,y)
        enddo

        do k = 1, 8
        fp(opp(k),x,y)= temp(k)
        gp(opp(k),x,y)= temp2(k)
        enddo

        endif
c---------------

    5  continue
c----------------------------------
        end
```

```
C---------------------------------------------------
      subroutine comp_rey(lx,ly,time,BT)
! Here is output some parameter into a file to keep a record.
      implicit none
      include "head.inc"
      integer  time,lx,ly
      real*8  rey
      character BT*10

      open(44,file='./out/parameter.dat')
      write(44,'("He_Chen_Doolen multiphase model")')
      write(44,'("Nwri=_",_i5)')   Nwri

      write(44,'("gforce:_",_f15.8)')   gforce
      write(44,'("viscosity_=_",f8.5,2X,"tau_f=",f17.8,2X,
     &___2X,"tau_g=",f17.8)_') c_squ*(tau_f-0.5), tau_f,tau_g
      write(44,'("@@@_mesh_size=_",i5,"X",i5)')   lx,   ly
      write(44,'("character_U_=_",f10.5)')  sqrt(-gforce*h_ref)
      write(44,'("Re_=_",f10.5)')
     &   h_ref*sqrt(-gforce*h_ref)/c_squ/(tau_f-0.5)
      write(44,'("Kappa_=_",f10.5)')   Kappa

      write(44,'(2X,"rho_h=",f17.7,2X,
     &___2X,"rho_l=",f17.7)_') rho_h,  rho_l
      write(44,'(2X,"psi_max=",f17.7,2X,
     &___2X,"psi_min=",f17.7)_') psi_max, psi_min
      close(44)
      return
      end




IO.for
C=============================================================
c Write a binary output file,
c for more detail, please refer to Chapter 6.
C-------------------------------------------------------------

      subroutine  write_results(lx,ly,obst,rho,upx,upy,p, n)
      implicit none
      include "head.inc"
      integer slicex2,lx,ly
      integer  x,y,i,j,n,obsval,jcen,k1,k,obst(lx,ly),slicex,slicey
      real*8   rho1,h(lx),rho(lx,ly),upx(lx,ly),upy(lx,ly)
     & , p(lx,ly)

      character*40 V1, V2, V3, V4, V5, V6, V7, V8, V9, V10, V11, V12
      REAL*4 ZONEMARKER, EOHMARKER
      character filename*16, B8*8
      integer*1 obob
      integer*2 ii, jj
      integer, parameter:: kmax=1
      character*40 Title,var, Zonename1

      write(B8,'(i8.8)') n

      open(41,file='./out/2D'//B8//'.plt',form="BINARY")
```

```
c-------------------------------------
      ZONEMARKER=              299.0
      EOHMARKER =              357.0

c      I. The header section.
      write(41) "#!TDV101"

c-----Integer value of 1.------------------
      write(41) 1

      Title="Rayleigh-Taylor"
      call dumpstring(Title)

c-----Number of variables (NumVar) in the datafile.
      write(41) 7
c------Variable names.
      V1='X'
        call dumpstring(V1)
      V2='Y'
        call dumpstring(V2)
      V3='u'
        call dumpstring(V3)
      V4='v'
        call dumpstring(V4)
      V5='rho1'
        call dumpstring(V5)
      V7='p'
        call dumpstring(V7)
      V8='obst'
        call dumpstring(V8)

      write(41) ZONEMARKER
      Zonename1="ZONE_001"
      call dumpstring(Zonename1)
c---------Zone Color
      write(41) -1
c---------ZoneType  !
      write(41) 0
c---------DataPacking 0=Block, 1=Point
      write(41) 1
c-Specify Var Location. 0 = Don't specify, all data is located at
! the nodes
      write(41) 0
c---------Number of user defined face neighbor connections
! (value >= 0)
      write(41) 0

      write(41) lx
      write(41) ly
      write(41) kmax

      write(41) 0
      write(41) EOHMARKER

c-----2.1 zone ----------------------------------
```

```
      write(41) Zonemarker

C--------variable data format, 1=Float, 2=Double,
c---3=LongInt, 4=ShortInt, 5=Byte, 6=Bit
      write(41) 4
      write(41) 4
      write(41) 1
      write(41) 1
      write(41) 1
      write(41) 1
      write(41) 5

      write(41) 0
      write(41) -1

    do k=1,kmax
     do j=1,ly
      do i=1,lx
        obob = obst(i,j)
        ii = i
        jj = j
        write(41) ii
        write(41) jj
        write(41) real(upx(i,j))
        write(41) real(upy(i,j))
        write(41) real(rho(i,j))
        write(41) real(p(i,j))
        write(41) obob
        end do
      end do
    end do

    close(41)
    end


C-----------------------------------------------------------

    subroutine dumpstring(instring)
     character(40) instring
     integer len

     len=LEN_TRIM(instring)

     do ii=1,len
       I=ICHAR(instring(ii:ii))
       write(41) I
     end do
     write(41) 0
    return
    end
C------------------------------
```

7.13 Appendix C: 3D code

```
c 1. This program is aimed to simulate a droplet or a bubble immersed
! in another fluid.
c 2. The fortran files: head.inc,  Main.for, Collision.for,
! Utility.for, Output.for.
c  should be put in a identical folder and compile together.
c 3. The main parameters control the flow are listed in subroutine
! "read_parameters".
c 4. Here output Tecplot files are Ascii files. Please refer to the
! code in Chapter 6
c  for writing binary files.

head.inc
C==========================================
      real*8 h_ref
      integer lx,ly,lz
      PARAMETER(lx=70,ly=70,lz=70)

      common/AA/  BUOYANCY,BD
      integer  BUOYANCY, BD

      common/AA/ ex(0:18),ey(0:18),ez(0:18)
      real*8 ex,ey,ez

      common/b/ error,vel,xc(0:18),yc(0:18),zc(0:18),t_k(0:18)
      real*8 error,vel,xc,yc,zc,t_k

      common/vel/ c_squ,cc,Nwri, opp(18)
      real*8 c_squ,cc
      integer Nwri,opp

      common/app/ psi_max,psi_min,rho_h,  rho_l,  t_0,t_1,t_2,xsten
      real*8 psi_max,psi_min,rho_h,  rho_l,  t_0,t_1,t_2,xsten

      common /prho/ dprho_x(lx,ly,lz), dprho_y(lx,ly,lz)
      real*8 dprho_x, dprho_y
      common /fai/ dfai_x(lx,ly,lz),dfai_y(lx,ly,lz),
     & dfai_z(lx,ly,lz)
      real*8 dfai_x, dfai_y,dfai_z
      common /fai2/ dprho_z(lx,ly,lz)
      real*8 dprho_z
      common /velo/ Ra,velS, Kappa, gforce, tau_f, tau_g, UU,DD
      real*8   Ra,velS, Kappa, gforce, tau_f, tau_g, UU,DD


Main.for
C==========================================
! This is 3D He-Chen-Zhang model code (He, Chen, Zhang,
! A Lattice Boltzmann scheme for incompressible multiphase flow
! and its application in simulation of Rayleigh-Taylor instability,
! JCP,152, 642, 1999)
      program D3Q19LBM
      implicit none
      include "head.inc"
```

```
      integer t_max,time,k
c   USE DFPORT

      integer obst(lx,ly,lz)
C-------------------------------------
! Author:  Haibo Huang, huanghb@ustc.edu.cn
C-------------------------------------
c    the lattice nodes are numbered as follows:
      real*8  u_x(lx,ly,lz),u_y(lx,ly,lz),rho(lx,ly,lz)
      real*8  rh(lx,ly,lz)
      real*8  u_z(lx,ly,lz),  p(lx,ly,lz)
      real*8  ff(0:18,lx,ly,lz),gp(0:18,lx,ly,lz),Fx(lx,ly,lz)
      &  ,Fy(lx,ly,lz), Fz(lx,ly,lz)
! Velocity vectors of the D3Q19 velocity model, xc, yc, zc are
! the x, y, z components of the vector, respectively. The order of
! the velocity vector is slightly different from that in Fig.1.1.
      data xc/0.d0, 1.d0, -1.d0, 0.d0, 0.d0, 0.d0, 0.d0, 1.d0, 1.d0,
      &  -1.d0, -1.d0, 1.d0, -1.d0, 1.d0, -1.d0, 0.d0, 0.d0,
      &  0.d0, 0.d0  /,
      &   yc/0.d0, 0.d0, 0.d0, 1.d0, -1.0d0, 0.d0, 0.d0, 1.d0, -1.d0,
      &  1.d0, -1.d0, 0.d0, 0.d0, 0.d0, 0.d0, 1.d0, 1.d0,
      &  -1.d0, -1.d0/,
      &   zc/0.d0, 0.d0, 0.d0, 0.d0, 0.d0, 1.d0, -1.d0, 0.d0, 0.d0,
      &  0.d0, 0.d0, 1.d0, 1.d0, -1.d0, -1.d0, 1.d0, -1.d0,
      &  1.d0, -1.d0/
      data ex/0, 1, -1, 0, 0, 0, 0, 1, 1,
      &  -1, -1, 1, -1, 1, -1, 0, 0,
      &  0, 0 /,
      &   ey/0, 0, 0, 1, -1, 0, 0, 1, -1,
      &  1, -1, 0, 0, 0, 0, 1, 1,
      &  -1, -1/,
      &   ez/0, 0, 0, 0, 0, 1, -1, 0, 0,
      &  0, 0, 1, 1, -1, -1, 1, -1,
      &  1, -1/
      data opp/2,1,4,3,6,5,10,9,8,7,14,13,12,11,18,17,16,15/
C-----------------

      BD = 0  ! It was not used at present code. It can be removed at
!        present stage.

! It turn on the buoyancy force if BUOYANCY=1
      BUOYANCY = 0

! Lattice speed
      cc = 1.d0
! Square of sound speed in the LBM
       c_squ = cc *cc / 3.d0
C-------------
! Weighting coefficients in the equilibrium distribution function for
! D3Q19 model.
       t_0 =  1.d0 / 3.d0
       t_1 =  1.d0 / 18.d0
       t_2 =  1.d0 / 36.d0

   t_k(0) = t_0
   do 1 k =1,6
```

```
        t_k(k) = t_1
    1 continue
      do 2 k =7,18
      t_k(k) = t_2
    2 continue
c-------------
      write (6,*) '@@@__LBM_for_3D_He_Chen_Zhang_LBM_starting.._@@@'
c---------------------------
c     Begin initialization
c---------------------------
! Specify how many steps to dump flow data file (flow field can be
! viewed by Tecplot)
      Nwri = 100

! Read important parameters in the flow we want to investigate
      call read_parameters(t_max)
! Initial each lattice node to be a fluid
! node (obst=0) or a solid node (obst=1)
      call read_obstacles(obst)
! Initial the flow field and f_i, g_i
      call init_density(obst,u_x,u_y,u_z,rh, rho,p, ff
     &   ,gp, Fx,Fy,Fz)

! Begin main loop
c--------------------------------------------------
      do 100 time = 1, t_max
      if ( mod(time, 50).eq. 0)  write(*,*) time
! Each "Nwri" time step dump a result
        if ( mod(time, Nwri) .eq. 0 .or. time. eq. 1) then
        call write_results2(obst,rho,u_x,u_y,u_z,p,time)
        end if

! Streaming step
        call stream(obst,ff )
        call stream(obst,gp )

! Obtained the macroscopic varibales
      call getuv(obst,u_x,u_y,u_z,rh, rho,ff, p,
     &   Fx, Fy,Fz)
      call geten(obst,u_x,u_y,u_z, rho,gp, p, Fx, Fy, Fz)
! Collision step
      call collision(obst,u_x,u_y,u_z,rh,
     &   rho,p,ff,gp, Fx,Fy,Fz)
c--------------------------------------------------
! End of main loop

  100 continue
  101   call write_results2(obst,rho,u_x,u_y,u_z,p,time)

  999 continue
      end


c--------------------------------------------------
      subroutine read_parameters(t_max)
      implicit none
      include "head.inc"
```

```
      real*8  Re1,  We
      integer  t_max

! This is the critical parameter to adjust surface tension
! independently.
      Kappa = 0.02d0
! Specify a characteristic velocity
      UU = 0.00d0
! Specify droplet's diameter
      DD = 30.d0

!  The \rho_h and \rho_l seems not able to choose randomly, it should
! be near to the values of  \psi_{max} and \psi_{min}.
! Usually, the maximum density ratio should not exceed the ratio
! between \psi_{max}
! and \psi_{min}.

      rho_h = 0.12d0
      rho_l = 0.04d0   ! Density ratio is 3
!  The \psi_{max} and \psi_{min} depend on the EOS we used.
      psi_max = 0.2516d0
      psi_min = 0.0241d0

      t_max  = 50000
! Relaxation time in the LBE for f_i.
      tau_f = 0.7d0
! Relaxation time in the LBE for g_i.
      tau_g = 0.7d0
! External force
      gforce = -0.0000d0
! Reynolds number
      Re1 = UU *DD/(1./3.)/tau_f
! Web number in the simulation
      We = rho_h *DD *UU*UU/(0.0011d0 *Kappa/0.1d0)
      write(*,'("tau_=_",_f13.7,_4X,_"gforce=",_f13.7)') tau_f,gforce
      write(*,'("UU=",_1f13.7,_10X,_"Re1=",1f13.7,_10X,_"We=",_
     & 1f13.7)')
     &  UU,Re1, We
       return
      end
C--------------------------------------------------

      subroutine read_obstacles(obst)
      implicit none
      include "head.inc"
      integer  x,y,z,obst(lx,ly,lz)
! Initially set all nodes to be fluid nodes
        do 5 z = 1, lz
         do 10 y = 1, ly
          do 40 x = 1, lx
            obst(x,y,z) =  0
40       continue
10     continue
5    continue

      end
```

```
c---------------------------------------------------


      subroutine init_density(obst,u_x,u_y,u_z,rh, rho,p, ff
     &  ,gp, Fx,Fy,Fz)
      implicit none
      include "head.inc"

      integer i,j,x,y,z,k,n,obst(lx,ly,lz)
      real*8  u_squ,u_n(0:18),fequi(0:18),u_x(lx,ly,lz),u_y(lx,ly,lz),
     & rho(lx,ly,lz),ff(0:18,lx,ly,lz),u_z(lx,ly,lz),rh(lx,ly,lz)
     & ,p(lx,ly,lz), Fx(lx,ly,lz), Fy(lx,ly,lz),Fz(lx,ly,lz),
     & gp(0:18,lx,ly,lz)

      do 12 z = 1, lz
        do 11 y = 1, ly
          do 10 x = 1, lx
          u_x(x,y,z) = 0.d0
          u_y(x,y,z) = 0.d0
          u_z(x,y,z) = 0.d0
          rh(x,y,z) = psi_min   !psi_max
          rho(x,y,z) = rho_l    !rho_h
cccc ---------------- phase-separate
c    call random_number (xx) !  random
c          rh(x,y) = 0.1d0 + 0.01* xx
c          rho(x,y) = rh(x,y)
cccc ---------------- phase-separate
          Fx(x,y,z) = 0.d0
          Fy(x,y,z) = 0.d0
          Fz(x,y,z) = 0.d0

  10      continue
  11    continue
  12 continue

      do 23 z = 1, lz
      do 22 y = 1, ly
      do 21 x = 1, lx

! Here a droplet is initialized.
      if( sqrt(float(x-lx/2)**2 + float(y-ly/2)**2 +
     &   float(z-lz/2)**2) .lt. DD/2.d0 ) then

          rh(x,y,z) = psi_max !psi_min  !psi_max  ! !
          rho(x,y,z) = rho_h  !rho_l   !rho_h
          u_z(x,y,z) = 0.d0   !UU/2.d0
      endif

! Here another droplet is initialized. For the case of droplet
! collision, it will be used.
c--------------
      if(0 .eq. 1)  then
        if( sqrt(float(x-lx/2)**2 + float(y-ly/2)**2 +
     &   float(z-62)**2) .lt. DD/2.d0 ) then !.or.
         rh(x,y,z) = psi_max !psi_min  !psi_max  ! !
         rho(x,y,z) = rho_h  !rho_l   !rho_h
```

```
          u_z(x,y,z) = -UU/2.d0
        endif
      endif
C---------------

    21 continue
    22 continue
    23 continue


       do 82 z = 1, lz
          do 81 y = 1, ly
             do 80 x = 1, lx
! Initially the hydrodynamic pressure is supposed to be equal to the
! thermodynamic one, which can be
! calculated from the C-S EOS (He, Chen, Zhang, JCP, 1999).
! In the C-S EOS: a=12RT=12c_s^2, b=4. The values are in lattice
! units
          p(x,y,z) =    rho(x,y,z)*rho(x,y,z)
     *          *c_squ*( 4.d0 - 2.d0*rho(x,y,z) )
     *                /(1.d0- rho(x,y,z))**3.d0
     *   - 12.d0* c_squ* rho(x,y,z)* rho(x,y,z) + rho(x,y,z)*c_squ

       if(obst(x,y,z) .eq. 1) then

       do 59 k=0, 18
         ff(k,x,y,z) = 0.d0
         gp(k,x,y,z) = 0.d0
    59 continue

       else
! Initial f_i is supposed to be the equilibrium value.
         call f_eq( rh(x,y,z),u_x(x,y,z),u_y(x,y,z),u_z(x,y,z),fequi)
         do 60 k = 0,18
            ff(k,x,y,z) = fequi(k)*rh(x,y,z)
    60  continue
! Initial g_i is supposed to be the equilibrium value.
         call g_eq(p(x,y,z),rho(x,y,z),u_x(x,y,z),u_y(x,y,z),
     &        u_z(x,y,z),fequi)
         do 61 k = 0,18
            gp(k,x,y,z) = fequi(k)
    61  continue
       endif

    80    continue
    81    continue
    82 continue
       end


Collision.for
C===============================================================
       subroutine stream(obst,f)
       implicit none
       include "head.inc"
       integer  k, obst(lx,ly,lz)
```

```
      real*8 f(0:18,lx,ly,lz),f_hlp(0:18,lx,ly,lz)
      integer  x,y,z,x_e,x_w,y_n,y_s,z_n,z_s

      do 12 z = 1, lz
       do 11 y = 1, ly
        do 10 x = 1, lx
! In x, y, z directions, periodic boundary conditions are applied
           z_n = mod(z,lz) + 1
           y_n = mod(y,ly) + 1
           x_e = mod(x,lx) + 1

           z_s = lz - mod(lz + 1 - z, lz)
           y_s = ly - mod(ly + 1 - y, ly)
           x_w = lx - mod(lx + 1 - x, lx)
c.........density propagation
           f_hlp(1 ,x_e,y  ,z  ) = f(1,x,y,z)
           f_hlp(2 ,x_w,y  ,z  ) = f(2,x,y,z)
           f_hlp(3 ,x  ,y_n,z  ) = f(3,x,y,z)
           f_hlp(4 ,x  ,y_s,z  ) = f(4,x,y,z)
           f_hlp(5 ,x  ,y  ,z_n) = f(5,x,y,z)
           f_hlp(6 ,x  ,y  ,z_s) = f(6,x,y,z)
           f_hlp(7 ,x_e,y_n,z  ) = f(7,x,y,z)
           f_hlp(8 ,x_e,y_s,z  ) = f(8,x,y,z)
           f_hlp(9 ,x_w,y_n,z  ) = f(9,x,y,z)
           f_hlp(10,x_w,y_s,z  ) = f(10,x,y,z)
           f_hlp(11,x_e,y  ,z_n) = f(11,x,y,z)
           f_hlp(12,x_w,y  ,z_n) = f(12,x,y,z)
           f_hlp(13,x_e,y  ,z_s) = f(13,x,y,z)
           f_hlp(14,x_w,y  ,z_s) = f(14,x,y,z)
           f_hlp(15,x  ,y_n,z_n) = f(15,x,y,z)
           f_hlp(16,x  ,y_n,z_s) = f(16,x,y,z)
           f_hlp(17,x  ,y_s,z_n) = f(17,x,y,z)
           f_hlp(18,x  ,y_s,z_s) = f(18,x,y,z)
   10    continue
   11    continue
   12 continue
c
      do 22 z = 1, lz
       do 21 y = 1, ly
        do 20 x = 1, lx
         do k = 1, 18
         f(k,x,y,z) = f_hlp(k,x,y,z)
         enddo
   20    continue
   21    continue
   22 continue
      return
      end

c----------------------------------------------------
      subroutine getuv(obst,u_x,u_y,u_z,rh, rho,fp, p,
     & Force1, Force2,Force3)
      implicit none
      include "head.inc"
      integer x,y,z,obst(lx,ly,lz),ip,jp,l,k,xp,yp, xn,yn,zp,zn
      real*8  temp(lx,ly,lz),
```

```
      &  u_x(lx,ly,lz),u_y(lx,ly,lz),rho(lx,ly,lz),rh(lx,ly,lz),
      &  fp(0:18,lx,ly,lz),  Force1(lx,ly,lz), fai(lx,ly,lz),
      &  Force2(lx,ly,lz), p(lx,ly,lz), prho(lx,ly,lz),
      &  Force3(lx,ly,lz),u_z(lx,ly,lz)

        do 10 z = 1, lz
          do 10 y = 1, ly
            do 10 x = 1, lx
            if(obst(x,y,z) .eq. 0) then
! Macroscopic variable \phi
            rh(x,y,z) = fp(0,x,y,z) +fp(1,x,y,z) +fp(2,x,y,z)
      &                    +fp(3,x,y,z) +fp(4,x,y,z) +fp(5,x,y,z)
      &                    +fp(6,x,y,z) +fp(7,x,y,z) +fp(8,x,y,z)
      &                    +fp(9,x,y,z) +fp(10,x,y,z) +fp(11,x,y,z)
      &                    +fp(12,x,y,z) +fp(13,x,y,z) +fp(14,x,y,z)
      &         +fp(15,x,y,z) +fp(16,x,y,z) +fp(17,x,y,z) +fp(18,x,y,z)
            endif
   10 continue

! From \phi to get density of fluid \rho
        call index_function( rh,rho)
C------------------------------------------------

        do 11 z = 1, lz
          do 11 y = 1, ly
            do 11 x = 1, lx
! To apply the EOS for thermodynamic pressure

          if(obst(x,y,z) .eq. 0) then
! \fai is (p_t- \rh* c_s^2), where p_t is the thermodynamic pressure
! (EOS is used)
! In the EOS: a=12RT=12c_s^2, b=4. The values are in lattice units
          fai(x,y,z) =
      &   rh(x,y,z)*rh(x,y,z) *c_squ*( 4.d0 - 2.d0*rh(x,y,z) )
      %               /(1.d0- rh(x,y,z))**3.d0
      %               - 12.d0* c_squ* rh(x,y,z)* rh(x,y,z)

!  prho means: p-\rho* c_s^2 , where p is the hydrodynamic pressure
          prho(x,y,z) = p(x,y,z) - rho(x,y,z) *c_squ

          endif
   11 continue


C----------------------------------------------------

        do 15 z = 1, lz
          do 15 y = 1, ly
            do 15 x = 1, lx
          xp = x+1
          yp = y+1
          zp = z+1
          xn = x-1
          yn = y-1
          zn = z-1
! Periodic boundary is applied here
        if (xp.gt.lx )    xp = 1
```

```
      if (xn.lt.1 )       xn = lx
      if (yp.gt.ly )       yp = 1
      if (yn.lt.1 )       yn = ly
      if (zp.gt.lz )       zp = 1
      if (zn.lt.1 )       zn = lz

! The value of Laplacian operator applied to the variable \rho
      temp(x,y,z)  =
   *   ( ( rh(xp,y,z) + rh(xn,y,z) + rh(x,yp,z)+ rh(x,yn,z)
   *        +rh(x,y,zp) + rh(x,y,zn) )*2.d0/6.d0
   *      +( rh(xp,yp,z )+ rh(xn,yn,z )
   *      +   rh(xp,yn,z )+ rh(xn,yp,z )
   *      +   rh(xp,y ,zp)+ rh(xn,y ,zn)
   *      +   rh(xp,y ,zn)+ rh(xn,y ,zp)
   *      +   rh(x ,yp,zp)+ rh(x ,yn,zn)
   *      +   rh(x ,yp,zn)+ rh(x ,yn,zp)        )/6.d0
   *      - 24.d0* rh(x,y,z)/6.d0
   *      )

  15 continue

c----------------------------------------
      do 20 z = 1, lz
        do 20 y = 1, ly
          do 20 x = 1, lx
          xp = x+1
          yp = y+1
          xn = x-1
          yn = y-1
          zp = z+1
          zn = z-1
! In x, y, z directions, periodic boundary conditions are applied
      if (xp.gt.lx )       xp = 1
      if (xn.lt.1 )        xn = lx
      if (yp.gt.ly )       yp = 1
      if (yn.lt.1 )        yn = ly
      if (zp.gt.lz )       zp = 1
      if (zn.lt.1 )        zn = lz

      Force1(x,y,z) = Kappa *rh(x,y,z)  *
   *   ( (temp(xp,y ,z )- temp(xn,y ,z ) )/6.d0
   *      +(temp(xp,yp,z )- temp(xn,yn,z ) )/12.d0
   *      +(temp(xp,yn,z )- temp(xn,yp,z ) )/12.d0
   *      +(temp(xp,y ,zp)- temp(xn,y ,zn) )/12.d0
   *      +(temp(xp,y ,zn)- temp(xn,y ,zp) )/12.d0
   *      )

      Force2(x,y,z) = Kappa *rh(x,y,z)  *
   &   ( (temp(x ,yp,z )- temp(x ,yn,z ) )/6.d0
   &      +(temp(xp,yp,z )- temp(xn,yn,z ) )/12.d0
   &      +(temp(xn,yp,z )- temp(xp,yn,z ) )/12.d0
   &      +(temp(x ,yp,zp)- temp(x ,yn,zn) )/12.d0
   &      +(temp(x ,yp,zn)- temp(x ,yn,zp) )/12.d0  )

      Force3(x,y,z) = Kappa *rh(x,y,z)  *
   &   ( (temp(x ,y ,zp)- temp(x ,y ,zn) )/6.d0
```

```
      &       +(temp(xp,y ,zp)- temp(xn,y ,zn) )/12.d0
      &       +(temp(xn,y ,zp)- temp(xp,y ,zn) )/12.d0
      &       +(temp(x ,yp,zp)- temp(x ,yn,zn) )/12.d0
      &       +(temp(x ,yn,zp)- temp(x ,yp,zn) )/12.d0  )

!  Buoyant force is added here
        if(BUOYANCY .eq. 1 ) then
        Force3(x,y,z) = Force3(x,y,z)
      &      + gforce *( rho(x,y,z)- rho_h )
        endif

   20 continue
C----------------------------------------

! To get the gradient (first derivative) of \fai and  p-\rho* c_s^2
C----------------------------------------
        do 30 z = 1, lz
          do 31 y = 1, ly
           do 32 x = 1, lx
        xp = x+1
        yp = y+1
        zp = z+1
        xn = x-1
        yn = y-1
        zn   = z-1
! In x, y, z directions, periodic boundary conditions are applied
        if (xp.gt.lx )     xp = 1
        if (xn.lt.1 )      xn = lx
        if (yp.gt.ly )     yp = 1
        if (yn.lt.1 )      yn = ly
        if (zp.gt.lz )     zp = 1
        if (zn.lt.1 )      zn = lz

! Gradient of \fai in x direction
        dfai_x(x,y,z) =
      *   ( (fai(xp,y ,z )- fai(xn,y ,z ) )/6.d0
      *      +(fai(xp,yp,z )- fai(xn,yn,z ) )/12.d0
      *      +(fai(xp,yn,z )- fai(xn,yp,z ) )/12.d0
      *      +(fai(xp,y ,zp)- fai(xn,y ,zn) )/12.d0
      *      +(fai(xp,y ,zn)- fai(xn,y ,zp) )/12.d0
      *      )

! Gradient of \fai in y direction
        dfai_y(x,y,z) =
      &   ( (fai(x ,yp,z )- fai(x ,yn,z ) )/6.d0
      &      +(fai(xp,yp,z )- fai(xn,yn,z ) )/12.d0
      &      +(fai(xn,yp,z )- fai(xp,yn,z ) )/12.d0
      &      +(fai(x ,yp,zp)- fai(x ,yn,zn) )/12.d0
      &      +(fai(x ,yp,zn)- fai(x ,yn,zp) )/12.d0   )

! Gradient of \fai in z direction
        dfai_z(x,y,z) =
      &   ( (fai(x ,y ,zp)- fai(x ,y ,zn) )/6.d0
      &      +(fai(xp,y ,zp)- fai(xn,y ,zn) )/12.d0
      &      +(fai(xn,y ,zp)- fai(xp,y ,zn) )/12.d0
```

```
      &       +(fai(x ,yp,zp)- fai(x ,yn,zn) )/12.d0
      &       +(fai(x ,yn,zp)- fai(x ,yp,zn) )/12.d0   )
c-------------------------------
      dprho_x(x,y,z) =
      *   ( (prho(xp,y ,z )- prho(xn,y ,z ) )/6.d0
      *       +(prho(xp,yp,z )- prho(xn,yn,z ) )/12.d0
      *       +(prho(xp,yn,z )- prho(xn,yp,z ) )/12.d0
      *       +(prho(xp,y ,zp)- prho(xn,y ,zn) )/12.d0
      *       +(prho(xp,y ,zn)- prho(xn,y ,zp) )/12.d0
      *       )

      dprho_y(x,y,z) =
      &   ( (prho(x ,yp,z )- prho(x ,yn,z ) )/6.d0
      &       +(prho(xp,yp,z )- prho(xn,yn,z ) )/12.d0
      &       +(prho(xn,yp,z )- prho(xp,yn,z ) )/12.d0
      &       +(prho(x ,yp,zp)- prho(x ,yn,zn) )/12.d0
      &       +(prho(x ,yp,zn)- prho(x ,yn,zp) )/12.d0   )

      dprho_z(x,y,z) =
      &   ( (prho(x ,y ,zp)- prho(x ,y ,zn) )/6.d0
      &       +(prho(xp,y ,zp)- prho(xn,y ,zn) )/12.d0
      &       +(prho(xn,y ,zp)- prho(xp,y ,zn) )/12.d0
      &       +(prho(x ,yp,zp)- prho(x ,yn,zn) )/12.d0
      &       +(prho(x ,yn,zp)- prho(x ,yp,zn) )/12.d0   )
 32    continue
 31    continue
 30  continue
      end

c------------------------------------------------------------------
      subroutine geten(obst,u_x,u_y,u_z,rho,gp, p,Fx,Fy, Fz)
      include "head.inc"
      integer x,y,obst(lx,ly,lz)
      real*8   u_x(lx,ly,lz),u_y(lx,ly,lz),rho(lx,ly,lz),
      & gp(0:18,lx,ly,lz),u_z(lx,ly,lz), Fx(lx,ly,lz),
      & Fy(lx,ly,lz), Fz(lx,ly,lz),p(lx,ly,lz)

      do 12 z = 1, lz
       do 11 y = 1, ly
        do 10 x = 1, lx
!  Initialization in each step
        u_x(x,y,z) = 0.d0
        u_y(x,y,z) = 0.d0
        u_z(x,y,z) = 0.d0

       if(obst(x,y,z) .eq. 0) then

        u_x(x,y,z)=(gp(1,x,y,z)+ gp(7,x,y,z)+ gp(8,x,y,z) +
      &             gp(11,x,y,z) + gp(13,x,y,z)
      &             -(gp(2,x,y,z) + gp(9,x,y,z) + gp(10,x,y,z)+
      &             gp(12,x,y,z) + gp(14,x,y,z) ))
      &       + 0.5d0 * c_squ *( Fx(x,y,z) +0.d0 )

c----------------------------------------------------------
        u_y(x,y,z) = (gp(3,x,y,z) + gp(7,x,y,z) + gp(9,x,y,z) +
      &             gp(15,x,y,z) + gp(16,x,y,z)
```

```
     &                  - (gp(4,x,y,z) + gp(8,x,y,z) + gp(10,x,y,z) +
     &                    gp(17,x,y,z) + gp(18,x,y,z) ))
     &              + 0.5d0 * c_squ *( Fy(x,y,z) +0.d0 )

c-------------------------------------------------------
       u_z(x,y,z) = (gp(5,x,y,z) + gp(11,x,y,z) + gp(12,x,y,z)+
     &                  gp(15,x,y,z) + gp(17,x,y,z)
     &                - (gp(6,x,y,z) + gp(13,x,y,z) + gp(14,x,y,z) +
     &                  gp(16,x,y,z) + gp(18,x,y,z) ))
     &              + 0.5d0 * c_squ *( Fz(x,y,z) +0.d0 )

       u_x(x,y,z) = u_x(x,y,z)/rho(x,y,z)/c_squ
       u_y(x,y,z) = u_y(x,y,z)/rho(x,y,z)/c_squ
       u_z(x,y,z) = u_z(x,y,z)/rho(x,y,z)/c_squ

        p(x,y,z) = gp(0,x,y,z) +gp(1,x,y,z) +gp(2,x,y,z)
     &                       +gp(3,x,y,z) +gp(4,x,y,z) +gp(5,x,y,z)
     &                       +gp(6,x,y,z) +gp(7,x,y,z) +gp(8,x,y,z)
     &                       +gp(9,x,y,z) +gp(10,x,y,z) +gp(11,x,y,z)
     &                       +gp(12,x,y,z) +gp(13,x,y,z) +gp(14,x,y,z)
     &   +gp(15,x,y,z) +gp(16,x,y,z) +gp(17,x,y,z) +gp(18,x,y,z)

        p(x,y,z) = p(x,y,z) - 0.5d0 * ( u_x(x,y,z)* dprho_x(x,y,z)
     %          + u_y(x,y,z)* dprho_y(x,y,z)
     %                  + u_z(x,y,z)* dprho_z(x,y,z)   )

       endif

 10    continue
 11   continue
 12 continue
      end

c-------------------------------------------------------
      subroutine collision(obst,ux,uy,uz,rh,
     & rho,p,ff,gp, Fx,Fy,Fz)
      implicit none
      include "head.inc"
      integer  l,obst(lx,ly,lz)
      real*8 ux(lx,ly,lz),uy(lx,ly,lz),ff(0:18,lx,ly,lz),rho(lx,ly,lz)
      real*8 Fx(lx,ly,lz),Fy(lx,ly,lz), rh(lx,ly,lz)
      real*8 Fz(lx,ly,lz),uz(lx,ly,lz),p(lx,ly,lz),gp(0:18,lx,ly,lz)

      integer x,y,z,k
      real*8  u_n(0:18),fequ(0:18),gequ(0:18),u_squ,temp(18),temp2(18)
      real*8  noteq(0:18),gneq(0:18)

      do 4 z = 1, lz
         do 5 y = 1, ly
          do 6 x = 1, lx
! Bounce back for solid nodes
            if(BD .eq. 1 .and. obst(x,y,z) .eq. 1) then
              do k=1, 18
              temp(k)   = ff(k,x,y,z)
              temp2(k)  = gp(k,x,y,z)
              enddo
```

```
                  do k=1, 18
                  ff(opp(k),x,y,z) = temp(k)
                  gp(opp(k),x,y,z) = temp2(k)
               enddo
            endif


            if(obst(x,y,z) .eq. 0) then

         call g_eq(p(x,y,z),rho(x,y,z),ux(x,y,z),uy(x,y,z),uz(x,y,z),gequ)
         call f_eq(rh(x,y,z),ux(x,y,z),uy(x,y,z),uz(x,y,z),fequ)

            do 60 k = 0,18
! Non-equilibrium distribution function
            noteq(k)= ff(k,x,y,z) - rh(x,y,z)*fequ(k)

            ff(k,x,y,z) = ff(k,x,y,z) -1.d0/tau_f * noteq(k)
! The source term in the LBE for f_i
     &          - (tau_f-0.5d0)/tau_f*
     &          ( (xc(k)-ux(x,y,z))* dfai_x(x,y,z)
     &          +(yc(k)-uy(x,y,z))* dfai_y(x,y,z)
     &             +(zc(k)-uz(x,y,z))* dfai_z(x,y,z) )*fequ(k)/c_squ
C----------------------------------------------------------------
! Non-equilibrium distribution function
            gneq(k)= gp(k,x,y,z) - gequ(k)

            gp(k,x,y,z) = gp(k,x,y,z) -1.d0/tau_g *gneq(k)
! The source term in the LBE for g_i
     &             + (tau_g-0.5d0)/tau_g *
     *          (
     &          (  (xc(k)-ux(x,y,z))*   Fx(x,y,z)
     &          +(yc(k)-uy(x,y,z))* (Fy(x,y,z) )
     &          +(zc(k)-uz(x,y,z))* (Fz(x,y,z) )
     &          )  *fequ(k)

     &             - (xc(k)-ux(x,y,z))*( fequ(k)-t_k(k) )*dprho_x(x,y,z)
     &             - (yc(k)-uy(x,y,z))*( fequ(k)-t_k(k) )*dprho_y(x,y,z)
     &             - (zc(k)-uz(x,y,z))*( fequ(k)-t_k(k) )*dprho_z(x,y,z)
     *          )
  60     continue

         endif

   6     continue
   5   continue
   4 continue

       return
       end
C------------------------------------------------------------


Utility.for
C==========================================================

! Linear map from index function \phi in JCP, 1999 (\rh) to density
! (\rho)
       subroutine index_function(rh,rho)
```

```
      implicit none
      include "head.inc"
      integer x,y,z
      real*8 rh(lx,ly,lz), rho(lx,ly,lz), psi_min0, psi_max0

!  The \psi_{max} and \psi_{min} depend on the EOS we used.

      psi_min0 = psi_min
      psi_max0 = psi_max

      do 1 x = 1,lx
      do 1 y = 1,ly
      do 1 z = 1,lz
      rho(x,y,z) = rho_l + ( rh(x,y,z)- psi_min0 )/(psi_max0 -
     & psi_min0)
     &   * (rho_h - rho_l)
    1 continue
      end

c----------------------------------------------------------
! Calculate the equilibrium distribution function of g_i, which is
! used to
! get pressure and velocity.
      subroutine g_eq(p,rh,u,v,w,gequ)
      implicit none
      include "head.inc"
      real*8 u,v,w,gequ(0:18),u_n(0:18),u_squ,rh,p
      integer k
          u_squ = u*u + v*v +w*w
      do 5 k = 0, 18
          u_n(k)   = xc(k)*u + yc(k)*v +zc(k)*w

          gequ(k) = t_k(k)* ( p + rh* c_squ* (
     &                u_n(k) / c_squ
     &              + u_n(k) *u_n(k) / (2.d0 * c_squ *c_squ)
     &              - u_squ / (2.d0 * c_squ)
     *              ) )
    5 continue
      end
cc‾‾‾‾‾‾‾‾‾‾‾‾‾‾‾‾‾‾‾‾‾‾‾‾‾‾‾‾‾‾‾‾‾‾‾‾‾‾‾‾‾‾‾
! Calculate the equilibrium distribution function f_i, which is used
! to get
! index function

      subroutine f_eq(rh,u,v,w,fequ)
      implicit none
      include "head.inc"
      real*8 u,v,w,fequ(0:18),u_n(0:18),rh,u_squ
      integer k
          u_squ = u*u + v*v +w*w
      do 65 k = 0, 18
          u_n(k)   = xc(k)*u + yc(k)*v +zc(k)*w

          fequ(k) = t_k(k)* ( 1.0d0+  u_n(k) / c_squ
     &              + u_n(k) *u_n(k) / (2.d0 * c_squ *c_squ)
     &              - u_squ / (2.d0 * c_squ))
   65 continue
      end
```

Output.for

```
C============================================================
! If you want the output tecplot file to be binary format, please
! change this output subroutine according to the binary file routine.
      subroutine  write_results2(obst,rho,upx,upy,upz,p, n)
      implicit none
      include "head.inc"
      integer  x,y,z,i,n,obsval,k1,k,obst(lx,ly,lz)
      real*8   rho1,h(lx),rho(lx,ly,lz),upx(lx,ly,lz),upy(lx,ly,lz)
      real*8   upz(lx,ly,lz),p(lx,ly,lz)
      character filename*16, B*7, C*1

      write(B,'(i7.7)') n
      filename='./out/3D'//B//'.plt'

      open(41,file=filename)

      write(41,*) 'variables_=_x,_y,_z,_rho,_p,_upx,_upy,_upz,_obst'
      write(41,*) 'zone_i=', lx, ',_j=', ly, ',_k=', lz, ',_f=point'
      do 10 z = 1, lz
         do 10 y = 1, ly
           do 10 x = 1, lx
C.........write results to file
           write(41,9) x, y, z,
     &        rho(x,y,z), p(x,y,z),
     &        upx(x,y,z), upy(x,y,z), upz(x,y,z), obst(x,y,z)
   10 continue

    9 format(3i4, 5f15.8, i4)

      close(41)
C--------------------------------
      end
```

CHAPTER 8

Axisymmetric multiphase HCZ model

8.1 Introduction

The model of He et al. (1999) has been extended to study axisymmetric multiphase flows, such as the two-droplet collision (Premnath and Abraham 2005b), break up of a liquid cylindrical column (Premnath and Abraham 2005a), and a drop impacting on wet walls (Mukherjee and Abraham 2007a).

This chapter is organized as the follows. First the macroscopic governing equation of our developed scheme is illustrated. Second, the axisymmetric multiphase HCZ LBM developed by Premnath and Abraham (2005a) is introduced. The MRT axisymmetric HCZ developed by McCracken and Abraham (2005) is also introduced briefly. In the third section, the examples of break up of a liquid cylindrical column and a two-droplet collision are illustrated.

Finally, our revised axisymmetric HCZ model is introduced. The scheme is used to simulate bubble rise and compared in detail with experimental data and other numerical results in the literature.

8.2 Methods

8.2.1 Macroscopic governing equations

The HCZ model is an index-function-based multiphase model. The governing equation for the index function ϕ in axisymmetric coordinates $((x, r)$ coordinates) is

$$\partial_t \phi + \partial_\beta (\phi u_\beta) + \frac{\phi u_r}{r} = \lambda \partial_\beta (\partial_\beta (p_{th} - c_s^2 \phi)). \tag{8.1}$$

It is a CH equation for tracking the interface and λ is the mobility (refer to Section 8.10.4). x and r are the axial and radial directions, respectively. p_{th} is the thermodynamic pressure calculated from the Carnahan–Starling EOS (He et al. 1999),

$$p_{th} = \phi RT \frac{1 + b\phi/4 + (b\phi/4)^2 - (b\phi/4)^3}{(1 - b\phi/4)^3} - a\phi^2, \tag{8.2}^*$$

Multiphase Lattice Boltzmann Methods: Theory and Application, First Edition.
Haibo Huang, Michael C. Sukop and Xi-Yun Lu.
© 2015 John Wiley & Sons, Ltd. Published 2015 by John Wiley & Sons, Ltd.
Companion Website: www.wiley.com/go/huang/boltzmann

where $a = b = 4$ and $RT = c_s^2$. For this parameter choice, through the Maxwell construction, we know that two phases with $\phi_g = 0.021$ and $\phi_l = 0.247$, where the subscripts g and l mean the gas and liquid phases respectively, would coexist.

For incompressible multiphase flow, the continuity equation is

$$\partial_t \rho + \partial_\beta(\rho u_\beta) = -\frac{\rho u_r}{r}. \tag{8.3}$$

The momentum governing equation (N–S equation) is:

$$\rho \partial_t u_\alpha + \rho \partial_\beta(u_\beta u_\alpha) = -\partial_\alpha p + \partial_\alpha[\eta(\partial_\alpha u_\beta + \partial_\beta u_\alpha)] + H_\alpha + (F_s)_\alpha + \mathcal{G}_\alpha, \tag{8.4}$$

where $(F_s)_\alpha$ is the α component of surface tension $\mathbf{F}_s$. In our simulation of bubble rise, $\mathcal{G}(\mathbf{x}) = -(\rho(\mathbf{x}) - \rho_l)\mathbf{g}$ is the buoyant force acting on the bubble, where $\mathbf{x}$ represents the position of a computational node in the computational domain.

Compared to 2D cases there are some source terms due to the axisymmetric effect on the right-hand side of the N–S equation, i.e.,

$$H_\alpha = \eta \left\{ \frac{\partial_r u_\alpha}{r} - \frac{u_r}{r^2}\delta_{\alpha r} + \partial_\alpha\left(\frac{u_r}{r}\right) \right\} - \rho\frac{u_r u_\alpha}{r}. \tag{8.5}$$

In the above expression η is the dynamic viscosity.

For 2D cases the surface tension $\mathbf{F}_s$ can be expressed in different forms (Kim 2005), for example in the study of He et al. (1999)

$$\mathbf{F}_s = \kappa\rho\nabla(\nabla^2\rho), \tag{8.6}$$

where κ is the surface tension coefficient. This is referred to as the "original" surface tension calculation.

For axisymmetric cases the surface tension is (Premnath and Abraham 2005a)

$$\mathbf{F}_s = \kappa\rho\partial_\alpha\left(\partial_\beta^2\rho + \frac{1}{r}\partial_r\rho\right). \tag{8.7}$$

In this expression $\kappa\rho\partial_\alpha\left(\frac{1}{r}\partial_r\rho\right)$ is the additional term in the surface tension that arises from axisymmetry.

Here, as suggested in Zhang et al. (2000) and as done in Chapter 7, we use the variable ϕ rather than ρ to calculate the surface tension, i.e., the surface tension in our axisymmetric model is

$$(F_s)_\alpha = \kappa\phi\partial_\alpha(\partial_\delta^2\phi) + \kappa\phi\partial_\alpha\left(\frac{1}{r}\partial_r\phi\right). \tag{8.8}$$

Here, all the source terms in the N–S equation are integrated as

$$\mathbf{F} = \mathbf{H} + \mathbf{F}_s + \mathcal{G} \tag{8.9}$$

or

$$F_\alpha = H_\alpha + (F_s)_\alpha + \mathcal{G}_\alpha. \tag{8.10}$$

8.2.2 Axisymmetric HCZ LBM (Premnath and Abraham 2005a)

This axisymmetric multiphase LBM is based on the HCZ model (He et al. 1999) in which the index function ϕ is used to track interfaces between liquid and gas. In the model, as in models described previously in Chapter 7, two distribution functions $\bar{f}_i$ and $\bar{g}_i$ are introduced, which are able to recover the CH equation for evolution of the index function and the N–S equations, respectively:

$$\bar{g}_i(\mathbf{x} + \mathbf{e}_i \Delta t, t + \Delta t) = \bar{g}_i(\mathbf{x}, t) - \frac{1}{\tau_1}(\bar{g}_i(\mathbf{x}, t) - \bar{g}_i^{eq}(\mathbf{x}, t)) + S_i(\mathbf{x}, t)\Delta t, \qquad (8.11)$$

$$\bar{f}_i(\mathbf{x} + \mathbf{e}_i \Delta t, t + \Delta t) = \bar{f}_i(\mathbf{x}, t) - \frac{1}{\tau_2}(\bar{f}_i(\mathbf{x}, t) - \bar{f}_i^{eq}(\mathbf{x}, t)) + S'_i(\mathbf{x}, t)\Delta t, \qquad (8.12)$$

where $\bar{f}_i(\mathbf{x}, t)$ is the density distribution function in the ith velocity direction at position $\mathbf{x}$ at time step t. τ_1 is a relaxation time and is related to the kinematic viscosity as $v = c_s^2(\tau_1 - 0.5)\Delta t$. τ_2 is related to mobility (refer to Section 8.10.4). Usually in our simulations

$$\tau_2 = \tau_1 \qquad (8.13)$$

is adopted. For the cases in which the kinematic viscosities of the liquid and gas are identical, $\tau_1 = \tau_l = \tau_g$. If the kinematic viscosities of the liquid and gas are different ($\tau_l \neq \tau_g$), then τ_1 should be a function of position. It should be interpolated through $\phi(\mathbf{x})$, i.e.,

$$\tau_1(\mathbf{x}) = \tau_g + \frac{\phi(\mathbf{x}) - \phi_g}{\phi_l - \phi_g}(\tau_l - \tau_g). \qquad (8.14)$$

$S_i(\mathbf{x}, t)$ and $S'_i(\mathbf{x}, t)$ are the source terms in Eqs (8.11) and (8.12), respectively. The equilibrium distribution functions $\bar{g}_i^{eq}(\mathbf{x}, t)$ and $\bar{f}_i^{eq}(\mathbf{x}, t)$ can be calculated as (He et al. 1999)

$$\bar{g}_i^{eq}(\mathbf{x}, t) = w_i \left[p_h + \rho c_s^2 \left(\frac{e_{i\alpha}u_\alpha}{c_s^2} + \frac{e_{i\alpha}u_\alpha e_{i\beta}u_\beta}{2c_s^4} - \frac{u_\alpha u_\alpha}{2c_s^2} \right) \right] \qquad (8.15)$$

and

$$\bar{f}_i^{eq}(\mathbf{x}, t) = w_i \phi \left[1 + \frac{e_{i\alpha}u_\alpha}{c_s^2} + \frac{e_{i\alpha}u_\alpha e_{i\beta}u_\beta}{2c_s^4} - \frac{u_\alpha u_\alpha}{2c_s^2} \right], \qquad (8.16)$$

respectively, where p_h and ρ are the hydrodynamic pressure and density of the fluid, respectively.

In Eqs (8.11) and (8.12) the $\mathbf{e}_i$s are the discrete velocities. For the D2Q9 model, $i = 0, 1, 2, \ldots, 8$ (see Chapter 1).

In Eqs (8.15) and (8.16), for the D2Q9 model, $w_i = 4/9$ ($i = 0$), $w_i = 1/9$, ($i = 1, 2, 3, 4$), $w_i = 1/36$, ($i = 5, 6, 7, 8$), and $c_s = \frac{c}{\sqrt{3}}$, where $c = \frac{\Delta x}{\Delta t}$ is the ratio of lattice spacing Δx and time step Δt.

The hydrodynamic pressure and the moments are obtained through

$$p_h = \sum \bar{g}_i + \frac{1}{2}u_\beta E_\beta \Delta t - \frac{1}{2}c_s^2 \rho \frac{u_r}{r}\Delta t, \qquad (8.17)$$

where $E_\beta = -\partial_\beta[\psi(\rho)]$ and $\psi(\rho) = p_h - c_s^2\rho$;

$$\rho c_s^2 u_\alpha = \sum \bar{g}_i e_{i\alpha} + \frac{1}{2}c_s^2 F_\alpha \Delta t. \qquad (8.18)$$

The above equation is correct while in the corresponding equation in Premnath and Abraham (2005a) there is a typographical error (missing coefficient RT or c_s^2 for the last two terms on the right-hand side in Eq. (69) in Premnath and Abraham (2005a)).

The index function is calculated from

$$\phi = \sum \bar{f}_i - \frac{\phi}{2}\frac{u_r}{r}. \qquad (8.19)$$

After ϕ is known, the corresponding density of the fluid can be calculated from

$$\rho(\phi) = \rho_g + (\phi - \phi_g)\frac{\rho_l - \rho_g}{\phi_l - \phi_g}, \qquad (8.20)$$

where ϕ_g and ϕ_l are the two equilibrium coexistence "densities" of the gas and liquid, respectively in Eq. (8.2).

The source terms, including the axisymmetric effect and surface tension, appearing in Eqs (8.11) and (8.12) are

$$S_i = \left(1 - \frac{1}{2\tau_1}\right)\left\{\left(e_{i\alpha} - u_\alpha\right)F_\alpha\Gamma_i(\mathbf{u}) + (e_{i\alpha} - u_\alpha)E_\alpha[\Gamma_i(\mathbf{u}) - \Gamma_i(0)] - w_i c_s^2\frac{\rho u_r}{r}\right\}, \qquad (8.21)$$

where $\Gamma_i(\mathbf{u}) = \bar{f}_i^{eq}/\phi$, and

$$S_i' = \left(1 - \frac{1}{2\tau_2}\right)\left\{\frac{\left(e_{i\alpha} - u_\alpha\right)F_\alpha'}{c_s^2}\Gamma_i(\mathbf{u}) - w_i\frac{\phi u_r}{r}\right\}, \qquad (8.22)$$

where $F_\alpha' = -\partial_\alpha[\psi(\phi)]$ and the function $\psi(\phi) = p_{th} - c_s^2\phi$.

It is noted that Eq. (8.21) includes all the terms in Eqs (61)–(63) in Premnath and Abraham (2005a) and Eq. (8.22) includes all terms in Eqs (66) and (67) in Premnath and Abraham (2005a).

8.2.3 MRT version of the axisymmetric LBM (McCracken and Abraham 2005)

As illustrated in the above section, the axisymmetric LBEs for $\bar{g}_i$ and $\bar{f}_i$ can be written in the following forms. Substituting Eq. (8.21) into Eq. (8.11) we have the explicit form

$$\bar{g}_i(\mathbf{x} + \mathbf{e}_i\Delta t, t + \Delta t) = \bar{g}_i(\mathbf{x}, t) - \frac{1}{\tau_1}(\bar{g}_i(\mathbf{x}, t) - \bar{g}_i^{eq}(\mathbf{x}, t)) + \left(1 - \frac{1}{2\tau_1}\right)(S_g)_i\Delta t, \qquad (8.23)$$

where

$$(S_g)_i = \left\{\left(e_{i\alpha} - u_\alpha\right)F_\alpha\Gamma_i(\mathbf{u}) + (e_{i\alpha} - u_\alpha)E_\alpha[\Gamma_i(\mathbf{u}) - \Gamma_i(0)] - w_i c_s^2\frac{\rho u_r}{r}\right\}. \qquad (8.24)$$

Substituting Eq. (8.22) into Eq. (8.12) we have the explicit form

$$\bar{f}_i(\mathbf{x} + \mathbf{e}_i\Delta t, t + \Delta t) = \bar{f}_i(\mathbf{x}, t) - \frac{1}{\tau_2}(\bar{f}_i(\mathbf{x}, t) - \bar{f}_i^{eq}(\mathbf{x}, t)) + \left(1 - \frac{1}{2\tau_2}\right)(S_f)_i\Delta t, \quad (8.25)$$

where

$$(S_f)_i = \left\{ \frac{(e_{i\alpha} - u_\alpha) F'_\alpha}{c_s^2}\Gamma_i(\mathbf{u}) - w_i\frac{\phi u_r}{r} \right\}. \quad (8.26)$$

As we introduced in Chapter 4, the difference between the MRT and BGK models is the collision term. The collision term $\Omega_i = -\frac{1}{\tau}(f_i - f_i^{eq})$ in BGK should be replaced by the MRT collision model (Lallemand and Luo 2000), that is,

$$\Omega_i = -M^{-1}\hat{S}\left[|m(\mathbf{x}, t)\rangle - |m^{(eq)}(\mathbf{x}, t)\rangle\right]. \quad (8.27)$$

For more detail, refer to Chapter 4.

The parameters in the MRT are chosen as $s_0 = s_3 = s_5 = 1.0$, $s_1 = 1.64$, $s_2 = 1.54$, $s_4 = s_6 = 1.7$, and $s_7 = s_8 = \frac{1}{\tau}$. The constants adopted here are identical to those in McCracken and Abraham (2005).

In the MRT scheme the collision process is usually executed in moment space, as illustrated in Eq. (8.27) (Lallemand and Luo 2000). Alternatively, however, it can also be executed in velocity space because the collision operator can be written as

$$\Omega_i = -M^{-1}\hat{S}M\left[|f(\mathbf{x}, t)\rangle - |f^{(eq)}(\mathbf{x}, t)\rangle\right]. \quad (8.28)$$

Compared with the BGK collision term $\Omega_i = -\frac{1}{\tau}(f_i - f_i^{eq})$ in the common LBE, we found there are two changes in this simplest MRT:
- $\frac{1}{\tau}$ is replaced with $M^{-1}\hat{S}M$ and
- $(f_i - f_i^{eq})$ is replaced with $|f(\mathbf{x}, t)\rangle - |f^{(eq)}(\mathbf{x}, t)\rangle$.

Analogously, in Eqs (8.23) and (8.25), if $\frac{1}{\tau}$ is replaced with $M^{-1}\hat{S}M$ we have

$$\bar{g}_i(\mathbf{x} + \mathbf{e}_i\Delta t, t + \Delta t) = \bar{g}_i(\mathbf{x}, t) - M^{-1}\hat{S}_1M\left[|\bar{g}\rangle - |\bar{g}^{(eq)}\rangle\right]$$

$$+ (S_g)_i\Delta t - \frac{1}{2}M^{-1}\hat{S}M|S_g\rangle\Delta t \quad (8.29)$$

and

$$\bar{f}_i(\mathbf{x} + \mathbf{e}_i\Delta t, t + \Delta t) = \bar{f}_i(\mathbf{x}, t) - M^{-1}\hat{S}_2M\left[|\bar{f}\rangle - |\bar{f}^{(eq)}\rangle\right]$$

$$+ (S_f)_i\Delta t - \frac{1}{2}M^{-1}\hat{S}M|S_f\rangle\Delta t. \quad (8.30)$$

These are the MRT formulae in McCracken and Abraham (2005).

It is noted that $\hat{S}_1$ and $\hat{S}_2$ are identical to $\hat{S}$, except $s_7 = s_8 = \frac{1}{\tau_1}$ in $\hat{S}_1$ and $s_7 = s_8 = \frac{1}{\tau_2}$ in $\hat{S}_2$. As we mentioned, in our simulations $\tau_2 = \tau_1$ is adopted. Hence $\hat{S}_2 = \hat{S}_1$. If $\tau_l \neq \tau_g$, $\hat{S}_1$ is a function of position as well as τ_1 (refer to Eq. (8.14)).

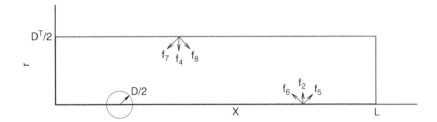

Figure 8.1   Geometry of computational domain $L \times \frac{D^T}{2}$ in the (x, r) coordinates. $r = 0$ is the axisymmetric axis.

8.2.4   Axisymmetric boundary conditions

Figure 8.1 shows the computational domain of size $L \times \frac{D^T}{2}$, where L is the length in the axisymmetric axis direction, i.e., the x axis, and $\frac{D^T}{2}$ is the radius of the domain. The computational domain is partitioned into a uniform Cartesian mesh with mesh spacing Δx.

In our simulations of droplet collision or elliptical droplet oscillation, the lower boundary where $r = 0$ represents the axisymmetric boundary and the slip boundary condition is applied. For the left and right boundaries (Figure 8.1) periodic boundary conditions are applied. For the upper boundary the slip boundary condition is applied (Mukherjee and Abraham 2007c).

In our simulations of bubble rise one boundary ($r = 0$) represents the axisymmetric axis and the other three boundaries ($x = 0$, $x = L$, and $r = \frac{D^T}{2}$) are walls. For the walls the no-slip wall boundary condition is imposed. In the LBM a simple bounce-back scheme is used to get the unknown f_i and g_i in the inward direction after the streaming step. In the scheme these unknown distribution functions are set to be the distribution functions in the corresponding reverse directions. For example, after the streaming step f_4, f_7, and f_8 are unknown on the upper wall boundary (refer to Figure 8.1); they are set to be $f_4 = f_2$, $f_7 = f_5$, and $f_8 = f_6$.

The source terms and boundary conditions perpendicular to the radial axis require additional attention. Source terms like $\rho \frac{u_r}{r}$ have to be handled carefully. According to l'Hôpital's rule, we have $\lim_{r \to 0} \rho \frac{u_r}{r} = \rho \partial_r u_r$. Hence, after we get $\partial_r u_r$ using the finite difference scheme, $\lim_{r \to 0} \rho \frac{u_r}{r}$ is obtained. For both f_i and g_i on the x axis the slip boundary condition should be applied. For example, on the x axis (refer to Figure 8.1) the unknown distribution functions are f_2, f_5, and f_6. After the streaming step these unknowns can be obtained through setting $f_2 = f_4$, $f_5 = f_8$, and $f_6 = f_7$.

8.3   The Laplace law

In this simulation the computational domain is 160×80. Initially half of a circle with radius $R = 30$ lu is filled with fluid $\rho = \rho_l$ ($\phi = \phi_l$) and the other region

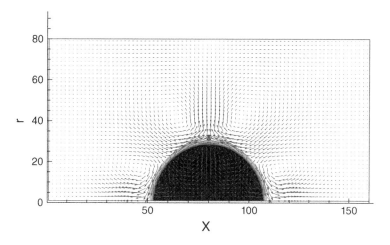

Figure 8.2 The pattern of spurious currents (case of $\kappa = 0.08$). The maximum velocity magnitude is about 0.0075 lu/ts.

is filled with gas. In the simulations $\rho_l = 0.12$, $\rho_g = 0.04$, and $\tau_g = \tau_f = 0.7$. The equilibrium state for a case with $\kappa = 0.08$ is shown in Figure 8.2. It is not a surprise that due to the curvature of the interface there is spurious current near the interface. The maximum magnitude of spurious currents is found to increase with κ. For example, for $\kappa = 0.02$ the maximum magnitude of the spurious currents is 0.0022 lu/ts. When $\kappa = 0.08$ (as shown in Figure 8.2), then the spurious current is larger at 0.0075 lu/ts.

The surface tension can be calculated from the pressure difference between the inside and outside of the drop, i.e., $\sigma = \frac{1}{2}(p_{in} - p_{out})R$. There is a coefficient $\frac{1}{2}$ because an axisymmetric case is a 3D case. The calculated surface tension as a function of κ for three simulations is shown in Figure 8.3.

From Figure 8.3 we obtained the slope of the line and hence

$$\sigma = 0.015\kappa. \qquad (8.31)$$

This formula is also found to be valid for the corresponding MRT version of the axisymmetric HCZ model.

In the above formula the coefficient 0.015 is independent of the relaxation time τ_g or τ_f. The coefficient also does not depend on the density ratio. That is easy to understand because in the surface tension formula (Eq. 8.8) ϕ appears rather than ρ.

8.4 Oscillation of an initially ellipsoidal droplet

An initially ellipsoidal drop suspended in gas will oscillate and, due to surface tension and viscous dissipation, the drop will eventually reach a spherical drop

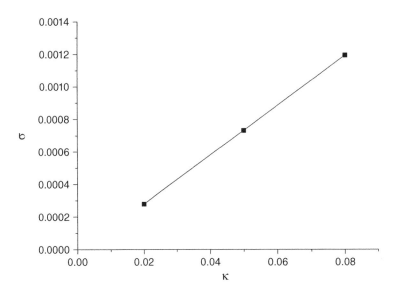

Figure 8.3   The surface tension as a function of κ in the axisymmetric HCZ model.

equilibrium state. There are many theoretical studies on this topic, for example Miller and Scriven (1968) and Lamb (1932). For the theoretical oscillation frequency of the ellipsoidal drop, refer to Section 7.10.

The boundary conditions and set up of the simulations are basically identical to those in Section 8.3. The difference is that here an ellipsoidal drop instead of a spherical drop is the initial condition. The mathematical description of the ellipsoid is

$$\frac{(x - \frac{L}{2})^2}{R_a^2} + \frac{r^2}{R_b^2} = 1,$$

(8.32)

where R_a and R_b are the maximum and minimum radii of the ellipsoid, respectively. It is noted that the equilibrium radius of the final spherial drop R_d can be calculated through mass conservation, i.e., $\frac{4}{3}\pi R_d^3 = \frac{4}{3}\pi R_a R_b^2$. The ellipsoid with $R_a = 50$ lu and $R_b = 40$ lu is initially placed in the center of the 181×91 computational domain. Here we carried out simulations of four cases. The parameters are listed in Table 8.1. Two of the cases (Cases III and IV) are simulated using the MRT version of the model, while Cases I and II are simulated using the BGK version.

Snapshots of the evolution of the drop for Case III are shown in Figure 8.4. In this case the density ratio is about 12. We found that this case can be simulated using the MRT version while the BGK version will blow up. Hence, simulating higher density ratios is one of the advantages of the MRT LBM.

From Figure 8.4 it can be seen that at $t = 8000$ ts the radius of the droplet in the x axis is shorter than that in the r axis; it is an oblate spheroid. At $t = 14,000$

Table 8.1   Theoretical and simulated oscillation period T of ellipsoidal drops ($R_a = 50$ lu and $R_b = 40$ lu).

Case	MRT or BGK	ρ_l	ρ_g	κ	τ	T (Eq. (7.64))	T (LBM)	Error of T
I	BGK	0.12	0.04	0.03	0.56	12,283	12,976	5.6%
II	BGK	0.12	0.03	0.02	0.53	14,303	14,717	2.9%
III	MRT	0.24	0.02	0.03	0.56	15,411	15,571	1.0%
IV	MRT	0.12	0.03	0.05	0.58	9,184	9,417	2.5%

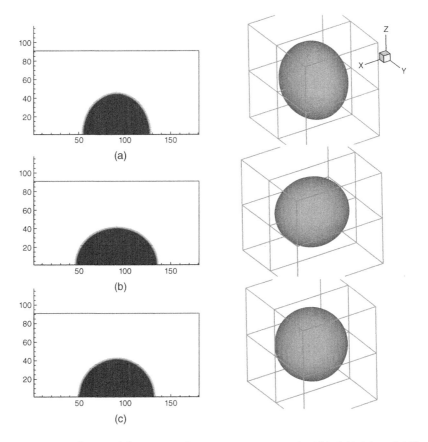

Figure 8.4   Drop shape at different times for Case III (parameters in Table 8.1). (a) $t = 8,000$ ts, (b) $t = 14,000$ ts, and (c) $t = 34,000$ ts. The horizontal and vertical radii for the times shown in (a), (b), and (c) are also labeled in Figure 8.5. The left-hand column shows the shape of the droplet in an axisymmetric plane; the right-hand column is the corresponding 3D view of the droplet.

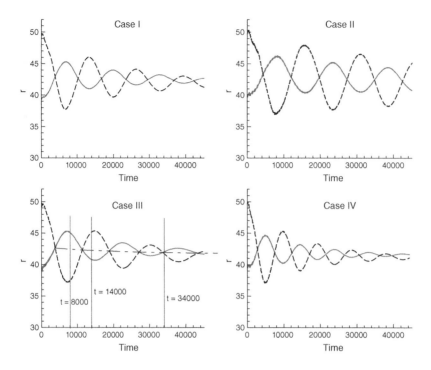

Figure 8.5 Radii of the oscillatory spheroid as a function of time for Cases I to IV. In each frame the dashed and solid lines represent the radius measured from the center of the droplet in the horizontal and vertical directions, respectively. Quantities are in lattice units.

ts it becomes a prolate spheroid. When the radii in the x axis and the r axis are equal (at $t = 34{,}000$ ts), the droplet is a perfect spheroid.

The periods of oscillation for the four cases are listed in Table 8.1. The period has unit ts. The theoretical period is obtained from Eq. (7.64), i.e., $T = \frac{2\pi}{\omega_n}$, where $n = 2$. From the table we can see that oscillation periods measured from both the BGK and MRT LBM are consistent with the theoretical predictions (with maximum error less than 6%).

The radii of the spheroid as a function of time are also shown in Figure 8.5. In each frame the dashed and solid lines represent the radius measured from the center of the droplet in the horizontal and vertical directions, respectively. Just as in the fully 3D case described in Chapter 7, the drop is spherical when the two curves (the dashed and solid lines) cross. The oscillatory period can be measured from these plots. As in Chapter 7, we can see that the amplitude of the oscillation decreases with time.

In Figure 8.5(c), if we connect the crossings of the two curves, the resulting line should be horizontal, indicating volume conservation. Although the connected line (the dash-dot line) is close to horizontal, there is a small discrepancy. That means the volume is not perfectly conserved. Similar behavior was also observed in simulations in Section 8.3. In that section the initial droplet radius is 30 lu, but eventually the radius at equilibrium state is less than 30 lu (refer to Figure 8.2). A possible reason for this is that in our initialization the coexisting ϕs are $\phi_l = 0.247$ and $\phi_g = 0.021$ but these values may not exactly represent the final/equilibrium ϕ and corresponding densities.

8.5   Cylindrical liquid column break

Through a linear stability analysis Rayleigh (1878) has shown that an inviscid column of cylindrical liquid of radius R_c is unstable if the axisymmetric wavelength of a disturbance λ_d is longer than its circumference, i.e., when $k^* = \frac{2\pi R_c}{\lambda_d} < 1$ the column may break (Premnath and Abraham 2005a).

   This test problem is illustrated in Premnath and Abraham (2005a). Here some simulations are performed to validate our code. Our simulations show that when $k^* > 1$, the liquid will not break. Four cases with $k^* < 1$ are tested. In our simulations the densities of the liquid and gas are $\rho_l = 0.12$ and $\rho_g = 0.03$ ($\phi_g = 0.021$ and $\phi_l = 0.247$), $\kappa = 0.05$, and the corresponding surface tension is $\sigma = 0.015\kappa$. The relaxation time constant $\tau_f = \tau_g = 0.7$. The computational domain is 300×100. The axisymmetric wavelength of a disturbance $\lambda_d = 300$ lu. The disturbance is initialized as $y(x) = R_c + A\cos(\frac{x}{2\pi\lambda_d})$, where A is the amplitude of the disturbance. In our simulations $A = 3$ lu. The cases of $R_c = 10$ lu, 20 lu, 30 lu, and 40 lu are simulated. The corresponding $k^* = 0.21, 0.42, 0.63$, and 0.84, respectively. The periodic boundary conditions are applied to the left and right boundaries (see Figure 8.6). The axisymmetric boundary condition is applied to the axis (bottom line). For the upper boundary, the macro-variables are extrapolated from the inner layer.

   Figure 8.6 shows the evolution of the liquid column. From Figure 8.6(a) to (c) we can see that at the beginning the liquid column will collapse along the slender segment and the fat segment becomes fatter. Figure 8.6(d) shows a snapshot when the liquid begins to break up into two droplets: one is the larger one, the other one is the satellite droplet. After the liquid breaks up into two parts both the main droplet and the satellite droplet oscillate until they reach equilibrium states. Figure 8.6(f) shows the equilibrium state. Some experimental and numerical works have measured the drop size distribution as a function of k^* (Lafrance 1975, Premnath and Abraham 2005a, Rutland and Jameson 1971). Here in Figure 8.7 the results of our four cases are illustrated. From Figure 8.7(a) it is found there are some smaller satellite drops in the axis. It is unclear at present whether that is physical or just numerical error. That is an open research question.

   We measured drop sizes and list them in Table 8.2. The drop size distribution as a function of wave number k^* is also plotted in Figure 8.8. In the figure the radii of drops are normalized by R_c. Our numerical result is compared with the analytical solution of Lafrance (1975). Considering that analytical work may adopt some

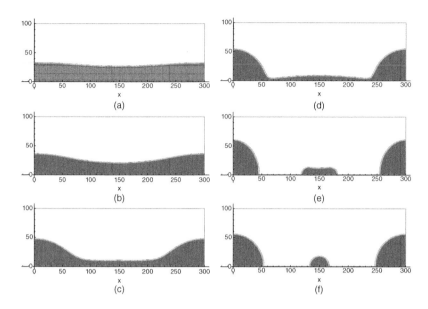

Figure 8.6   The column of cylindrical liquid breaks up into two drops under an axisymmetric disturbance (case of $R_c = 30$ lu). (a) $t = 0$, (b) $t = 12{,}000$ ts, (c) $t = 21{,}000$ ts, (d) $t = 24{,}000$ ts, (e) $t = 27{,}000$ ts, and (f) $t = 60{,}000$ ts.

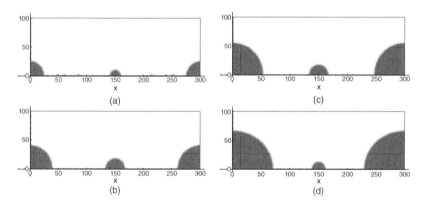

Figure 8.7   The final equilibrium states of liquid break up under an axisymmetric disturbance for four different cases. (a) $R_c = 10$ lu, (b) $R_c = 20$ lu, (c) $R_c = 30$ lu, and (d) $R_c = 40$ lu.

assumptions that are not consistent with reality (e.g., inviscid liquid), our results agree well with the analytical expectation except for a large discrepancy for satellite drop size at $k^* = 0.21$. A possible reason for this is that some of the smaller satellite drops are split from the largest satellite drop. Another possible reason is that our present computational domain is not sufficient for the case $k^* = 0.21$, which means larger domain size with larger R_c is required for accurate simulation.

Table 8.2 Measured equilibrium drop sizes for cases of cylindrical liquid break up with $\lambda_d = 300$ lu.

Case	R_c	$k^* \left(\frac{2\pi R_c}{\lambda_d} \right)$	R_1 (main drop)	R_2 (satellite drop)	$\frac{R_1}{R_c}$	$\frac{R_2}{R_c}$
a	10	0.21	24.6	9.2	2.46	0.92
b	20	0.42	39.6	18.5	1.98	0.93
c	30	0.63	54.2	18.2	1.81	0.61
d	40	0.84	68.5	12.7	1.71	0.32

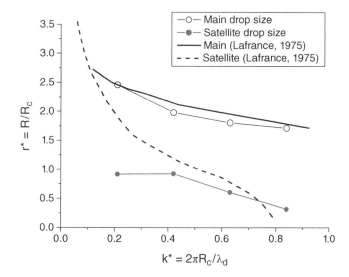

Figure 8.8 Theoretical and simulated (plotted as points from Table 8.2) main and satellite drop sizes as function of wave number k^*.

8.6 Droplet collision

Droplet collision was simulated in Premnath and Abraham (2005a,b) using the axisymmetric HCZ model. In the context of the axisymmetric model we are only interested in the head-on collision, in which the separation distance between the droplet centers is parallel to the relative velocity **U**.

In this section the Reynolds number is defined as

$$Re = \frac{\rho_l DU}{\eta},$$ (8.33)

the Weber number is

$$We = \frac{\rho_l DU^2}{\sigma},$$ (8.34)

and the Ohnesorge number Oh is

$$Oh = \frac{16\eta}{\sqrt{\rho_l D\sigma/2}}.$$ (8.35)

From those definitions we immediately found that

$$Oh = \frac{16\sqrt{2We}}{Re},$$ (8.36)

which means that only two of the above non-dimensional numbers are independent.

Ashgriz and Poo (1990) investigated water droplets colliding in one atmosphere pressure air. They found that the head-on collision may result in permanent coalescence or separation. The permanent coalescence occurs when We is smaller than a critical value, which may depend on the impact parameters (density ratio, viscosity ratio). At larger We the initially merged droplets would eventually split apart and, simultaneously, smaller satellite droplets may appear (Qian and Law 1997).

First, a case of permanent coalescence ($Re = 171$ and $We = 21$) was simulated. The computational domain is 640×180. The diameters of the two droplets are $D = 80$ lu. The densities of the liquid and gas are $\rho_l = 0.12$ and $\rho_g = 0.03$, respectively. Initially, the two droplets collide head-on with relative velocity $U = 0.06$ lu/ts, which means the left droplet initially moves with velocity $\frac{U}{2}$ while the right one moves with $\frac{-U}{2}$. In our simulation $\tau_l = \tau_g = 0.584$ and $\kappa = 0.11$. Figure 8.9 shows the evolution of the coalescence. The time is normalized by $t^* = \frac{tU}{D}$. From Figure 8.9 (a) to (d), i.e., $t^* = 0$ to $t^* = 2.25$, the droplets merged into a larger deformed disk-like drop. After that the large deformed drop undergoes an oscillatory behavior. At $t^* = 9$ the large droplet is still oscillating. Finally, we can imagine that due to the viscous damping the large drop will reach an equilibrium state (a spherical drop).

A case in the separation regime was also simulated. The computational domain size, droplet size, and the densities of fluid and gas are all identical to those in the above case. The other parameters are $\kappa = 0.03$, $\tau_f = \tau_g = 0.543$, and $U = 0.07$ lu/ts. Hence, the important non-dimensional parameters in this case are $Re = 391$ and $We = 105$.

The evolution of the droplet collision is shown in Figure 8.10 and the solution is rotated about the axisymmetric axis to create 3D images of the same simulation in Figure 8.11. Here we can see that initially the two droplets merge into a larger deformed droplet. The droplet then deforms into a much thinner disk-like drop (compared to Figure 8.9). In Figure 8.10 (e) it is stretching and at $t^* = 19.25$ (Figure 8.10(f)) the drop begins to split into three parts. Later, two main droplets are separated simultaneously with the creation of a small satellite droplet between them (Figure 8.10(h)). After the snapshot of (h) the

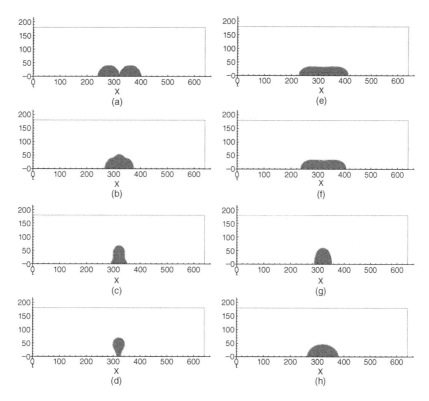

Figure 8.9 Evolution of coalescence of two droplets (case of $Re = 171$ and $We = 21$). (a) $t^* = 0$, (b) $t^* = 0.75$, (c) $t^* = 1.5$, (d) $t^* = 2.25$, (e) $t^* = 3.75$, (f) $t^* = 6$, (g) $t^* = 6.75$, and (h) $t^* = 9$. $t^* = \frac{tU}{D}$. Axes are in lattice units.

three droplets are still oscillating to reach their equilibrium states (three spherical droplets).

This case is almost identical to those illustrated in Figure 11 in Premnath and Abraham (2005a) or Figure 7 in Premnath and Abraham (2005b), which were simulated using the impact parameters $We = 100$, $Oh = 0.589$, $Re = 384.5$, and density ratio 4. Here our predicted separation time is about $t^* = 19.25$, which is consistent with the time a little bit earlier than $t^* = 19.9$ in Figure 11 in Premnath and Abraham (2005a) and $t^* = 19.59$ in Figure 7 in Premnath and Abraham (2005b).

8.6.1   Effect of gradient and Laplacian calculation

In our simulations we found that the finite difference scheme applied for gradient and Laplacian calculation plays a very important role in interface evolution.

Let us review where should we calculate the gradient and Laplacian. For the Laplacian, the only term is $\partial_\delta^2 \phi$ in Eq. (8.8). For the gradient there are several terms:

- $\kappa\phi\partial_\alpha(\partial_\delta^2\phi)$ in Eq. (8.8),

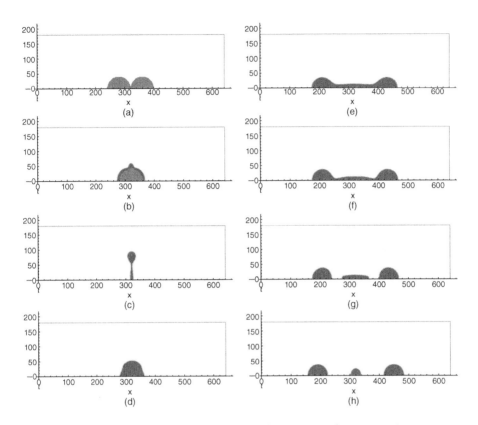

Figure 8.10 Evolution of droplet collision (case of $Re = 391$ and $We = 105$, in separation regime). The MRT version is used in the simulation. (a) $t^* = 0$, (b) $t^* = 0.875$, (c) $t^* = 3.5$, (d) $t^* = 7.875$, (e) $t^* = 17.5$, (f) $t^* = 19.25$, (g) $t^* = 21$, and (h) $t^* = 31.5$.

- $\kappa\phi\partial_\alpha\left(\frac{1}{r}\partial_r\phi\right)$ in Eq. (8.8),
- $E_\beta = -\partial_\beta[\psi(\rho)]$ in Eq. (8.17),
- $F'_\alpha = -\partial_\alpha[\psi(\phi)]$ in Eq. (8.22),
- $\partial_r u_\alpha$ and $\partial_\alpha\left(\frac{u_r}{r}\right)$ in Eq. (8.5).

First, we would like to see how Premnath and Abraham (2005a) handle the gradient calculation. In Premnath and Abraham (2005a) a fourth-order finite difference scheme was used to calculate the gradient:

$$\partial_\alpha\zeta = \frac{1}{36\Delta x}\sum_{i=1}^{8}[8\zeta(\mathbf{x}+\mathbf{e}_i\Delta t) - \zeta(\mathbf{x}+2\mathbf{e}_i\Delta t)]\frac{e_{i\alpha}}{c}, \qquad (8.37)$$

where ζ represents a macro-variable. In McCracken and Abraham (2005) gradients are calculated using a hybrid method where the sixth-order central difference method gets 75% weighting and is given by

$$\partial_x\zeta = \frac{1}{60\Delta x}[-\zeta_{(i-3,j)} + 9\zeta_{(i-2,j)} - 45\zeta_{(i-1,j)} + 45\zeta_{(i+1,j)} - 9\zeta_{(i+2,j)} + \zeta_{(i+3,j)}]. \qquad (8.38)$$

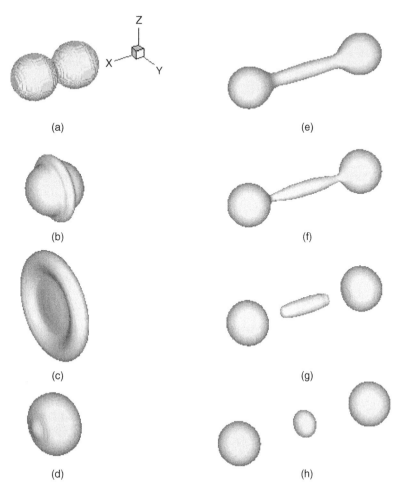

Figure 8.11   3D view of Figure 8.10 (case of $Re = 391$ and $We = 105$). (a) $t^* = 0$, (b) $t^* = 0.875$, (c) $t^* = 3.5$, (d) $t^* = 7.875$, (e) $t^* = 17.5$, (f) $t^* = 19.25$, (g) $t^* = 21$, and (h) $t^* = 31.5$.

A 25% weighting is given to a fourth-order method in Eq. (8.37).

For the LaplacianindexLaplacian, a second-order finite difference scheme was used (Premnath and Abraham 2005a):

$$\partial_\alpha^2 \zeta = \frac{1}{3\Delta x^2} \sum_{i=1}^{8} [\zeta(\mathbf{x} + \mathbf{e}_i \Delta t) - \zeta(\mathbf{x})].$$ (8.39)

In McCracken and Abraham (2005) the Laplacian is calculated through

$$\partial_\alpha^2 \zeta = \frac{1}{6\Delta x^2} [\zeta_{(i+1,j+1)} + \zeta_{(i+1,j-1)} + \zeta_{(i-1,j+1)} + \zeta_{(i-1,j-1)} - 20\zeta_{(i,j)}]$$

$$+ \frac{4}{6\Delta x^2} [(\zeta_{i,j+1} + \zeta_{i,j-1} + \zeta_{(i+1,j)} + \zeta_{(i-1,j)})],$$ (8.40)

which can also be written concisely as:

$$\partial_\alpha^2 \zeta = \frac{c^2}{c_s^2 \Delta x^2} \sum_{i=1}^{8} w_i [\zeta(\mathbf{x} + \mathbf{e}_i \Delta t) - 2\zeta(\mathbf{x}) + \zeta(\mathbf{x} - \mathbf{e}_i \Delta t)]$$

$$= \frac{2c^2}{c_s^2 \Delta x^2} \sum_{i=1}^{8} w_i [\zeta(\mathbf{x} + \mathbf{e}_i \Delta t) - \zeta(\mathbf{x})], \tag{8.41}$$

where w_i is the weighting coefficient in the equilibrium distribution function
(Eq. (8.15) or (8.16)). Premnath and Abraham (2005b) only cited the above
two references (Premnath and Abraham 2005a, McCracken and Abraham 2005).
Hence, we believe that the finite difference calculation of gradient in their simulations is at least fourth-order. It is reasonable to assume that Eq. (8.39) is used
to calculate the Laplacian in Premnath and Abraham (2005b) since McCracken
and Abraham (2005) and Premnath and Abraham (2005b) are by the same
authors.

In our simulations we adopt Eqs (8.39) and (8.37) for the Laplacian and
gradient calculations, respectively. The above collision cases (Figures 8.9 and
8.10) are all simulated using this strategy. If the gradient calculation adopts
higher-order finite difference schemes as suggested in McCracken and Abraham
(2005), i.e., the hybrid method including 75% of the sixth-order central difference plus 25% of the fourth-order difference, the result is almost identical to
the result with only fourth-order difference (Figure 8.10). Results show that the
calculation of the Laplacian using Eq. (8.39) or Eq. (8.40) makes no difference.
That is easy to understand because both Laplacian calculations are second-order
accurate.

Here we are curious about why at least a fourth-order finite difference scheme
has to be adopted in the simulations since the LBM is a well-known second-order
spatial method. Hence, a second-order finite difference scheme may be more
compatible with the LBM. Usually, for the second-order finite difference calculation, the following formula can be adopted:

$$\partial_x \zeta = \frac{1}{3\Delta x}[\zeta_{(i+1,j)} - \zeta_{(i-1,j)}]$$

$$+ \frac{1}{12\Delta x}[\zeta_{(i+1,j+1)} - \zeta_{(i-1,j-1)} + \zeta_{(i+1,j-1)} - \zeta_{(i-1,j+1)}]. \tag{8.42}$$

It can also be written concisely as:

$$\partial_\alpha \zeta = \frac{c}{2c_s^2 \Delta x} \sum_{i=1}^{8} w_i e_{i\alpha} [\zeta(\mathbf{x} + \mathbf{e}_i \Delta t) - \zeta(\mathbf{x} - \mathbf{e}_i \Delta t)]$$

$$= \frac{1}{c_s^2 \Delta t} \sum_{i=1}^{8} w_i e_{i\alpha} \zeta(\mathbf{x} + \mathbf{e}_i \Delta t). \tag{8.43}$$

It is worth mentioning that using the concise form of the first derivative,
the truncation error of the finite difference is easy to determine using Taylor
expansion and the lattice tensor properties of the D2Q9 model. For example, the

calculation of the first derivative in Eq. (8.37) can be analyzed as

$$
\partial_\alpha \zeta = \frac{8}{36\Delta x} \sum_{i=1}^{8} \frac{e_{i\alpha}}{c} \left[\zeta + (\Delta t) e_{i\beta} \partial_\beta \zeta + \frac{(\Delta t)^2}{2} e_{i\beta} e_{i\gamma} \partial_\beta \partial_\gamma \zeta \right]
$$

$$
+ \frac{8}{36\Delta x} \sum_{i=1}^{8} \frac{e_{i\alpha}}{c} \left[\frac{(\Delta t)^3}{6} e_{i\beta} e_{i\gamma} e_{i\delta} \partial_\beta \partial_\gamma \partial_\delta \zeta + \frac{(\Delta t)^4}{24} e_{i\beta} e_{i\gamma} e_{i\delta} e_{i\theta} \partial_\beta \partial_\gamma \partial_\delta \partial_\theta \zeta + \dots \right]
$$

$$
- \frac{1}{36\Delta x} \sum_{i=1}^{8} \frac{e_{i\alpha}}{c} \left[\zeta + (2\Delta t) e_{i\beta} \partial_\beta \zeta + \frac{(2\Delta t)^2}{2} e_{i\beta} e_{i\gamma} \partial_\beta \partial_\gamma \zeta \right]
$$

$$
- \frac{1}{36\Delta x} \sum_{i=1}^{8} \frac{e_{i\alpha}}{c} \left[\frac{(2\Delta t)^3}{6} e_{i\beta} e_{i\gamma} e_{i\delta} \partial_\beta \partial_\gamma \partial_\delta \zeta + \frac{(\Delta t)^4}{24} e_{i\beta} e_{i\gamma} e_{i\delta} e_{i\theta} \partial_\beta \partial_\gamma \partial_\delta \partial_\theta \zeta + \dots \right]
$$

$$(8.44)$$

According to the definition of the D2Q9 velocity model, due to symmetry the summation of an odd number of $\mathbf{e}_i$ is zero, for example

$$
\sum_i e_{i\alpha} = 0, \quad \sum_i e_{i\alpha} e_{i\beta} e_{i\gamma} = 0. \tag{8.45}
$$

The following summations of lattice tensors can also be obtained:

$$
\sum_{i=1}^{4} e_{i\alpha} e_{i\beta} = 2c^2 \delta_{\alpha\beta}, \quad \sum_{i=5}^{8} e_{i\alpha} e_{i\beta} = 4c^2 \delta_{\alpha\beta},
$$

$$
\sum_{i=1}^{4} e_{i\alpha} e_{i\beta} e_{i\gamma} e_{i\gamma} = 2c^4 \delta_{\alpha\beta\gamma\delta},
$$

$$
\sum_{i=5}^{8} e_{i\alpha} e_{i\beta} e_{i\gamma} e_{i\gamma} = 4c^4 (\delta_{\alpha\beta} \delta_{\gamma\beta} + \delta_{\gamma\beta} \delta_{\alpha\delta} + \delta_{\delta\beta} \delta_{\gamma\alpha}) - 8c^4 \delta_{\alpha\beta\gamma\delta}. \tag{8.46}
$$

Hence, from Eq. (8.44) we have

$$
\partial_\alpha \zeta = \frac{8}{36c\Delta x} \sum_{i=1}^{8} [0 + (\Delta t) e_{i\alpha} e_{i\beta} \partial_\beta \zeta + 0]
$$

$$
+ \frac{8}{36c\Delta x} \sum_{i=1}^{8} \left[\frac{(\Delta t)^3}{6} e_{i\alpha} e_{i\beta} e_{i\gamma} e_{i\delta} \partial_\beta \partial_\gamma \partial_\delta \zeta + 0 + O(c^6 \Delta t^5) + \dots \right]
$$

$$
- \frac{1}{36c\Delta x} \sum_{i=1}^{8} [0 + (2\Delta t) e_{i\alpha} e_{i\beta} \partial_\beta \zeta + 0]
$$

$$
- \frac{1}{36c\Delta x} \sum_{i=1}^{8} \left[\frac{(2\Delta t)^3}{6} e_{i\alpha} e_{i\beta} e_{i\gamma} e_{i\delta} \partial_\beta \partial_\gamma \partial_\delta \zeta + 0 + O(c^6 \Delta t^5) + \dots \right]
$$

$$
= \partial_\alpha \zeta + O(\Delta x^4). \tag{8.47}
$$

We can see that the finite difference formula (Eq. (8.37)) is fourth-order accurate. Through a similar analysis we know that Eq. (8.43) has second-order accuracy.

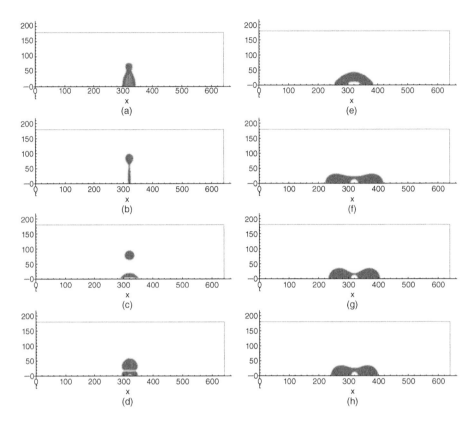

Figure 8.12   Evolution of droplet collision (case of $Re = 391$ and $We = 105$). In the MRT simulation the gradient is calculated using a second-order finite difference scheme. (a) $t^* = 1.75$, (b) $t^* = 3.5$, (c) $t^* = 7$, (d) $t^* = 12.25$, (e) $t^* = 14$, (f) $t^* = 19.25$, (g) $t^* = 24.5$, and (h) $t^* = 26.25$.

If the gradient calculation is not the above fourth-order formula Eq. (8.37), but rather the above second-order formula, what will happen? Figure 8.12 shows the result of a case where $Re = 391$ and $We = 105$ with identical parameters to those in Figure 8.10. Through comparison between Figure 8.10 and Figure 8.12 it is found that the evolutions are very different. From Figure 8.12(c) and (d) we can see that gas is trapped in the drop. In frame (g) the separation is most likely to appear at time $t^* = 24.5$ because the neck is then the narrowest. However, the expected separation does not happen. After $t^* = 26.25$ the coalesced drop oscillates to reach an equilibrium spherical state. Hence, in Figure 8.12 there is only coalescence, and separation is not observed, which is very different from Figure 8.10 and experimental observation (Ashgriz and Poo 1990). The fourth-order accuracy code for the required first-order derivatives is provided in the subroutine below.

```
      subroutine fourthorder_derivative(lx,ly,lam, lam_x, lam_y)
! 'lam' is a macro-variable: an array with dimension lx \times ly.
```

```
! to calculate the gradient $\partial_x lam$, and $\partial_y lam$
      implicit none
      include "head.inc"
      integer lx,ly, x, y, k, xp,yp, xn, yn, xd, yd
      integer xp2, xn2, yp2, yn2, xd2, yd2 , ex(0:8),ey(0:8)
      real*8 lam(lx,ly), lam_x(lx,ly), lam_y(lx,ly)
      data ex/0, 1, 0, -1, 0, 1, -1, -1, 1/,   ! e_ix
     &   ey/0, 0, 1, 0, -1, 1, 1, -1, -1/    ! e_iy

      do 20 y = 1, ly
      do 21 x = 1, lx
      xp = x+1
      yp = y+1
      xn = x-1
      yn = y-1
      xp2 = x+2
      yp2 = y+2
      xn2 = x-2
      yn2 = y-2
      if (xp.gt.lx ) xp = 1
      if (xn.lt.1  ) xn = lx
      if (xp2.gt.lx) xp2 = xp2-lx
      if (xn2.lt.1 ) xn2 = xn2+lx

c--- fourth-order accuracy
      if(y.eq. 1 .or. y.eq. ly) then
          lam_x(x,y) = (8.d0*( lam(xp,y)- lam(xn,y) ) -
     &    ( lam(xp2,y)- lam(xn2,y) )    ) /12.d0
      elseif(y.eq. 2 .or. y.eq. ly-1) then
          lam_x(x,y) = (8.d0*( lam(xp,y)- lam(xn,y) ) -
     &    ( lam(xp2,y)- lam(xn2,y) )    ) /12.d0
      else
        lam_x(x,y) = 0.d0
        do k =1, 8
        xd = x+ex(k)
        yd = y+ey(k)
        xd2 = x+2*ex(k)
        yd2 = y+2*ey(k)
        if(xd  .gt. lx) xd = xd-lx
        if(xd2 .gt. lx) xd2= xd2-lx
        if(xd  .lt. 1)  xd = xd+lx
        if(xd2 .lt. 1)  xd2= xd2+lx

        lam_x(x,y) = lam_x(x,y)+ xv(k)*(8.d0* lam(xd,yd)
     &  -lam(xd2,yd2)    )
! Fourth-order finite difference scheme in x direction
        enddo
        lam_x(x,y) =1.d0/36.d0*lam_x(x,y)
      endif

c------------------------
      if(y.eq. 1) then
        lam_y(x,y) = lam(x,2)-lam(x,1)
      elseif (y.eq. 2 .or. y.eq. ly-1) then
        lam_y(x,y) =
     &  ( (lam(x ,yp)- lam(x ,yn) )/3.d0
```

```
      &      +(lam(xp,yp)- lam(xn,yn) )/12.d0
      &      +(lam(xn,yp)- lam(xp,yn) )/12.d0   )
! Here locally for layer y=2, and y=ly-1, the finite difference is
! second-order.
          elseif (y.eq. ly) then
            lam_y(x,y) = lam_y(x,y-1)
          else
            lam_y(x,y) = 0.d0
            do k =1, 8
              xd = x+ex(k)
              yd = y+ey(k)
              xd2 = x+2*ex(k)
              yd2 = y+2*ey(k)
              if(xd .gt.  lx) xd = xd-lx
              if(xd2 .gt. lx) xd2= xd2-lx
              if(xd .lt.   1) xd = xd+lx
              if(xd2 .lt.  1) xd2= xd2+lx

              lam_y(x,y) = lam_y(x,y)+ yv(k)*(8.d0* lam(xd,yd)
      &              -lam(xd2,yd2)     )
! Fourth-order finite difference scheme in y direction
            enddo
              lam_y(x,y) =1.d0/36.d0*lam_y(x,y)
        endif
   21 continue
   20 continue

      end
```

8.6.2 Effect of BGK and MRT

Premnath and Abraham (2005b) used the MRT version of the axisymmetric HCZ model to illustrate the case where $Re = 391$ and $We = 105$. Premnath and Abraham (2005a) show the same case ($Re = 391$ and $We = 105$), but do not mention if the simulation result is coming from the MRT version. However, Premnath and Abraham (2005a) only mentioned the BGK model and that suggests that only the BGK model is used. Comparing Figure 11 in Premnath and Abraham (2005a) and Figure 7 in Premnath and Abraham (2005b), we find that the evolutions are almost identical. Hence, it seems that in these studies (Premnath and Abraham (2005a) and Premnath and Abraham (2005b)) results from the BGK and MRT versions are consistent.

However, we found that in our simulations the BGK results in a significantly different evolution path for the case for $Re = 391$ and $We = 105$. The result from the BGK simulation is shown in Figure 8.13. It is seen that although the initially coalesced drop separates at $t^* = 24.5$ (Figure 8.13(g)), the separated droplets quickly and permanently coalesced. The evolution is significantly different from that obtained from the MRT simulation (Figure 8.10). From Figure 8.13(c) and (d) we can see that gas bubbles are trapped in the drop. The entrapment is understood as follows: from (b) to (c) the disk-like drop splits into two parts at the neck. One part becomes a torus and the other part is a small satellite drop crossing the

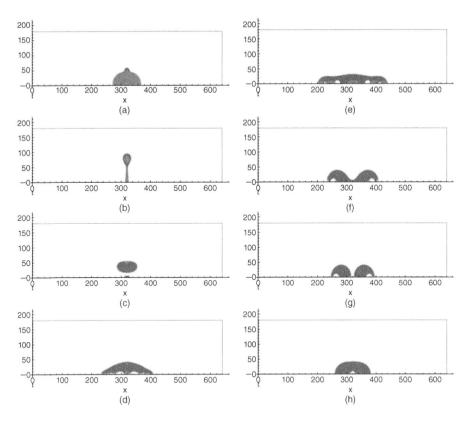

Figure 8.13 Evolution of droplet collision (case of $Re = 391$ and $We = 105$, BGK simulation). (a) $t^* = 0.875$, (b) $t^* = 3.5$, (c) $t^* = 7.875$, (d) $t^* = 9.625$, (e) $t^* = 12.25$, (f) $t^* = 21$, (g) $t^* = 24.5$, and (h) $t^* = 43.75$.

axisymmetric axis. The toroidal part then contracts and coalesces with the satellite one. The entrapment of gas obviously occurs when the toroidal part coalesces with the satellite one.

Premnath and Abraham (2005b) argue that the bubble entrapment is a physical effect and not a numerical artifact because it has been observed experimentally by Ashgriz and Poo (1990). Premnath and Abraham (2005b) only mentioned the entrapment of gas due to the formation of curved interfaces on the approaching sides of the drops. However, from Figure 6 in Premnath and Abraham (2005b) frame (d) to (e) we can see that the entrapment of gas is likely due to the mechanism described in the last paragraph rather than entrapment inside curved interfaces on the approaching sides of the drops.

We are curious as to whether the mechanism we described above is physical. At the present time this is an open question. More experimental and numerical studies need to be carried out to answer it.

The above question is also relative to understanding why the BGK result is not consistent with the MRT one. Effectively, the two Lattice Boltzmann equations (equations about $\bar{g}_i$ and $\bar{f}_i$) solve the macroscopic N–S equations and the CH equation, respectively. Using the MRT Lattice Boltzmann equation to solve the N–S equations should work. However, it is an unresolved question if using the MRT Lattice Boltzmann equation to solve the CH equation is superior than the BGK Lattice Boltzmann equation.

8.7   A revised axisymmetric HCZ model (Huang et al. 2014a)

In the following, our axisymmetric model (Huang et al. 2014a) based on He et al. (1999) will be introduced in detail. Our revision of the axisymmetric model is based on two improvements. One is the surface tension calculation, the other is mass conservation.

Basically, formulae in this revised model are similar to those in Premnath and Abraham (2005a) except that there are some typographical errors in Premnath and Abraham (2005a). The typographical errors are a term $-\eta\frac{u_r}{r^2}$ is missing in the right-hand side of Eq. (2) in Premnath and Abraham (2005a), and the viscous term on the right-hand side of both Eq. (16) and Eq. (55) in Premnath and Abraham (2005a) should be $\eta\left\{\frac{\partial_r u_\alpha}{r} - \frac{u_r}{r^2}\delta_{\alpha r} + \partial_\alpha\left(\frac{u_r}{r}\right)\right\}$ instead of $\frac{\eta}{r}[\partial_r u_i + \partial_i u_r]$.

8.7.1   MRT collision
In higher Re cases, for example $Re > 50$, smaller τ_1 is preferred. However, if the above BGK collision is used, numerical simulation may become unstable. Here, the MRT collision model of McCracken and Abraham (2005) and Lallemand and Luo (2000) is used in the simulations.

The collision terms $-\frac{1}{\tau_1}(\bar{g}_i - \bar{g}_i^{eq})$ and $-\frac{1}{\tau_2}(\bar{f}_i - \bar{f}_i^{eq})$ in Eqs (8.11) and (8.12) should be replaced by the MRT collision model (McCracken and Abraham 2005), that is,

$$\bar{g}_i(\mathbf{x} + \mathbf{e}_i\Delta t, t + \Delta t) = \bar{g}_i(\mathbf{x}, t) - \sum_j \bar{\mathbf{S}}_{1ij}(\bar{g}_j - \bar{g}_j^{eq}) + Q_i\Delta t - \frac{\Delta t}{2}\sum_j \bar{\mathbf{S}}_{1ij}Q_j, \qquad (8.48)$$

where $\bar{\mathbf{S}}_1 = \mathbf{M}^{-1}\hat{\mathbf{S}}_1\mathbf{M}$, $Q_i = S_i/(1 - \frac{1}{2\tau_1})$, and

$$\bar{f}_i(\mathbf{x} + \mathbf{e}_i\Delta t, t + \Delta t) = \bar{f}_i(\mathbf{x}, t) - \sum_j \bar{\mathbf{S}}_{2ij}(\bar{f}_j - \bar{f}_j^{eq}) + Q_i'\Delta t - \frac{\Delta t}{2}\sum_j \bar{\mathbf{S}}_{2ij}Q_j', \qquad (8.49)$$

where $\bar{\mathbf{S}}_2 = \mathbf{M}^{-1}\hat{\mathbf{S}}_2\mathbf{M}$ and $Q_i' = S_i'/(1 - \frac{1}{2\tau_2})$.

The matrix $\mathbf{M}$ is illustrated in Appendix A of Chapter 4. The parameters in the MRT model are chosen as $s_0 = s_3 = s_5 = 1.0$, $s_1 = 1.64$, $s_2 = 1.54$, $s_4 = s_6 = 1.7$, and $s_7 = s_8 = \frac{1}{\tau}$. In the collision matrices $\hat{\mathbf{S}}_1$ and $\hat{\mathbf{S}}_2$, s_0 to s_6 are identical to those illustrated in the above except $s_7 = s_8 = \frac{1}{\tau_1}$ in $\hat{\mathbf{S}}_1$ and $s_7 = s_8 = \frac{1}{\tau_2}$ in $\hat{\mathbf{S}}_2$.

8.7.2 Calculation of the surface tension

The surface tension form, i.e., Eq. (8.7), is applicable to simulate some multiphase flows. However, this original surface tension calculation may change its direction across the bubble interface in the HCZ model. A wiggle over the interface region may appear and cause numerical instability at large density ratio (Chao et al. 2011, Kim 2005). The following improved surface tension calculation is suggested (Chao et al. 2011, Kim 2005).

$$\mathbf{F}_s = -k\nabla \cdot \left(\frac{\mathbf{n}}{|\mathbf{n}|} \right) |\nabla\phi|\nabla\phi, \tag{8.50}$$

where the normal vector $\mathbf{n}$ is defined to be $\mathbf{n} \equiv \nabla\phi$.

For axisymmetric cases there is an extra term in $\mathbf{F}_s$, that is,

$$\mathbf{F}_s = -k\nabla \cdot \left(\frac{\nabla\phi}{|\nabla\phi|} \right) |\nabla\phi|\nabla\phi - k\frac{\partial_r\phi}{r} \cdot \nabla\phi. \tag{8.51}$$

Evaluation of the terms in Eq. (8.51) using finite difference approximation is illustrated in the following.

The computational domain is partitioned into a uniform Cartesian mesh with mesh spacing Δx. Suppose the center of each cell, $\Omega_{i,j}$, is located at $(x_i, r_j) = ((i - 0.5)\Delta x, (j - 0.5)\Delta x)$. The cell vertices are located at $(x_{i+\frac{1}{2}}, r_{j+\frac{1}{2}}) = (i\Delta x, j\Delta x)$.

The discretized form of Eq. (8.51) can be written as

$$\mathbf{F}_s = -k\nabla_d \cdot \left(\frac{\nabla\phi}{|\nabla\phi|} \right)_{ij} |\nabla_d\phi_{ij}|\nabla_d\phi_{ij} - k\frac{\partial_r\phi}{r} \cdot \nabla_d\phi_{ij}, \tag{8.52}$$

where ∇_d is a finite difference approximation to the divergence operator.

The normal vector at a vertex can be obtained by differentiating the phase field over the centers of the four surrounding cells (Kim 2005). For example, the normal vector $\mathbf{n}_{i+\frac{1}{2},j+\frac{1}{2}} \equiv (\nabla\phi)_{i+\frac{1}{2},j+\frac{1}{2}}$ at the top right vertex of cell $\Omega_{i,j}$ is

$$\mathbf{n}_{i+\frac{1}{2},j+\frac{1}{2}} = \left(n^x_{i+\frac{1}{2},j+\frac{1}{2}}, n^r_{i+\frac{1}{2},j+\frac{1}{2}} \right)$$

$$= \left(\frac{\phi_{i+1,j} + \phi_{i+1,j+1} - \phi_{i,j} - \phi_{i,j+1}}{2\Delta x}, \frac{\phi_{i,j+1} + \phi_{i+1,j+1} - \phi_{i,j} - \phi_{i+1,j}}{2\Delta x} \right). \tag{8.53}$$

The curvature of the interface $s(\phi_{i,j})$ at a cell center (i, j) is calculated from the above vertex-centered normals:

$$s(\phi_{i,j}) = \nabla_d \cdot \left(\frac{\mathbf{n}}{|\mathbf{n}|} \right)_{i,j} = \frac{1}{2\Delta x} \frac{n^x_{i+\frac{1}{2},j+\frac{1}{2}} + n^r_{i+\frac{1}{2},j+\frac{1}{2}}}{\left| \mathbf{n}_{i+\frac{1}{2},j+\frac{1}{2}} \right|}$$

$$+ \frac{1}{2\Delta x} \left(\frac{n^x_{i+\frac{1}{2},j-\frac{1}{2}} - n^r_{i+\frac{1}{2},j-\frac{1}{2}}}{\left| \mathbf{n}_{i+\frac{1}{2},j-\frac{1}{2}} \right|} - \frac{n^x_{i-\frac{1}{2},j+\frac{1}{2}} - n^r_{i-\frac{1}{2},j+\frac{1}{2}}}{\left| \mathbf{n}_{i-\frac{1}{2},j+\frac{1}{2}} \right|} - \frac{n^x_{i-\frac{1}{2},j-\frac{1}{2}} + n^r_{i-\frac{1}{2},j-\frac{1}{2}}}{\left| \mathbf{n}_{i-\frac{1}{2},j-\frac{1}{2}} \right|} \right). \tag{8.54}$$

The cell-centered normal is the average of the vertex normals,

$$\nabla_d \phi_{ij} = \frac{1}{4}\left(\mathbf{n}_{i+\frac{1}{2},j+\frac{1}{2}} + \mathbf{n}_{i+\frac{1}{2},j-\frac{1}{2}} + \mathbf{n}_{i-\frac{1}{2},j+\frac{1}{2}} + \mathbf{n}_{i-\frac{1}{2},j-\frac{1}{2}}\right). \tag{8.55}$$

8.7.3 Mass correction

In the original HCZ model without mass correction the mass of a rising bubble may increase with time, which is an unfavorable behavior (Chao et al. 2011). Of course, the volume of a rising bubble is expected to increase as lower confining pressures are encountered. The following corrections assume that the density of gas is constant and any pressure differences are negligible.

In the axisymmetric simulations of Huang et al. (2014a) similar unfavorable behavior in bubble rise is also observed. Hence, a mass correction step is necessary to ensure mass conservation. The volume of the bubble can be corrected using the scheme of Son (2001),

$$\frac{\partial \phi}{\partial \tau_3} = (V - V_0)|\nabla\phi|, \tag{8.56}$$

where V is the bubble volume before the correction. $V_0 = \frac{1}{6}\pi D^3$ is the initial volume of the bubble, where D is the initial diameter of the bubble. τ_3 is an artificial time. The equation is computed after each streaming step until the steady state $V = V_0$ is reached.

In the implementation V is determined in the following way. For a computational node $\mathbf{x}$, if $\rho(\mathbf{x}) < \frac{\rho_l + \rho_g}{2}$, the node is supposed to be occupied by gas and it is labeled as $\mathbf{x}_g$. The formula calculation of the volume (integration scheme) is

$$V = \sum_{\mathbf{x}_g} 2\pi r \Delta r \Delta x, \tag{8.57}$$

where $\Delta x = \Delta r = 1$ lu and r is the radius from $\mathbf{x}_g$ to $r = 0$ in lattice units (refer to Figure 8.1). Here V has units of lu^3.

In the implementation we take the following form (Chao et al. 2011),

$$\frac{\partial \phi}{\partial \tau_3} + \mathbf{u} \cdot \nabla\phi = 0, \tag{8.58}$$

where $\mathbf{u} = (u_x, u_r) = -(V - V_0)\nabla\phi/|\nabla\phi|$. Although a high-order scheme can be used to solve this equation (Chao et al. 2011), here for simplicity the first-order upwind finite difference scheme is used to discretize the convection term, i.e., $\mathbf{u} \cdot \nabla\phi$. The first-order Euler scheme is used to discretize the time derivative. For example, if $u_x > 0$, $\partial_x^- \phi = \frac{\phi_{(i,j)} - \phi_{(i-1,j)}}{\Delta x}$, if $u_x < 0$, $\partial_x^+ \phi = \frac{\phi_{(i+1,j)} - \phi_{(i,j)}}{\Delta x}$. In our simulations solving this equation the time step $d\tau_3 \sim \frac{0.15}{V_0}$ is adopted. If $u_x^+ = \max(u_x, 0)$ and $u_x^- = \min(u_x, 0)$ are defined in the x axis and the corresponding variables in the r direction are defined similarly, then the discretized form can be written as

$$\phi_{(i,j)}^{\tau+d\tau_3} = \phi_{(i,j)}^{\tau} - (u_x^+ \partial_x^- \phi + u_x^- \partial_x^+ \phi)d\tau_3 - (u_r^+ \partial_r^- \phi + u_r^- \partial_r^+ \phi)d\tau_3. \tag{8.59}$$

Huang et al. (2014a) has shown that the mass conservation property of this model is much better than that in Premnath and Abraham (2005a). With the mass conservation step, the bubble volume is kept almost constant with very small oscillation. However, without the mass correction step the bubble continuously grows (Huang et al. 2014a).

The filter technique mentioned in Chao et al. (2011) seems not to be critical in the scheme (Huang et al. 2014a). The key point of the technique is to smoothly remove local extremes, which may be helpful for increased numerical stability. Without the technique most simulations we performed are still stable.

8.8 Bubble rise

Sankaranarayanan et al. (2002) and Gupta and Kumar (2008) used the SC model to study a rising bubble. However, in both studies the parameters are limited to a very narrow range. Besides that, in the 2D study of Gupta and Kumar (2008), the comparison between the LBM simulation and experimental data is poor in terms of bubble shape. Srivastava et al. (2013) developed the SC multiphase model for axisymmetric flows. However, it has never been tested for complex flow phenomena, e.g., bubble rise, and the mass conservation aspects are unknown.

Frank et al. (2005) used the free-energy-based LBM to simulate bubble rise. However, only cases with very small Reynolds number were simulated and corresponding terminal spherical and oblate ellipsoidal bubbles were observed (Frank et al. 2005). Cheng et al. (2010) used a free-energy-based model (Zheng et al. 2006) to study 3D bubble–bubble interactions. However, because the model they used is not able to include the density contrast effect, the result is limited to density-matched cases.

Recently, a model (Amaya-Bower and Lee 2010) based on the model of He et al. (1999) has been further developed to handle higher density ratio bubble rise. The parameter range study (Amaya-Bower and Lee 2010) is much wider than the other related LBM studies.

However, in terms of terminal bubble shape, the results of Amaya-Bower and Lee (2010) have large discrepancies with the experimental data. For example, Figure 8.14 shows some comparisons of experimental and simulated bubbles from Table 2 of Amaya-Bower and Lee (2010). For case A4, the bubble is a spherical cap in the experiment (Bhaga and Weber 1981). However, the simulated result is a toroidal bubble, which is very different from the spherical cap bubble. They attributed the discrepancy to grid resolution (Amaya-Bower and Lee 2010), but it is difficult to explain why the aspect ratio of the simulated toroidal bubble is very different from the experimental one. For the cases A5 and A6, basically the terminal bubble shapes are skirted with a rounded lower edge. However, for cases with high *Re*, the rounded lower edge should become sharper (refer to Figure 8.16).

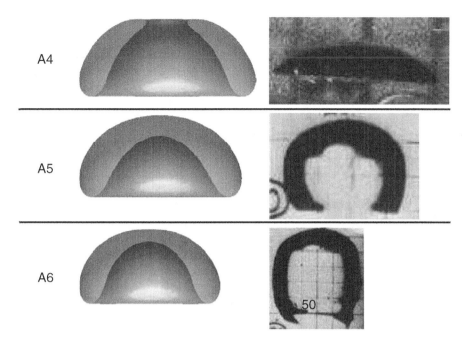

Figure 8.14 Comparison of LBM results (Amaya-Bower and Lee (2010), left-hand column) and experimental data (Bhaga and Weber 1981, right-hand column). The figure is copied from Table 2 of Amaya-Bower and Lee (2010). The Eötvös number and Morton number of cases A4, A5, and A6 (upper, middle, and lower rows) are listed in Table 8.3. Source: Amaya-Bower and Lee (2010). Reproduced with permission of Elsevier.

Although some other studies based on the HCZ model or Lee–Lin model (Lee and Lin 2005) for the axisymmetric two-phase flows have been carried out (McCracken and Abraham 2005, Mukherjee and Abraham 2007c, Premnath and Abraham 2005a), the validation cases are mainly focussed on very simple droplet flow problems, for example droplet oscillation and droplet collision. In Section 8.8.1 simulation results using the model (Premnath and Abraham 2005a) are poor when they are compared with experimental data. In terms of mass conservation, these models (including the Lee–Lin model) based on the HCZ model are believed to be not so satisfactory for bubble rise as they are developed from the original HCZ model, in which the mass conservation has been questioned (Chao et al. 2011).

Based on the HCZ model and a technique to ensure mass conservation, Huang et al. (2014a) developed an axisymmetric HCZ model to simulate the bubble rise problem. As we mentioned previously, the revised model is based on two aspects. One is the calculation of surface tension, the other is mass conservation. In Huang et al. (2014a) all typical bubble shapes observed in experiments (Bhaga and Weber 1981) are compared in detail with simulation results. The effect of the initial bubble shape reported in the literature is reproduced correctly.

Table 8.3 Cases for bubble rising ($\sigma = 0.011\kappa$, $v = (\tau - 0.5)c_s^2 \Delta t$, $D = 100$ lu, tube diameter $D_T = 5D$).

Case	Eo	Mo	Re.*	τ_l	κ	g	Re. (LBM)
A1	17.7	711	0.232	0.862	0.0005	4.06×10^{-8}	0.288
A2	32.2	8.2×10^{-4}	55.3	0.54081	0.008	1.18×10^{-6}	48.4
A3	243	266	7.77	0.9155	0.004	4.46×10^{-6}	6.93
A4	115	4.63×10^{-3}	94.	0.536	0.005	2.64×10^{-6}	83.7
A5	339	43.1	18.3	0.710	0.003	4.66×10^{-6}	17.0
A6	641	43.1	30.3	0.646	0.002	5.88×10^{-6}	27.7
A7	116	5.51	13.3	0.712	0.005	2.66×10^{-6}	12.0
A8	114	8.6×10^{-4}	151.	0.5238	0.005	2.61×10^{-6}	131.3
A9	116	0.103	42.2	0.578	0.005	2.66×10^{-6}	36.0
B1	94.3	4.85×10^{-3}	77.9	0.5487	0.008	3.46×10^{-6}	70.2
B2	61.9	8.2×10^{-4}	99.5	0.5274	0.005	1.42×10^{-6}	82.6
B3	292	26.7	22.1	0.750	0.005	6.69×10^{-6}	16.6

*Source: Huang et al. 2014. Reproduced with permission of Elsevier.

8.8.1 Numerical validation

In Huang et al. (2014a) all typical terminal bubble shapes are observed. The terminal bubble shapes and velocities are compared with the experimental data in detail. The mass conservation property of the model, the effect of different surface tension calculations, the density effects on the terminal shape, and velocity are investigated (Huang et al. 2014a). The axisymmetric model (Huang et al. 2014a) is demonstrated to be able to reproduce the effect of initial bubble shape in Hua and Lou (2007). In this section we briefly introduce some cases and conclusions.

In the literature usually two independent non-dimensional numbers, the Eötvös number and the Morton number, are defined for bubble rise problems (Bhaga and Weber 1981):

$$Eo = \frac{gD^2 \rho}{\sigma}, \tag{8.60}$$

$$Mo = \frac{g\eta^4}{\rho\sigma^3}. \tag{8.61}$$

The Reynolds number is based on terminal bubble rise velocity U_t and initial bubble diameter D, i.e., $Re = \frac{U_t D}{v}$, where v is the kinematic viscosity of the liquid surrounding the bubble.

Table 8.3 lists the main parameters used in our simulations. In these simulations the typical terminal bubble shapes, such as spherical, oblate ellipsoidal, disk, spherical cap, and skirted, are all observed. To compare with the experimental data, the tube diameter in our simulation is $D_T = 5D$. All of the initial bubbles are supposed to be spherical. The wobbling bubbles that usually appear at high

Reynolds number and high Eo number are not considered because wobbling is not an axisymmetric flow.

In the LBM simulations, to ensure better incompressibility conditions, the maximum velocity magnitude should not exceed 0.1 lu/ts. In Table 8.3 all terminal Res are known from experiment (Bhaga and Weber 1981), so we can calculate the expected terminal velocity in the LBM simulations. From the calculation the expected rise velocities are all less than 0.02 lu/ts, which sufficiently satisfies the incompressibility condition. For this unsteady flow problem a smaller time step is preferred. Hence, smaller τ is preferred in the simulations. On the other hand, smaller τ may induce numerical instability. The parameters are chosen according to a balance between the two constraints.

In all of the simulations in Table 8.3 we set $\rho_l = 0.247$ mu/lu^3 and $\rho_g = 0.016$ mu/lu^3, which means the maximum density ratio in our simulations is $\frac{\rho_l}{\rho_g} = 15.5$. In Table 8.3 first we can set $\kappa \approx O(0.001)$, then the desired surface tension is $\sigma = 0.011\kappa$ (the relationship is measured from the Laplace law). The acceleration of gravity is determined from

$$g = \frac{\sigma Eo}{D^2 \rho_l}. \tag{8.62}$$

The kinematic viscosity of the liquid is determined through

$$\nu_l = \left(\frac{D^2 \sigma^2 Mo}{\rho_l^2 Eo} \right)^{\frac{1}{4}}. \tag{8.63}$$

The relaxation time for liquid is $\tau_l = \frac{\nu_l}{c_s^2 \Delta t} + 0.5$. In our simulations the velocity of the top interface can be considered as the velocity of the bubble. For simplicity, the precise vertical position of the front interface is recorded and through the time derivative of the position the velocity can be determined.

Although the relaxation time for gas τ_g does not have to be identical to τ_l, here $\tau_2 = \tau_g = \tau_l$ is adopted in simulations in Table 8.3. This assumption has been demonstrated to be acceptable in terms of bubble shape and rise velocity in Huang et al. (2014a).

Our LBM results for terminal bubble rise Res are listed in the far right column of Table 8.3 and compared to the Res measured from experiments (Bhaga and Weber 1981). The measured Res from LBM simulation agree well with the experimental Res from Bhaga and Weber (1981).

For cases with very small Mo (high Re), such as A2, A4, A8, B1, and B2 in Table 8.3, numerical instability may appear if the BGK collision model is used and τ_1 and τ_2 are small. For these cases the simulations with density ratio $\frac{\rho_l}{\rho_g} = 15.5$ are not stable using the BGK collision model. Hence, the MRT collision is used for these simulations. The Res from our MRT LBM simulations are listed in Table 8.3 and have small discrepancies (about 10%) with the experimental data because the density ratio is not as high as that in the experiment (Bhaga and Weber 1981).

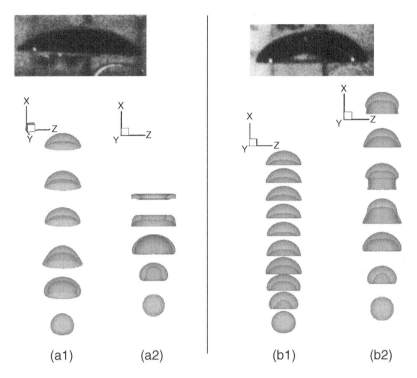

Figure 8.15 The top left and top right photographs are the experiment results (Bhaga and Weber 1981) for Cases A4 and A9, respectively. (a1) Bubble shape evolution of Case A4 with *revised* surface tension calculation, (a2) Case A4 with *original* surface tension calculation, (b1) Case A9 with *revised* surface tension calculation, and (b2) Case A9 with *original* surface tension calculation. The parameters in Cases A4 and A9 are listed in Table 8.3.

8.8.2 Surface-tension calculation effect

After revising the typographical error of Mukherjee and Abraham (2007c) it is expected that the axisymmetric model is able to simulate the bubble rise correctly. However, without the revised surface tension calculation mentioned in Section 8.7.2 the axisymmetric HCZ model (Premnath and Abraham 2005a) can only give poor results compared to the experimental data (Huang et al. 2014a).

Cases A4 and A9 with density ratio of 3 ($\rho_l = 0.24$, $\rho_g = 0.08$) were simulated to compare the effect of surface-tension calculation schemes. Results designed to simulate bubble evolution during the rise are shown in Figure 8.15. The bubble shapes are obtained through sweeping the 2D bubble shape 360° along the axisymmetric axis.

For Case A4 it is seen that with the original surface calculation the bubble becomes toroidal (refer to (a2)), which is very different from the experimental photograph shown in the top left of Figure 8.15. For Case A9 the bubble shapes in (b2) still have some discrepancies with the experimental photograph. On the contrary, the results with revised surface tension calculation ((a1) and (b1)) are all very consistent with the experiment data (Bhaga and Weber 1981).

8.8.3 Terminal bubble shape

In principle, the surface tension force tends to maintain the bubble in a spherical shape. For cases with a particular density, bubble diameter, and acceleration of gravity a high Eo number (Eq. 8.60) means low surface tension, which allows large deformation (Hua and Lou 2007). On the other hand, higher Re would induce larger deformation in the bubble's vertical direction (Hua and Lou 2007). The final bubble shape is determined by the combined effect of Eo and Re.

Figure 8.16 shows the comparison of simulated terminal bubble shapes with the experimental data (Bhaga and Weber 1981) in nine cases. The descriptions for the bubble shapes are also listed in the far right column. For Case A1, due to small Eo and Re (refer to Table 8.3), the bubble remains almost spherical when it rises. The final bubble shape in Case A2 is oblate ellipsoidal (disk- like). In Case A3 the bubble shape is an oblate ellipsoidal cap. At the lower part of the bubble there is a rounded lower edge. Because of high Re, the bubble shape in Case A4 is a spherical cap. Because of large Eo, both terminal bubble shapes in Cases A5 and A6 are skirted (cases with density ratio 3). Cases A7 and A9 are described as oblate ellipsoidal caps and Case A8 is a spherical cap (Bhaga and Weber 1981). The simulated shapes all agree well with the experimental ones (Bhaga and Weber 1981).

Hua and Lou (2007) and Huang et al. (2014a) found that the effect of density ratio on terminal velocity is more significant than it is on terminal shape. Huang et al. (2014a) shows in Case A7 that for low Eo the bubble shapes are almost unaffected by density ratios. The simulated terminal rise velocities with density ratio 15.5 are very close to the experimental ones (density ratio 1000). See Re for simulations and experiments in Table 8.3.

For the larger Eo cases the effect of density ratio is found to be more significant. Figure 8.17 illustrates the effect. For cases A5 ($Eo = 339$) and A6 ($Eo = 641$) simulation results with density ratio 3 are more consistent with the experimental data than those with density ratio 15.5. In the experiment (Bhaga and Weber 1981) the density ratio is approximately 1000.

8.8.4 Wake behind the bubble

To further validate the axisymmetric model, wakes behind the bubble are compared with experimental observations (Bhaga and Weber 1981). In the experiment H_2 tracer was used to obtain the flow visualization. Figure 8.18 shows the terminal bubble wakes in three cases: B1, B2, and B3. The main parameters used in the simulations are listed in Table 8.3. In each case the closed toroidal wake predicted by our LBM simulation agrees well with the experiment. For Case B1, which is shown in Figure 8.18(a), there are some bright spots at the lower outside of the bubble rim in the experiment photograph. These may be

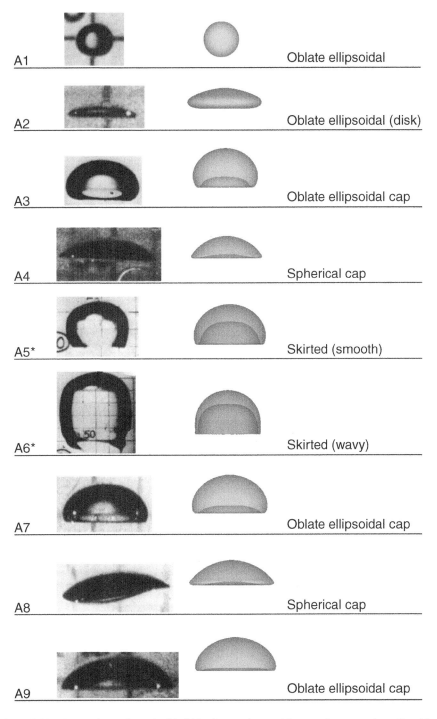

Figure 8.16 Comparison of terminal bubble shapes observed in experiments and predicted by LBM (A1–A9, for various Eo and Mo, Table 8.3, density ratios are 15.5). Left, middle, and right columns represent experiment results (Bhaga and Weber 1981), LBM results, and the descriptions of the bubble shape (Bhaga and Weber 1981), respectively. *Density ratio 3.

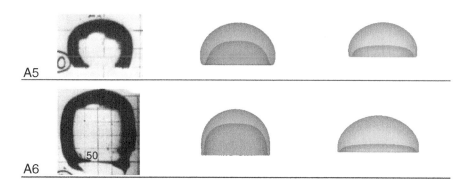

Figure 8.17 Comparison of terminal bubble shapes observed in Cases A5 and A6 with different density ratios. The left column represents experiment results (Bhaga and Weber 1981); the middle and right columns are LBM results with density ratio 3 and density ratio 15, respectively.

due to the secondary wake circulations that occur just behind the bubble rim. In our LBM simulation the two small secondary wakes are captured. In Case B2 (Figure 8.18(b)) the secondary wake is not observed. In Case B3 (Figure 8.18(c)) the secondary circulation area in the skirt bubble wake becomes larger. The LBM results are consistent with both previous numerical (Hua and Lou 2007) and experimental (Bhaga and Weber 1981) studies.

The effect of initial bubble shape was discussed in (Hua and Lou 2007). For bubble rise with lower Reynolds number, the terminal bubble shape and rise velocity are almost not affected by the initial bubble shape. However, for the high-*Re* cases the initial bubble shape significantly affects the terminal bubble shape (Hua and Lou 2007). Huang et al. (2014a) successfully reproduced the effect observed in Hua and Lou (2007) in Case A8 (density ratio 3). In the investigation (Huang et al. 2014a) two different initial bubble shapes are set for Case A8: one is an oblate bubble (aspect ratio between height and width of the initial bubble is 0.716) and the other is spherical. The result of the case with initial oblate bubble is consistent with experimental data (Bhaga and Weber 1981) and the possible reason is given by Hua and Lou (2007).

8.9 Conclusion

In this chapter axisymmetric models (McCracken and Abraham 2005, Premnath and Abraham 2005a) that are based on the HCZ model are introduced and applied to simulate some axisymmetric two-phase flows. A mass-conserving axisymmetric HCZ model (Huang et al. 2014a) is also presented. Because of the revised surface tension calculation, bubble rise shapes are consistent with experimental results and the effects of initial bubble shape were reproduced successfully. Quantitatively, the wakes behind the bubble agree well with the experimental ones and those of Hua and Lou (2007).

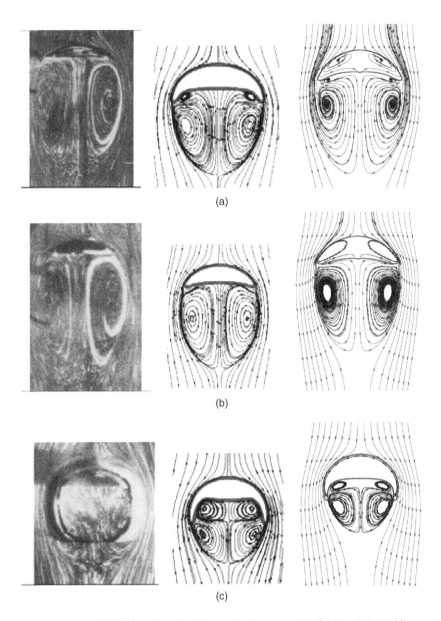

(a)

(b)

(c)

Figure 8.18 Terminal bubble wakes in (a) Case B1 (upper row), (b) Case B2 (middle row), and (c) Case B3 (bottom row). The left, middle, and right columns are results from experiment (Bhaga and Weber 1981), numerical work (Hua and Lou 2007), and our LBM, respectively. Source: Huang et al. 2014. Reproduced with permission of Elsevier.

Through adopting the mass correction step in our scheme and using a revised surface tension calculation, the axisymmetric multiphase Lattice Boltzmann model (Huang et al. 2014a) is superior to the previous one (Mukherjee and Abraham 2007c) for bubble rise simulations.

For high-density ratios we will introduce other models in Chapter 10.

8.10   Appendix A: Chapman–Enskog analysis

Applying the Taylor expansion to Eq. (8.11) and adopting the Chapman–Enskog expansion $\partial_t = \partial_{t_1} + \varepsilon\partial_{t_2} + \dots$ and $\bar{g}_i = \bar{g}_i^{(0)} + \varepsilon\bar{g}_i^{(1)} + \varepsilon^2\bar{g}_i^{(2)}$, where $\varepsilon = \Delta t$, we have

$$\varepsilon(\partial_\alpha + \varepsilon\partial_\alpha + e_{i\alpha}\partial_\alpha)\left(\bar{g}_i^{(0)} + \varepsilon\bar{g}_i^{(1)} + \varepsilon^2\bar{g}_i^{(2)}\right)$$

$$+ \frac{\varepsilon^2}{2}(\partial_\alpha + \varepsilon\partial_\alpha + e_{i\alpha}\partial_\alpha)^2\left(\bar{g}_i^{(0)} + \varepsilon\bar{g}_i^{(1)}\right)$$

$$= -\frac{1}{\tau_1}\left(\bar{g}_i^{(0)} + \varepsilon\bar{g}_i^{(1)} + \varepsilon^2\bar{g}_i^{(2)} - \bar{g}_i^{(eq)}\right) + S_i\Delta t. \tag{8.64}$$

Retaining terms to $O(\varepsilon^2)$, in a different time scale, Eq. (8.64) yields

$$O(\varepsilon): (\bar{g}_i^{(0)} - \bar{g}_i^{eq})/\tau_1 = 0, \tag{8.65}$$

$$O(\varepsilon^1): (\partial_{t_1} + e_{i\alpha}\partial_\alpha)\bar{g}_i^{(0)} + \frac{1}{\tau_1}\bar{g}_i^{(1)} - S_i = 0, \tag{8.66}$$

$$O(\varepsilon^2): \partial_{t_2}\bar{g}_i^{(0)} + \left(1 - \frac{1}{2\tau_1}\right)(\partial_{t_1} + e_{i\alpha}\partial_\alpha)\bar{g}_i^{(1)} + \frac{1}{2}(\partial_{t_1} + e_{i\alpha}\partial_\alpha)S_i + \frac{1}{\tau_1}\bar{g}_i^{(2)} = 0. \tag{8.67}$$

8.10.1   Preparation for derivation

We note Eq. (8.17) yields

$$\sum\bar{g}_i = \sum\bar{g}_i^{(0)} + \Delta t\sum\bar{g}_i^{(1)} + \Delta t^2\sum\bar{g}_i^{(2)} + \dots = \left(p_h - \frac{\Delta t}{2}u_\beta E_\beta\right) + c_s^2\frac{\rho u_r}{2r}\Delta t. \tag{8.68}$$

From Eqs (8.65) and (8.15) we have

$$\sum\bar{g}_i^{(0)} = \sum\bar{g}_i^{eq} = p_h. \tag{8.69}$$

Hence, from the above two equations we can derive

$$\sum\bar{g}_i^{(1)} = -\frac{1}{2}u_\beta E_\beta + c_s^2\frac{\rho u_r}{2r}, \quad \sum\bar{g}_i^{(2)} = 0. \tag{8.70}$$

Similarly, from Eq. (8.18) we can obtain $\sum e_{i\alpha}\bar{g}_i = \sum e_{i\alpha}\bar{g}_i^{(0)} + \Delta t\sum e_{i\alpha}\bar{g}_i^{(1)} = c_s^2(\rho u_\alpha - \frac{\Delta t}{2}F_\alpha)$. Hence, the first moments of $\bar{g}_i^{(0)}$ and $\bar{g}_i^{(1)}$ are

$$\sum e_{i\alpha}\bar{g}_i^{(0)} = c_s^2\rho u_\alpha, \quad \sum e_{i\alpha}\bar{g}_i^{(1)} = -\frac{1}{2}F_\alpha c_s^2. \tag{8.71}$$

The zeroth and first moments of the source term S_i take the form

$$\sum_i S_i = \left(1 - \frac{1}{2\tau_1}\right)\left\{u_\alpha E_\alpha - c_s^2 \frac{\rho u_r}{r}\right\}, \quad \sum_i e_{i\beta}S_i = c_s^2\left(1 - \frac{1}{2\tau_1}\right)F_\beta. \quad (8.72)$$

To be well prepared for further derivation, we also write down the second moment of $\bar{g}_i^{(0)}$ and S_i. From Eq. (8.15) we have

$$\sum_i g_i^{(0)} e_{i\alpha} e_{i\beta} = c_s^2(p_h \delta_{\alpha\beta} + \rho u_\alpha u_\beta). \quad (8.73)$$

Using Eq. (8.21) and omitting higher-order terms of $O(u^3)$ we have

$$\sum_i e_{i\alpha}e_{i\beta}S_i = c_s^2\left(1 - \frac{1}{2\tau_1}\right)\left\{E_\gamma\left(u_\alpha\delta_{\beta\gamma} + u_\beta\delta_{\alpha\gamma} + u_\gamma\delta_{\beta\alpha}\right)\right.$$

$$\left. + (F_\beta u_\alpha + F_\alpha u_\beta) - c_s^2 \frac{\rho u_r}{r}\delta_{\alpha\beta}\right\}. \quad (8.74)$$

8.10.2   Mass conservation

Summing both sides of Eq. (8.66) over i and using Eqs (8.69), (8.70), and (8.72) gives

$$\partial_{t_1} p_h + c_s^2\partial_\alpha(\rho u_\alpha) - u_\alpha E_\alpha + c_s^2 \frac{\rho u_r}{r} = (\partial_{t_1} + u_\alpha\partial_\alpha)p_h + c_s^2\left(\rho\partial_\alpha u_\alpha + \frac{\rho u_r}{r}\right) = 0. \quad (8.75)$$

We then proceed to $O(\varepsilon^2)$. Using Eqs (8.69), (8.70), (8.71), and (8.72) and summing both sides of Eq. (8.67) over i gives

$$\partial_{t_2} p_h = 0. \quad (8.76)$$

From Eqs (8.75) and (8.76) we obtain

$$(\partial_t + u_\alpha\partial_\alpha)p_h + c_s^2\left(\rho\partial_\alpha u_\alpha + \frac{\rho u_r}{r}\right) = 0. \quad (8.77)$$

Here we can see that only when

$$(\partial_t + u_\alpha\partial_\alpha)p_h = 0 \quad (8.78)$$

can the incompressible condition $\rho\partial_\alpha u_\alpha + \frac{\rho u_r}{r} = 0$ be satisfied.

8.10.3   Momentum conservation

Multiplying Eq. (8.66) by $e_{i\beta}$, summing over i, and using Eqs (8.71), (8.72) and (8.73) gives

$$O(\varepsilon) : \partial_{t_1}(\rho u_\beta) + \partial_\alpha(p_h\delta_{\alpha\beta} + \rho u_\alpha u_\beta) - F_\beta = 0. \quad (8.79)$$

Multiplying Eq. (8.67) by $e_{i\beta}$ and summing over i gives

$$O(\varepsilon^2) : c_s^2\partial_{t_2}(\rho u_\beta) + \left(1 - \frac{1}{2\tau_1}\right)\left[\partial_{t_1}\left(\sum_i e_{i\beta}\bar{g}_i^{(1)}\right) + \partial_\alpha\left(\sum_i e_{i\alpha}e_{i\beta}\bar{g}_i^{(1)}\right)\right]$$

$$+ \frac{1}{2}\left[\partial_{t_1}\left(\sum_i e_{i\beta}S_i\right) + \partial_\alpha\sum_i(e_{i\beta}e_{i\alpha}S_i)\right] = 0. \quad (8.80)$$

From Eqs (8.71) and (8.72) we know that

$$\left(1 - \frac{1}{2\tau_1}\right)\partial_{t_1}\left(\sum e_{i\beta}\bar{g}_i^{(1)}\right) = c_s^2\left(1 - \frac{1}{2\tau_1}\right)\partial_{t_1}\left(-\frac{1}{2}F_\alpha\right) = -\frac{1}{2}\partial_{t_1}\left(\sum e_{i\beta}S_i\right).$$

(8.81)

Hence, Eq. (8.80) can be simplified as

$$c_s^2\partial_{t_2}(\rho u_\beta) + \left(1 - \frac{1}{2\tau_1}\right)\partial_\alpha\sum\left(e_{i\alpha}e_{i\beta}\bar{g}_i^{(1)}\right) + \frac{1}{2}\partial_\alpha\sum(e_{i\beta}e_{i\alpha}S_i) = 0.$$

(8.82)

To perform the derivation, now our task is to get $\sum(e_{i\alpha}e_{i\beta}\bar{g}_i^{(1)})$. We note from Eq. (8.66) that $\bar{g}_i^{(1)} = -\tau_1(\partial_{t_1} + e_{i\alpha}\partial_\alpha)\bar{g}_i^{(0)} + \tau_1 S_i$. Substituting this equation into Eq. (8.82) further simplifies it to

$$c_s^2\partial_{t_2}(\rho u_\beta) - (\tau_1 - 0.5)\partial_\alpha\sum\left(e_{i\alpha}e_{i\beta}\left(\partial_{t_1} + e_{i\gamma}\partial_\gamma\right)\bar{g}_i^{(0)}\right) + \tau_1\partial_\alpha\sum(e_{i\beta}e_{i\alpha}S_i) = 0.$$

(8.83)

Eq. (8.83) includes the third moments of $\bar{g}_i^{(0)}$. From Eq. (8.15) we have

$$\partial_\gamma\sum\left(e_{i\alpha}e_{i\beta}e_{i\gamma}\bar{g}_i^{(0)}\right) = c_s^4\partial_\gamma[\rho(u_\alpha\delta_{\beta\gamma} + u_\beta\delta_{\alpha\gamma} + u_\gamma\delta_{\beta\alpha})].$$

(8.84)

Substituting Eqs (8.73), (8.74), and (8.84) into Eq. (8.83) gives

$$\partial_{t_2}(\rho u_\beta)$$

$$- (\tau_1 - 0.5)\left\{\partial_\alpha\partial_{t_1}\left(p_h\delta_{\alpha\beta} + \underbrace{\rho u_\alpha u_\beta}\right) + c_s^2\partial_\alpha\partial_\gamma\left[\rho\left(u_\alpha\delta_{\beta\gamma} + u_\beta\delta_{\alpha\gamma}\right) + \underline{\rho u_\gamma\delta_{\beta\alpha}}\right]\right\}$$

$$+ (\tau_1 - 0.5)\left\{\partial_\alpha\underbrace{\left[F_\beta u_\alpha + F_\alpha u_\beta\right]} + \partial_\alpha\left[E_\gamma\left(u_\alpha\delta_{\beta\gamma} + u_\beta\delta_{\alpha\gamma}\right) + \overbrace{E_\gamma u_\gamma\delta_{\alpha\beta}} - c_s^2\overbrace{\frac{\rho u_r\delta_{\alpha\beta}}{r}}\right]\right\}$$

$$= 0.$$

(8.85)

In the appendix to Chapter 7 we shown that the underbraced terms and terms of $\partial_\alpha(u_\alpha\partial_\beta p + u_\beta\partial_\alpha p)$ in Eq. (8.85) can be canceled. From Eq. (8.75) we know

$$\partial_{t_1}(p_h\delta_{\alpha\beta}) = -c_s^2\left[\partial_\gamma\left(\rho u_\gamma\right) + \overbrace{\frac{\rho u_r}{r}}\right]\delta_{\alpha\beta} + \overbrace{u_\gamma E_\gamma\delta_{\alpha\beta}}.$$

(8.86)

Substituting this equation we can see that the overbraced terms in Eqs (8.86) and (8.85) can be canceled. The underlined terms in Eqs (8.86) and (8.85) can also be canceled. The widetilde term in Eq. (8.86) can be canceled with the widetilde term in Eq. (8.85).

Substituting $E_\gamma = -\partial_\gamma(p_h - c_s^2\rho)$ into Eq. (8.85), and noticing the terms $u_\alpha\partial_\beta p_h$, $u_\beta\partial_\alpha p_h$ will be canceled with their counterparts in $\partial_{t1}(\rho u_\alpha u_\beta)$ (see the appendix to Chapter 7), yields

$$\partial_{t_2}(\rho u_\beta) - c_s^2(\tau_1 - 0.5)\partial_\alpha\{\rho[\partial_\beta u_\alpha + \partial_\alpha u_\beta]\} = 0.$$

(8.87)

From Eqs (8.79) and (8.87), using $\partial_t(\rho u_\beta) = \partial_{t_1}(\rho u_\beta) + \Delta t \partial_{t_2}(\rho u_\beta)$ and $\nu = c_s^2(\tau_1 - 0.5)\Delta t$, we have the N–S equations,

$$\partial_t(\rho u_\beta) + \partial_\alpha(\rho u_\alpha u_\beta) = -\partial_\beta p_h + \nu \partial_\alpha\{\rho[\partial_\beta u_\alpha + \partial_\alpha u_\beta]\} + F_\beta. \tag{8.88}$$

If Eq. (8.88) is consistent with Eq. (8.4), the following formula should be satisfied.

$$\partial_t \rho + u_\alpha \partial_\alpha \rho = 0. \tag{8.89}$$

That means the material derivative $\frac{D\rho}{Dt} = 0$. Hence, as a conclusion, only if Eqs (8.78) and (8.89) are satisfied, i.e., the material derivatives of both p_h and ρ are zero, can the model be valid for incompressible two-phase flow.

8.10.4 CH equation

To obtain the equation tracking the interface, we start from the equation analogous to Eqs (8.65)–(8.67) for the distribution functions $\bar{f}_i$. From Eqs (8.19) and (8.64) we know that $\sum_i \bar{f}_i^{(0)} = \phi$, $\sum_i \bar{f}_i^{(1)} = \frac{\phi u_r}{2r}$, and $\sum_i \bar{f}_i^{(n)} = 0$, for $n > 1$. Summing both sides of Eq. (8.66) over i gives

$$\partial_{t_1}\phi + \partial_\alpha(\phi u_\alpha) + \frac{\phi u_r}{r} = 0. \tag{8.90}$$

Summing Eq. (8.67) over i and substituting $\sum_i S_i' = -\left(1 - \frac{1}{2\tau_2}\right)\frac{\phi u_r}{r}$ yields

$$\partial_{t_2}\phi + \left(1 - \frac{1}{2\tau_2}\right)\left[\partial_\alpha\left(\sum_i e_{i\alpha}\bar{f}_i^{(1)}\right) + \partial_{t_1}\left(\frac{\phi u_r}{2r}\right)\right]$$

$$+ \frac{1}{2}\left[\partial_\alpha\left(\sum_i e_{i\alpha}S_i'\right) - \left(1 - \frac{1}{2\tau_2}\right)\partial_{t_1}\left(\frac{\phi u_r}{r}\right)\right] = 0. \tag{8.91}$$

In the simulation, $\sum_i \bar{f}_i e_{i\alpha}$ is unknown or not calculated; although $\sum_i \bar{f}_i^{eq} e_{i\alpha}$ is known, it is impossible to determine $\sum_i e_{i\alpha}\bar{f}_i^{(1)}$ in the LBM. If this term is assumed to be omitted, substituting $\sum_i e_{i\alpha}S_i' = \left(1 - \frac{1}{2\tau_2}\right)F_\alpha'$, Eq. (8.91) yields

$$\partial_{t_2}\phi + \frac{1}{2}\left(1 - \frac{1}{2\tau_2}\right)\partial_\alpha F_\alpha' = 0. \tag{8.92}$$

Substituting $F_\alpha' = -\partial_\alpha(p_{th} - c_s^2\phi)$ into the above equation we have

$$\partial_{t_2}\phi = \frac{1}{2}\left(1 - \frac{1}{2\tau_2}\right)\partial_\alpha(\partial_\alpha(p_{th} - c_s^2\phi)). \tag{8.93}$$

Combining Eqs (8.90) and (8.93) together gives

$$\partial_t\phi + \partial_\alpha(\phi u_\alpha) + \frac{\phi u_r}{r} = \lambda\partial_\alpha^2(p_{th} - c_s^2\phi), \tag{8.94}$$

where $\lambda = \frac{1}{2}\left(1 - \frac{1}{2\tau_2}\right)\Delta t$ is the mobility. This is the macro phase-tracking equation, which is a CH-like equation in our axisymmetric model based on the HCZ model (He et al. 1999).

CHAPTER 9

Extensions of the HCZ model for high-density ratio two-phase flows

9.1 Introduction

As described in Chapter 7, He et al. (1999) proposed an incompressible multiphase LBM. In the model two distribution functions are used. One is used to recover the incompressibility condition and N–S equations, the other is used to track the interfaces between different phases and is able to recover a macroscopic equation similar to the CH equation. The CH equation is used to model the creation, evolution, and dissolution of phase-field interfaces (Jacqmin 1999). The equation can be written as $\frac{\partial \phi}{\partial t} + \mathbf{u} \cdot \nabla \phi = \nabla \cdot (\lambda \nabla \mu)$, where $\phi = \frac{\rho - \rho_g}{\rho_l - \rho_g}$ is the index function, μ is the chemical potential, and λ is the mobility.

Recently, a series of models (Amaya-Bower and Lee 2005; Lee and Lin 2010; Lee and Liu 2010) based on the model of He et al. (1999) have been further developed to handle higher-density ratio multiphase flows. Lee and Lin (2005) proposed the idea of directional derivatives (Kikkinides et al. 2008). Later, Lee and Fischer (2006), Lee and Liu (2010), and Amaya-Bower and Lee (2010) applied an isotropic discretization scheme in their models as well as the directional derivatives. The models have been used to study Rayleigh–Taylor instability (Chiappini et al. 2010), etc.

The difference between Lee and Fischer (2006) and the papers of Amaya-Bower and Lee (2010) and Lee and Liu (2010) is that the model in Lee and Fischer (2006) uses only one distribution function and the compressibility effect is larger than that in the two-distribution-function models (Amaya-Bower and Lee 2010; Lee and Liu 2010). Here we do not discuss this compressible model (i.e. Lee and Fischer (2006)). The other models (Amaya-Bower and Lee 2010; Lee and Lin 2005; Lee and Liu 2010) seem able to simulate density ratios as high as 1000. For a droplet splashing a thin liquid film, the results show good agreement with some experimental data (Lee and Lin 2005). The results of rising bubble simulations are also consistent with experimental data (Amaya-Bower and Lee 2010).

Multiphase Lattice Boltzmann Methods: Theory and Application, First Edition.
Haibo Huang, Michael C. Sukop and Xi-Yun Lu.
© 2015 John Wiley & Sons, Ltd. Published 2015 by John Wiley & Sons, Ltd.
Companion Website: www.wiley.com/go/huang/boltzmann

The theoretical differences between the original model of He et al. (1999) and the high-density models (Amaya-Bower and Lee 2010; Lee and Lin 2005; Lee and Liu 2010) and how the differences between them affect numerical results has not been described. Here, we try to identify the differences between the models in terms of the macroscopic equations that the models recover. We study the Galilean invariance property of each model. We also propose a simple way to modify the models so that they correctly recover the CH equation. Numerical simulations are carried out to confirm our theoretical conclusions.

You can find an introduction to the HCZ model (He et al. 1999) in Chapter 7. Here we describe four models which we designate as I, II, III, and IV, corresponding to Lee and Lin (2005), Amaya-Bower and Lee (2010), Lee and Liu (2010), and our revision.

9.2   Model I (Lee and Lin 2005)

9.2.1   Stress and potential form of intermolecular forcing terms

As we know, the common Lattice Boltzmann equation is able to recover the momentum equations (Lee and Lin 2005)

$$\partial_t \rho u_\alpha + \partial_\beta (\rho u_\alpha u_\beta) = \partial_\beta \sigma_{\alpha\beta}^{ig} + F_\alpha, \tag{9.1}$$

where the ideal-gas stress tensor is

$$\sigma_{\alpha\beta}^{ig} = -\rho c_s^2 \delta_{\alpha\beta} + \eta(\partial_\beta u_\alpha + \partial_\alpha u_\beta). \tag{9.2}$$

F_α is the intermolecular forcing term.

For non-ideal gases the stress tensor is

$$\sigma_{\alpha\beta} = -P\delta_{\alpha\beta} + \sigma_{\alpha\beta}^{(v)} + \sigma_{\alpha\beta}^{(1)}, \tag{9.3}$$

where $P(\rho)$ is the thermodynamic pressure for the isothermal fluid and the viscous stress tensor is

$$\sigma_{\alpha\beta}^{(v)} = \eta(\partial_\beta u_\alpha + \partial_\alpha u_\beta) - \frac{2}{3}\eta\partial_\delta u_\delta \delta_{\alpha\beta}. \tag{9.4}$$

The stress $\sigma_{\alpha\beta}^{(1)}$ takes the form (Lee and Lin 2005)

$$\sigma_{\alpha\beta}^{(1)} = \kappa \left[\left(\frac{1}{2}\partial_\delta\rho\partial_\delta\rho + \rho\partial_\delta^2\rho \right) \delta_{\alpha\beta} - \partial_\alpha\rho\partial_\beta\rho \right]. \tag{9.5}$$

Hence, if we use the standard Lattice Boltzmann equation to simulate two-phase flow, we should add some forcing terms to it. Correspondingly, there is a force term F_α on the right-hand side of Eq. (9.1). The expression for F_α can be easily obtained through comparison of $\sigma_{\alpha\beta}^{(ig)}$ (Eq. (9.2)) and $\sigma_{\alpha\beta}$ (Eq. (9.3)). Hence,

$$F_\alpha = \partial_\beta(\rho c_s^2 - P)\delta_{\alpha\beta} + \kappa\left[\left(\frac{1}{2}\partial_\delta\rho\partial_\delta\rho + \rho\partial_\delta^2\rho\right)\delta_{\alpha\beta} - \partial_\alpha\rho\partial_\beta\rho\right]. \tag{9.6}$$

For the intermolecular force F_α, Lee and Lin (2005) suggested two forms for the two Lattice Boltzmann equations f_i and g_i: the potential form and the stress form. In the Lattice Boltzmann equation recovering the momentum equations, the stress form should be used (Lee and Lin 2005), i.e.,

$$F_\alpha = \partial_\beta\left(\rho c_s^2 - p\right)\delta_{\alpha\beta} + \kappa\partial_\beta(\partial_\delta\rho\partial_\delta\rho\delta_{\alpha\beta} - \partial_\alpha\rho\partial_\beta\rho), \tag{9.7}$$

in which the modified pressure is defined as

$$p = P - \kappa\rho\partial_\delta^2\rho + \frac{\kappa}{2}\partial_\delta\rho\partial_\delta\rho. \tag{9.8}$$

For the Lattice Boltzmann equation for the order parameter, which is able to recover the CH equation macroscopically, the intermolecular force is suggested to take the potential form (Lee and Lin 2005)

$$F_\alpha = \partial_\beta\left(\rho c_s^2 - P\right)\delta_{\alpha\beta} + \kappa\rho\partial_\alpha\partial_\beta^2\rho. \tag{9.9}$$

It is argued that the potential form is more suitable than the stress form (Eq. (9.7)) when phase separation is important.

Substituting Eq. (9.8) into Eq. (9.7) we have

$$F_\alpha = \partial_\beta\left(\rho c_s^2 - P + \kappa\rho\partial_\delta^2\rho - \frac{\kappa}{2}\partial_\delta\rho\partial_\delta\rho\right)\delta_{\alpha\beta}$$
$$+ \kappa\partial_\beta(\partial_\delta\rho\partial_\delta\rho\delta_{\alpha\beta} - \partial_\alpha\rho\partial_\beta\rho)$$
$$= \partial_\beta\left(\rho c_s^2 - P\right)\delta_{\alpha\beta} + \underbrace{\kappa\partial_\alpha\left(\rho\partial_\delta^2\rho\right) + \frac{\kappa}{2}\partial_\alpha(\partial_\delta\rho\partial_\delta\rho)} - \underline{\kappa\partial_\beta(\partial_\alpha\rho\partial_\beta\rho)}$$
$$= \partial_\beta\left(\rho c_s^2 - P\right)\delta_{\alpha\beta} + \underbrace{\kappa\rho\partial_\alpha\left(\partial_\delta^2\rho\right) + \kappa\partial_\alpha\rho\partial_\delta^2\rho + \kappa\partial_\delta\rho\partial_\alpha\partial_\delta\rho}$$
$$\underline{-\kappa\partial_\alpha\rho\partial_\beta^2\rho - \kappa\partial_\beta\rho\partial_\alpha\partial_\beta\rho}$$
$$= \partial_\beta\left(\rho c_s^2 - P\right)\delta_{\alpha\beta} + \kappa\rho\partial_\alpha\partial_\beta^2\rho. \tag{9.10}$$

In the above derivation, the underbraced terms are equivalent and underlined terms are equivalent. It is also noted that $\frac{\kappa}{2}\partial_\alpha(\partial_\delta\rho\partial_\delta\rho) = \kappa\partial_\delta\rho\partial_\alpha\partial_\delta\rho$. Hence, theoretically, the stress form (Eq. (9.7)) and the potential form (Eq. (9.9)) are identical.

9.2.2 Model description

Lee and Lin (2005) developed the HCZ model through adoption of the stress and potential forms of the intermolecular forcing terms in the momentum equation and the equation of the order parameter, respectively. The model of Lee and Lin (2005) applied the pre-stream and post-stream collision steps. Theoretically, the two steps can be combined to a single collision step, as illustrated in Eqs (7.1) and (7.2). Lee and Lin (2005) proposed using different discretization schemes in each collision step. Only in that way can cases with high-density ratio be achieved. The above strategies may improve the numerical stability.

In the description of He et al. (2005) the density instead of the index function is used. Here we follow their description.

In their model, instead of using the non-ideal gas equation as in He et al. (1999), they used the free-energy scheme. The free energy takes the form (Lee and Lin 2005)

$$E_f(\rho) \approx \beta(\rho - \rho_g)^2(\rho - \rho_l)^2. \tag{9.11}$$

The corresponding analogous functions are

$$E_\alpha = \partial_\alpha \rho, \tag{9.12}$$

$$F_\alpha = \kappa[\partial_\alpha(\partial_\delta \rho \partial_\delta \rho) - \partial_\delta(\partial_\alpha \rho \partial_\delta \rho)], \tag{9.13}$$

and

$$F'_\alpha = \partial_\alpha \rho c_s^2 - \rho \partial_\alpha \mu, \tag{9.14}$$

where

$$\mu(\rho) = \mu_0(\rho) - \kappa \partial_\delta^2 \rho. \tag{9.15}$$

The chemical potential $\mu_0(\rho)$ is (Lee and Lin 2005)

$$\mu_0(\rho) = \frac{\partial E_f}{\partial \rho} \approx 4\beta(\rho - \rho_g)(\rho - \rho_l)\left(\rho - \frac{\rho_l + \rho_g}{2}\right), \tag{9.16}$$

where

$$\beta = \frac{12\sigma}{W(\rho_l - \rho_g)^4} \tag{9.17}$$

is a constant that can be calculated from a given surface tension σ (Lee and Lin 2005). The parameter κ is also related to surface tension by

$$\kappa = 1.5\frac{\sigma W}{(\rho_l - \rho_g)^2}, \tag{9.18}$$

where W is the width of the interface.

We also note that from the free energy a simplified EOS can be derived (Lee and Lin 2005) once the function $E_f(\rho)$ is given. Here we give the simplified equation of state explicitly as (Lee and Lin 2005)

$$P = \rho\left(\frac{\partial E_f}{\partial \rho}\right)_T - E_f = \beta(\rho - \rho_l)(\rho - \rho_g)(3\rho^2 - \rho(\rho_l + \rho_g) - \rho_l\rho_g), \tag{9.19}$$

where P is the thermodynamic pressure. The pressure P as a function of $\frac{1}{\rho}$ is shown in Figure 9.1. To show the curve more clearly, here the log-scale is used for the $\frac{1}{\rho}$ axis. From the figure we can see that there are three roots for the equation $P = 0$. Two of them are $\rho = \rho_l$ and $\rho = \rho_g$. The third root between them represents an unstable state. The Maxwell construction is also satisfied. Obviously, this simplified EOS is similar to the van der Waals EOS.

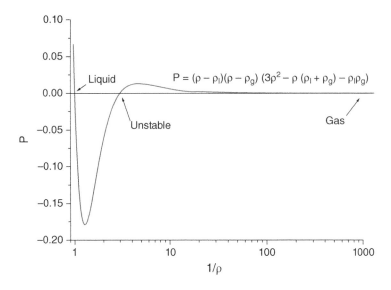

Figure 9.1 The pressure as a function of $\frac{1}{\rho}$ (Eq. (9.19) with $\beta = 1$, $\rho_l = 1$, and $\rho_g = 0.001$).

From Eq. (7.22) we have $\bar{f}_i^{(1)} = -\tau_2(\partial_{t_1} + e_{i\gamma}\partial_{\gamma})\bar{f}_i^{(0)} + \tau_2 S_i'$. Substituting this equation with $\sum_i \bar{f}_i^{(0)} e_{i\alpha} = \rho u_\alpha$ and $\sum_i \bar{f}_i^{(0)} e_{i\alpha} e_{i\gamma} = \rho c_s^2 \delta_{\alpha\gamma} + \rho u_\alpha u_\gamma$ into Eq. (7.45) yields

$$\partial_{t_2}\rho - (\tau_2 - 0.5)\partial_\alpha[\partial_{t_1}(\rho u_\alpha) + \partial_\gamma(\rho c_s^2 \delta_{\alpha\gamma} + \rho u_\alpha u_\gamma)] + \tau_2\partial_\alpha\left(\sum_i e_{i\alpha}S_i'\right) = 0. \quad (9.20)$$

Using the Euler equation derived from summation of $\mathbf{e}_i$ times Eq. (7.22), and substituting $\sum_i e_{i\alpha}S_i' = \left(1 - \frac{1}{2\tau_2}\right)F_\alpha'$, Eq. (9.20) yields

$$\partial_{t_2}\rho - (\tau_2 - 0.5)\partial_\alpha[F_\alpha + \partial_\alpha(\rho c_s^2 - p)] + (\tau_2 - 0.5)\partial_\alpha F_\alpha' = 0. \quad (9.21)$$

Substituting F_α and F_α' into the above equation we have the CH-like equation

$$\partial_t\rho + \partial_\alpha(\rho u_\alpha) = (\tau_2 - 0.5)\Delta t \left\{\kappa\left[\partial_\alpha^2\partial_\delta^2\rho - \partial_\alpha\partial_\delta\left(\partial_\alpha\rho\partial_\delta\rho\right)\right] - \left[\partial_\alpha^2 p - \partial_\alpha\left(\rho\partial_\alpha\mu\right)\right]\right\}. \quad (9.22)$$

Here we can see that this CH-like equation is not $\partial_t\rho + u_\alpha\partial_\alpha\rho = \partial_\gamma[\lambda\partial_\gamma(P - p)]$ (i.e., Eq. (36) in Lee and Lin (2005)) or $\partial_t\rho + u_\alpha\partial_\alpha\rho = \partial_\gamma(\lambda\partial_\gamma\mu_0)$ (i.e., Eq. (37) in Lee and Lin (2005)), where λ is the mobility in the CH equation.

Substituting the F_β into Eq. (7.43), the corresponding N–S equations derived from this model are $\partial_\alpha(u_\beta\partial_\alpha\rho)$ or $\partial_\alpha(u_\alpha\partial_\beta\rho)$

$$\partial_t(\rho u_\beta) + \partial_\alpha(\rho u_\alpha u_\beta) = -\partial_\beta p + v\partial_\alpha\{\rho[\partial_\beta u_\alpha + \partial_\alpha u_\beta]\} + \kappa[\partial_\beta(\partial_\delta^2\rho) - \partial_\delta(\partial_\beta\rho\partial_\delta\rho)]. \quad (9.23)$$

This equation is identical to Eq. (29) in Lee and Lin (2005).

It is noted that the equilibrium distribution function in Lee and Lin (2005) for $\bar{g}_i^{eq}$ is slightly different from Eq. (7.4) in Chapter 7:

$$\bar{g}_i^{eq}(\mathbf{x}, t) = w_i \left[\frac{p}{c_s^2} + \rho \left(\frac{e_{i\alpha} u_\alpha}{c_s^2} + \frac{e_{i\alpha} u_\alpha e_{i\beta} u_\beta}{2c_s^4} - \frac{u_\alpha u_\alpha}{2c_s^2} \right) \right]. \tag{9.24}$$

The macroscopic variable p should be given as

$$p = c_s^2 \sum \bar{g}_i + \frac{\Delta t}{2} u_\alpha \frac{\partial \rho c_s^2}{\partial x_\alpha} \tag{9.25}$$

instead of $p = \sum \bar{g}_i + \frac{\Delta t}{2} u_\alpha \frac{\partial \rho c_s^2}{\partial x_\alpha}$ (Eq. (49) in Lee and Lin (2005)). This is a typographical error in Eq. (49) in Lee and Lin (2005).

9.2.3 Implementation

Model I is implemented in three steps: the pre-streaming collision step, the streaming step, and the post-streaming collision step (Lee and Lin 2005).

Pre-streaming collision step

$$\bar{g}_i(\mathbf{x}, t) = g_i(\mathbf{x}, t) - \frac{1}{2\tau_1 - 1}(g_i - \bar{g}_i^{eq})$$

$$+ \frac{\Delta t}{2} \frac{(e_{i\alpha} - u_\alpha)\partial_\alpha \rho c_s^2}{c_s^2}[\Gamma_i(\mathbf{u}) - \Gamma_i(0)]$$

$$+ \frac{\Delta t}{2} \frac{(e_{i\alpha} - u_\alpha)[\kappa \partial_\alpha(\partial_\beta \rho \partial_\beta \rho) - \kappa \partial_\gamma(\partial_\alpha \rho \partial_\gamma \rho)]}{c_s^2}\Gamma_i(\mathbf{u}) \tag{9.26}$$

$$\bar{f}_i(\mathbf{x}, t) = f_i(\mathbf{x}, t) - \frac{1}{2\tau_2 - 1}(f_i - \bar{f}_i^{eq})$$

$$+ \frac{\Delta t}{2} \frac{(e_{i\alpha} - u_\alpha)[\partial_\alpha \rho c_s^2 - \rho \partial_\alpha \mu]}{c_s^2}\Gamma_i(\mathbf{u}) \tag{9.27}$$

Streaming step

$$\bar{g}_i(\mathbf{x} + \mathbf{e}_i \Delta t, t + \Delta t) = \bar{g}_i(\mathbf{x}, t) \tag{9.28}$$

$$\bar{f}_i(\mathbf{x} + \mathbf{e}_i \Delta t, t + \Delta t) = \bar{f}_i(\mathbf{x}, t) \tag{9.29}$$

Post-streaming collision step

$$g_i(\mathbf{x}, t) = \bar{g}_i(\mathbf{x}, t) - \frac{1}{2\tau_1}(\bar{g}_i - \bar{g}_i^{eq})$$

$$+ \frac{\Delta t}{2} \left(1 - \frac{1}{2\tau_1} \right) \frac{(e_{i\alpha} - u_\alpha)\partial_\alpha \rho c_s^2}{c_s^2}[\Gamma_i(\mathbf{u}) - \Gamma_i(0)]$$

$$+ \frac{\Delta t}{2} \left(1 - \frac{1}{2\tau_1} \right) \frac{(e_{i\alpha} - u_\alpha)[\kappa \partial_\alpha(\partial_\beta \rho \partial_\beta \rho) - \kappa \partial_\gamma(\partial_\alpha \rho \partial_\gamma \rho)]}{c_s^2}\Gamma_i(\mathbf{u}) \tag{9.30}$$

$$f_i(\mathbf{x}, t) = \bar{f}_i(\mathbf{x}, t) - \frac{1}{2\tau_2}(\bar{f}_i - \bar{f}_i^{eq})$$

$$+\frac{\Delta t}{2}\left(1 - \frac{1}{2\tau_1}\right)\frac{(e_{i\alpha} - u_\alpha)[\partial_\alpha \rho c_s^2 - \rho \partial_\alpha \mu]}{c_s^2}\Gamma_i(\mathbf{u}) \qquad (9.31)$$

In Lee and Lin (2005), in the Lattice Boltzmann equations for both f_i and g_i, only one uniform relaxation time τ is used. We have shown that is not necessary to use uniform τ in the Lattice Boltzmann equations. The relationship between τ_1, τ_2, here and the τ in Lee and Lin (2005) is $\tau = \tau_1 - 0.5 = \tau_2 - 0.5$. Here $\tau_1 > 0.5$ and $\tau_2 > 0.5$. For the forcing term discretization the second-order biased/mixed difference is recommended for the pre-streaming collision step while the standard central difference is suggested for the post-streaming collision step (Lee and Lin 2005).

9.2.4 Directional derivative

There are some first and second derivatives in the forcing term in the Lattice Boltzmann equations (Eqs (9.26), (9.27), (9.30), and (9.31)). Lee and Lin (2005) proposed the idea of using a "directional derivative", which means $e_{i\alpha}$ times a derivative, for example $e_{i\alpha}\Delta t \frac{d\zeta}{dx_\alpha}$. In this model we should pay more attention to the directional derivatives of μ and ρ in the above equations.

The second-order central difference (CD) for the directional derivative is (Lee and Lin 2005)

$$e_{i\alpha}\Delta t \partial_\alpha^{CD}\zeta|_{\mathbf{x}} = \frac{1}{2}[\zeta(\mathbf{x} + \mathbf{e}_i\Delta t) - \zeta(\mathbf{x} - \mathbf{e}_i\Delta t)], \qquad (9.32)$$

where ζ denotes the variable μ or the density of fluid ρ (Lee and Lin 2005).

In the following equations, the superscripts CD and BD denote the central difference and the biased difference, respectively. MD means a mixture of CD and BD (Lee and Lin 2005).

The second-order BD for the directional derivative is (Lee and Lin 2005)

$$e_{i\alpha}\Delta t \partial_\alpha^{BD}\zeta|_{\mathbf{x}} = \frac{1}{2}[-\zeta(\mathbf{x} + 2\mathbf{e}_i\Delta t) + 4\zeta(\mathbf{x} + \mathbf{e}_i\Delta t) - 3\zeta(\mathbf{x})]. \qquad (9.33)$$

In Model I, " ... it is reasonable to apply the second-order biased/mixed difference in the pre-streaming collision step while the standard central difference in the post-streaming collision step ... " (Lee and Lin 2005). Actually, this proposed approach is critical for this high-density ratio model. Without this, the model is unable to handle high-density ratio cases. The mixed finite difference stencil is (Lee and Lin 2005)

$$e_{i\alpha}\Delta t \partial_\alpha^{MD}\zeta|_{\mathbf{x}} = \begin{cases} e_{i\alpha}\Delta t \partial_\alpha^{BD}\zeta|_{\mathbf{x}}, & \text{if } \left(e_{i\alpha}\Delta t \partial_\alpha^{BD}\zeta|_{\mathbf{x}}\right) \times \left(e_{i\alpha}\Delta t \partial_\alpha^{CD}\zeta|_{\mathbf{x}+\mathbf{e}\Delta t}\right) \geq 0, \\ e_{i\alpha}\Delta t \partial_\alpha^{CD}\zeta|_{\mathbf{x}}, & \text{if } \left(e_{i\alpha}\Delta t \partial_\alpha^{BD}\zeta|_{\mathbf{x}}\right) \times \left(e_{i\alpha}\Delta t \partial_\alpha^{CD}\zeta|_{\mathbf{x}+\mathbf{e}\Delta t}\right) \leq 0. \end{cases} \qquad (9.34)$$

For the derivatives other than the directional derivatives, the common formulae for first and second derivatives can be used (refer to Section 8.6.1).

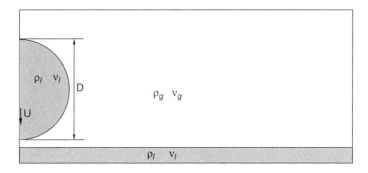

Figure 9.2   Schematic diagram of a droplet splashing on a thin liquid film.

Theoretically, combining the pre-streaming and post-streaming collision steps into one single collision step is fine. However, " ... combining the pre-streaming and post-streaming collision steps into one single collision step poses problems. If the second-order central difference is used in the single collision step, the solution becomes unstable as the density ratio increases and the interface thickness decreases. If the second-order biased or mixed difference is exclusively used in the single collision step, the interface tends to smear due to numerical diffusion and asymmetry" (Lee and Lin 2005).

9.2.5   Droplet splashing on a thin liquid film

The case of a droplet splashing on a thin liquid film at $Re = 500$ and $We = 8000$ is simulated as an example of the capabilities of this model. The schematic diagram for the example is shown in Figure 9.2. The computational domain is 800×400. The Reynolds number and Weber number are defined as

$$Re = \frac{UD}{v_l} \quad \text{and} \quad We = \frac{\rho_l U^2 D}{\sigma}. \tag{9.35}$$

We can see that due to symmetry this setup is efficient when the symmetric boundary condition is applied to the left boundary. The symmetric boundary is easy to implement in the LBM. It is identical to the slip wall boundary condition. For more details refer to Section 8.2.4. The whole droplet/splash slice image in Figure 9.3 is obtained through mirroring our result across this boundary.

For the right and upper boundaries, second-order extrapolation of the distribution functions is applied (Chen et al. 1996). For the bottom wall boundary, the simple bounce-back scheme is imposed. The densities of liquid and gas are $\rho_l = 1.0$ and $\rho_g = 0.001$. The non-dimensional relaxation parameters in pure liquid and gas regimes are $\tau_l = 0.506$ and $\tau_g = 0.65$, respectively. The corresponding kinematic viscosity of the liquid and gas are $v_l = 0.002$ and $v_g = 0.05$. The dynamic viscosity ratio is 40. The thickness of the interface is specified as $W = 4$ lu, surface tension $\sigma = 6.25 \times 10^{-7}$, $\beta = 2 \times 10^{-6}$, the diameter of the droplet

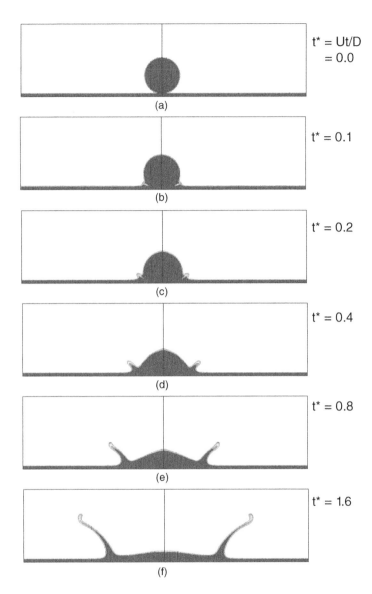

Figure 9.3   Snapshots of a droplet splashing on a thin liquid film with $Re = 500$, $We = 8000$, and $\frac{\rho_l}{\rho_g} = 1000$ using Model I. The labeled non-dimensional time $t^* = \frac{tU}{D}$.

is $D = 200$ lu, and its downward velocity before impact is $U = 0.005$ lu/ts. Initially, the droplet contacts the thin film, as shown in Figure 9.3. Snapshots of the splashing are shown in Figure 9.3.

The result is highly consistent with Figure 15 in Lee and Lin (2005). The log–log plot of the spread factor $\frac{r}{D}$ as a function of $t^* = \frac{Ut}{D}$ also agrees well with the power law (not shown) (Lee and Lin 2005).

9.3   Model II (Amaya-Bower and Lee 2010)

Recently, Amaya-Bower and Lee (2010) revised Model I according to Lee and Fischer (2006) to simulate rising bubble phenomena. Hereafter, this model is referred to as Model II. In this model they not only used the directional derivatives (Lee and Lin 2005) but also applied isotropic discretization (Lee and Fischer 2006). For consistency, in the following description our $\bar{f}_i$ is the $\bar{h}_i$ in Amaya-Bower and Lee (2010).

In Model II the corresponding

$$E_\alpha = c_s^2 \partial_\alpha \rho, \tag{9.36}$$

$$F_\alpha = -\phi \partial_\alpha \mu, \tag{9.37}$$

and

$$F'_\alpha = \partial_\alpha(\phi c_s^2 - p) - (\partial_\phi p_2)\partial_\alpha \phi - \phi \partial_\alpha \mu. \tag{9.38}$$

We also note that in the model there is an extra term

$$\left(1 - \frac{1}{2\tau_2}\right) \partial_\gamma(\lambda \partial_\gamma \mu)\Gamma_i(\mathbf{u}) \tag{9.39}$$

in the LBE for $\bar{f}_i$ (refer to Eq. (7.19)), i.e.,

$$S'_i = \left(1 - \frac{1}{2\tau_2}\right) \frac{(e_{i\alpha} - u_\alpha)F'_\alpha}{c_s^2 \rho} \bar{f}_i^{eq} + \left(1 - \frac{1}{2\tau_2}\right) \partial_\gamma(\lambda \partial_\gamma \mu)\Gamma_i(\mathbf{u}). \tag{9.40}$$

It should be noted that the modified distribution function illustrated in Appendix A in Amaya-Bower and Lee (2010) will lead to the coefficient $\left(1 - \frac{1}{2\tau_2}\right)$ before the extra term. Hence, we have

$$\sum S'_i = \left(1 - \frac{1}{2\tau_2}\right) \partial_\alpha(\lambda \partial_\alpha \mu) \tag{9.41}$$

and

$$\sum S'_i e_{i\alpha} = \left(1 - \frac{1}{2\tau_2}\right) [u_\alpha \partial_\gamma(\lambda \partial_\gamma \mu) + F'_\alpha]. \tag{9.42}$$

In Amaya-Bower and Lee (2010) the density or the index function is redefined as

$$\phi = \sum \bar{f}_i + \frac{\Delta t}{2} \partial_\alpha(\lambda \partial_\alpha \mu). \tag{9.43}$$

In the above equation, because μ is a function of ϕ, Amaya-Bower and Lee (2010) suggested using an implicit method to solve the equation. However, this may be not necessary because the numerical result is almost not improved even using the implicit method. Hence, we have $\sum \bar{f}_i^{(1)} = -\frac{1}{2}\partial_\alpha(\lambda \partial_\alpha \mu)$. Repeating the derivation procedure in Section 7.3.2 here, the corresponding Eq. (7.44) becomes

$$\partial_{t_1}\phi + \partial_\alpha(\phi u_\alpha) = \partial_\alpha(\lambda \partial_\alpha \mu). \tag{9.44}$$

Finally, the CH-like equation in the model is (refer to Eq. (7.47))

$$\partial_t \phi + \partial_\alpha(\phi u_\alpha) = \partial_\alpha(\lambda \partial_\alpha \mu) + \frac{1}{2}\left(1 - \frac{1}{2\tau_2}\right)\Delta t \partial_\alpha[F_\alpha' + u_\alpha \partial_\gamma(\lambda \partial_\gamma \mu)]. \tag{9.45}$$

The term $(\partial_\phi p_2)\partial_\alpha \phi$ in F_α' is an "artificial gradient" in the model to avoid negative ϕ (Amaya-Bower and Lee 2010). The meaning of "p_2" in Amaya-Bower and Lee (2010) is unclear; it only serves the purpose of avoiding negative ϕ. Obviously, this derived CH-like equation is different from the one claimed in Amaya-Bower and Lee (2010).

The corresponding N–S equations derived from this model are:

$$\partial_t(\rho u_\beta) + \partial_\alpha(\rho u_\alpha u_\beta) = -\partial_\beta p + \nu \partial_\alpha\{\rho[\partial_\beta u_\alpha + \partial_\alpha u_\beta]\} - \phi \partial_\beta \mu. \tag{9.46}$$

We can see that these N–S equations look different from the N–S equations that are recovered from the HCZ model (Eq. (7.43)) and Model I (Eq. (9.23)). The difference lies in the different way of incorporating the surface tension effect in (Kim 2005). For example, in the HCZ model (Eq. (7.43)), the surface tension term is supposed to be $\kappa \rho \partial_\alpha(\partial_\delta^2 \rho)$ while in Model II the corresponding term is $-\phi \partial_\beta \mu$. It is an open question which surface tension model is better (Kim 2005). Hence, at present we are still unable to conclude which one is superior.

9.3.1 Implementation
In the implementation, the forcing terms ($E_\alpha, F_\alpha, F_\alpha'$ illustrated in Eqs (9.36) to (9.38)) should be discretized in different ways (finite central difference or biased difference).

The equilibrium distribution functions for g_i and f_i are

$$\hat{g}_i^{eq}(\mathbf{x}, t) = w_i\left[p + c_s^2\rho\left(\frac{e_{i\alpha}u_\alpha}{c_s^2} + \frac{e_{i\alpha}u_\alpha e_{i\beta}u_\beta}{2c_s^4} - \frac{u_\alpha u_\alpha}{2c_s^2}\right)\right] \tag{9.47}$$

and

$$\hat{f}_i^{eq}(\mathbf{x}, t) = w_i\phi\left(1 + \frac{e_{i\alpha}u_\alpha}{c_s^2} + \frac{e_{i\alpha}u_\alpha e_{i\beta}u_\beta}{2c_s^4} - \frac{u_\alpha u_\alpha}{2c_s^2}\right). \tag{9.48}$$

respectively.

In our LBEs (Eqs (7.1) and (7.2)), the distribution functions are $\bar{g}_i$ and $\bar{f}_i$, respectively. In our implementation, the forcing terms are discretized in the following way (Amaya-Bower and Lee 2010):

$$\bar{g}_i(\mathbf{x} + \mathbf{e}_i\Delta t, t + \Delta t) = \bar{g}_i(\mathbf{x}, t) - \frac{1}{\tau_1}(\bar{g}_i - \bar{g}_i^{eq})$$
$$+ \Delta t(e_{i\alpha} - u_\alpha)\left\{-\phi\partial_\alpha^{MD}\mu\Gamma_i(\mathbf{u}) + c_s^2\partial_\alpha^{MD}\rho[\Gamma_i(\mathbf{u}) - \Gamma_i(0)]\right\}, \tag{9.49}$$

$$\bar{f}_i(\mathbf{x} + \mathbf{e}_i\Delta t, t + \Delta t) = \bar{f}_i(\mathbf{x}, t) - \frac{1}{\tau_2}(\bar{f}_i - \bar{f}_i^{eq})$$
$$+ \frac{\Delta t}{c_s^2(\rho_l - \rho_g)}(e_{i\alpha} - u_\alpha)\{\partial_\alpha^{MD}(\phi c_s^2 - p) - (\partial_\phi p_2)\partial_\alpha^{MD}\phi - \phi\partial_\alpha^{MD}\mu\}. \tag{9.50}$$

The equilibrium distribution functions $\bar{g}_i^{eq}$ and $\bar{f}_i^{eq}$ are

$$\bar{g}_i^{eq} = \hat{g}_i^{eq} - \frac{\Delta t}{2}(e_{i\alpha} - u_\alpha)\left\{-\phi\partial_\alpha^{CD}\mu\Gamma_i(\mathbf{u}) + c_s^2\partial_\alpha^{CD}\rho[\Gamma_i(\mathbf{u}) - \Gamma_i(0)]\right\} \tag{9.51}$$

and

$$\bar{f}_i^{eq} = \hat{f}_i^{eq} - \frac{\Delta t}{2c_s^2(\rho_l - \rho_g)}(e_{i\alpha} - u_\alpha)\left\{\partial_\alpha^{CD}(\phi c_s^2 - p) - (\partial_\phi p_2)\partial_\alpha^{CD}\phi - \phi\partial_\alpha^{CD}\mu\right\}. \tag{9.52}$$

Here we can see that the evolution equation for $\bar{f}_i$ is basically consistent with Eq. (7.2) except $\frac{1}{\rho}$ in Eq. (7.2) is replaced with $\frac{1}{\rho_l - \rho_g}$.

In the above equations,

$$\mu = \mu_0 - \kappa\partial_\alpha^2\phi. \tag{9.53}$$

The chemical potential is

$$\mu_0 = \partial_c E_0 = 4\beta\phi(\phi - 1)(\phi - 0.5), \tag{9.54}$$

where $E_0 = \beta\phi^2(1 - \phi^2)$ is the bulk energy.

The above directional derivatives can be obtained through the finite difference scheme illustrated in Section 9.2.4. For example, the mixed direction derivative for variable ϕ can be written as

$$e_{i\alpha}\Delta t\partial_\alpha^{MD}\phi|_{\mathbf{x}} = \frac{1}{2}\left[e_{i\alpha}\Delta t\partial_\alpha^{CD}\phi|_{\mathbf{x}} + e_{i\alpha}\Delta t\partial_\alpha^{BD}\phi|_{\mathbf{x}}\right]$$

$$= \frac{1}{4}[-\phi(\mathbf{x} + 2e_{i\alpha}\Delta t) + 5\phi(\mathbf{x} + e_{i\alpha}\delta t) - 3\phi(\mathbf{x}) - \phi(\mathbf{x} - e_{i\alpha}\Delta t)]. \tag{9.55}$$

It is noted that there is a typographical error in Eq. (23) in Sun et al. (2013), which reads:

$$e_{i\alpha}\Delta t\partial_\alpha^{CD}\phi|_{\mathbf{x}} = \frac{1}{2}[\phi(\mathbf{x} + e_{i\alpha}\Delta t) + \phi(\mathbf{x} - e_{i\alpha}\Delta t)]. \tag{9.56}$$

Obviously, this is wrong and the equation should be

$$e_{i\alpha}\Delta t\partial_\alpha^{CD}\phi|_{\mathbf{x}} = \frac{1}{2}[\phi(\mathbf{x} + e_{i\alpha}\Delta t) - \phi(\mathbf{x} - e_{i\alpha}\Delta t)]. \tag{9.57}$$

Using Eq.(9.57), the other derivatives can also be calculated through

$$\partial_\alpha^{CD}\phi|_{\mathbf{x}} = \frac{1}{c_s^2\Delta t}\sum_i w_i e_{i\alpha}(e_{i\beta}\Delta t\partial_\beta^{CD}\phi)|_{\mathbf{x}}. \tag{9.58}$$

For $c_s^2 = \frac{1}{3}$, the equation can be written as

$$\partial_x^{CD}\phi|_{\mathbf{x}} = \frac{1}{3}[\phi(\mathbf{x} + \mathbf{e}_1\Delta t) - \phi(\mathbf{x} + \mathbf{e}_3\Delta t)]$$

$$+ \frac{1}{12}[\phi(\mathbf{x} + \mathbf{e}_5\Delta t) - \phi(\mathbf{x} + \mathbf{e}_7\Delta t)] + \frac{1}{12}[\phi(\mathbf{x} + \mathbf{e}_8\Delta t) - \phi(\mathbf{x} + \mathbf{e}_6\Delta t)]. \tag{9.59}$$

Similarly, the biased derivative can be calculated by

$$\partial_\alpha^{BD}\phi|_{\mathbf{x}} = \frac{1}{c_s^2\Delta t}\sum_i w_i e_{i\alpha}(e_{i\beta}\Delta t\partial_\beta^{BD}\phi)|_{\mathbf{x}}. \tag{9.60}$$

For example, $\partial_x^{BD}\phi|_{\mathbf{x}}$ can be written as

$$
\begin{aligned}
\partial_x^{BD}\phi|_{\mathbf{x}} = & \frac{1}{6}[-\phi(\mathbf{x}+2\mathbf{e}_1\Delta t)+4\phi(\mathbf{x}+\mathbf{e}_1\Delta t)] \\
& -\frac{1}{6}[-\phi(\mathbf{x}+2\mathbf{e}_3\Delta t)+4\phi(\mathbf{x}+\mathbf{e}_3\Delta t)] \\
& +\frac{1}{24}\{-\phi(\mathbf{x}+2\mathbf{e}_5\Delta t)+4\phi(\mathbf{x}+\mathbf{e}_5\Delta t)-[-\phi(\mathbf{x}+2\mathbf{e}_7\Delta t)+4\phi(\mathbf{x}+\mathbf{e}_7\Delta t)]\} \\
& +\frac{1}{24}\{-\phi(\mathbf{x}+2\mathbf{e}_8\Delta t)+4\phi(\mathbf{x}+\mathbf{e}_8\Delta t)-[-\phi(\mathbf{x}+2\mathbf{e}_6\Delta t)+4\phi(\mathbf{x}+\mathbf{e}_6\Delta t)]\}.
\end{aligned}
$$

$$(9.61)$$

A 2D code for this model is presented in Appendix B of this chapter. For the bubble rise problem, the term $\partial_\phi p_2$ (see Eq. (9.38)) may be necessary for numerical stability (Amaya-Bower and Lee 2010). Here in the code the simplest flow around one droplet or bubble immersed in another fluid is simulated. In this simple flow problem the term $\partial_\phi p_2$ is not considered. In this code the extra body force, e.g., the gravity force, is also not considered.

9.4   Model III (Lee and Liu 2010)

Lee and Liu (2010) applied the multiphase Model II (Amaya-Bower and Lee 2010) with some minor changes to simulate micro-scale drop impact on a dry surface. Here this method is referred to as Model III. Chiappini et al. (2009) also applied this model to simulate the phenomenon of two-droplet coalescence. However, the simulation result is not consistent with experimental data (Chiappini et al. 2009). Actually this conclusion is not only applicable to Model III, but also to Models I and II. We may address this issue in future research.

In Model III the corresponding

$$E_\alpha = c_s^2 \partial_\alpha \rho, \tag{9.62}$$

$$F_\alpha = -\phi \partial_\alpha \mu, \tag{9.63}$$

and

$$F_\alpha' = c_s^2 \partial_\alpha \phi - \frac{\phi}{\rho}\partial_\alpha p - \frac{\phi^2}{\rho}\partial_\alpha \mu. \tag{9.64}$$

In this model there is no artificial gradient such as the one that appears in Model II. The index function is defined as $\phi = \sum \bar{f}_i$, and hence, $\sum \bar{f}_i^{(n)} = 0$ for $n \geq 1$ in this model.

We note that in the model there is an extra term

$$
\frac{1}{2}(\lambda \partial_\gamma^2 \mu)\Gamma_i(\mathbf{u})|_{(\mathbf{x},t)} + \frac{1}{2}(\lambda \partial_\gamma^2 \mu)\Gamma_i(\mathbf{u})|_{(\mathbf{x}+\mathbf{e}_i\Delta t,t)}
\tag{9.65}
$$

in the LBE for $\bar{f}_i$ (refer to Eq. (7.19)). Through Taylor expansion with truncated error $O(\Delta t^2)$ this extra term can be rewritten as

$$(2 + \Delta t e_{i\alpha} \partial_\alpha) \frac{1}{2} (\lambda \partial_\gamma^2 \mu) \Gamma_i(\mathbf{u})|_{(\mathbf{x},t)}. \tag{9.66}$$

Hence, we have

$$\sum S_i' = \lambda \partial_\gamma^2 \mu + \frac{\Delta t}{2} \partial_\alpha [u_\alpha (\lambda \partial_\gamma^2 \mu)] \tag{9.67}$$

and keeping the truncation error to $O(\Delta t)$,

$$\sum S_i' e_{i\alpha} = \left(1 - \frac{1}{2\tau_2}\right) F_\alpha' + u_\alpha \lambda \partial_\gamma^2 \mu. \tag{9.68}$$

The corresponding Eqs (7.44) and (7.46) become

$$\partial_{t_1} \phi + \partial_\alpha (\phi u_\alpha) = \lambda \partial_\alpha^2 \mu + \frac{\Delta t}{2} \partial_\alpha [u_\alpha (\lambda \partial_\gamma^2 \mu)] \tag{9.69}$$

and

$$\partial_{t_2} \phi = -\frac{1}{2} \partial_\alpha \left[\left(1 - \frac{1}{2\tau_2}\right) F_\alpha' + u_\alpha \lambda \partial_\gamma^2 \mu \right], \tag{9.70}$$

respectively.

Combining Eqs (9.69) and (9.70) together, the CH-like equation for this model is

$$\partial_t \phi + \partial_\alpha (\phi u_\alpha) = \lambda \partial_\gamma^2 \mu - \frac{\Delta t}{2} \left(1 - \frac{1}{2\tau_2}\right) \partial_\alpha F_\alpha'. \tag{9.71}$$

Here again we can see that there are extra terms in Eq. (9.71) compared to the CH equation. We note that in Lee and Liu (2010) the authors chose $\tau_2 = 1.0$ for all cases because here $\tau_2 = \tau_2' + 0.5$, where $\tau_2' = 0.5$ is fixed in Lee and Liu (2010). Even with this choice, the CH equation is still only approximately recovered. The N–S equations derived from this model are identical to those obtained from Model II in Eq. (9.46).

9.5 Model IV

In Models II, and III we can see that to recover the diffusion term on the right-hand side of the CH equation correctly, τ_2 must be 0.5. However, when the relaxation time approaches 0.5, numerical instability appears and the simulation may blow up.

Here, we propose a simple way to incorporate the diffusion term in the CH equation into Model I. This means Model IV is identical to Model I except for the following different choices for functions F_α, F_α', and E_α. It is much easier to implement than the above models and without the limitation of $\tau_2 = 0.5$. In our model,

$$F_\alpha = \kappa \rho \partial_\alpha (\partial_\delta^2 \rho), \tag{9.72}$$

$$F_\alpha' = \partial_\alpha (\rho c_s^2 - p) + \kappa \rho \partial_\alpha (\partial_\delta^2 \rho) \underline{- \lambda \partial_\alpha \mu}, \tag{9.73}$$

and

$$E_\alpha = \partial_\alpha \rho. \tag{9.74}$$

The underlined term incorporates the diffusion term in the CH equation into the LBM (refer to Section 9.2.2). In this model there are no other extra terms in Eq. (7.19). The density or the index function is defined in the usual way: $\rho = \sum \bar{f}_i$. The derived CH equation takes the form

$$\partial_t \rho + \partial_\alpha (\rho u_\alpha) = (\tau_2 - 0.5)(\Delta t) \partial_\alpha (\lambda \partial_\alpha \mu). \tag{9.75}$$

If we choose a constant τ_2 for the collision step in Eq. (7.2), then Eq. (9.75) is exactly the CH equation. The derived N–S equations are identical to those of the original incompressible HCZ model in Eq. (7.43), which are slightly different from those from Models II and III in Eq. (9.46). The resulting model is referred to as Model IV_a.

If we want to recover the N–S equation (Eq. (9.46)) in these models, we should choose:

$$F_\alpha = -\rho \partial_\alpha \mu, \tag{9.76}$$

$$F_\alpha' = \partial_\alpha (\rho c_s^2 - p) - \rho \partial_\alpha \mu \underline{- \lambda \partial_\alpha \mu}, \tag{9.77}$$

and

$$E_\alpha = \partial_\alpha \rho. \tag{9.78}$$

This model is referred to as Model IV_b.

9.6   Numerical tests for different models

To compare the performance of the above five models we carry out simulations that have theoretical solutions. The typical benchmark problems are a drop inside a box with periodic boundary conditions and two-phase layered flow in a channel.

9.6.1   A drop inside a box with periodic boundary conditions

We compared the surface tension measured using the Laplace law from the converged simulation results with the theoretical solution for the surface tension we specified as part of the model input. We also investigate the behavior of the spurious current and the pressure field.

In our simulations the computational domain is 100×100. Initially we set a circle with density of liquid $\rho_l = 1.0$ and radius about 20 lu. Outside the circle, the density of gas is $\rho_g = 0.001$. The pressure is initially set as zero everywhere in the computational domain. The relaxation parameters for liquid and gas are $\tau_l = 0.53$ and $\tau_g = 0.65$, respectively. The relaxation time τ_1 is a function of position and is interpolated as

$$\tau_1(\mathbf{x}) = \phi(\mathbf{x})\tau_l + (1 - \phi(\mathbf{x}))\tau_g. \qquad (9.79)$$

The parameter τ_2 can be specified as $\tau_2 = \tau_1$ or as a constant because τ_2 is not relevant to viscosity and only affects the phase-field equation. The surface tension is initially set to $\sigma = 6.25 \times 10^{-6}$ mu/ts^2 and $W = 4$ lu.

The surface tension σ in our simulation results is measured with the Laplace law. The Laplace law is $\Delta p = \frac{\sigma}{R}$, where Δp is the pressure difference between the inside of the drop and a point outside the drop but far away from the interface vicinity, and R is the radius of the drop.

The convergence of the model is defined by the stabilization of the surface tension σ

$$E_\sigma = \left| \frac{\sigma_{LBM}(t) - \sigma_{LBM}(t - \Delta t_E)}{\sigma_{LBM}(t)} \right|. \qquad (9.80)$$

In all cases, $\Delta t_E = 10^4 ts$. The convergence criterion is $E_\sigma < 10^{-5}$.

It is found that in Case 1 of Table 9.1 using the finite difference stencil (Eq. (9.34)) the drop will drift away from its initial location at the center of the box due to asymmetries in the spurious current after about 6.5×10^5 ts. It may be due to the stencil used in Model I. Case 2 confirmed this conjecture. We can

Table 9.1 Cases of an initial stationary drop inside a box with theoretical $\sigma = 6.25 \times 10^{-6}$ mu/ts^2 and $\tau_l = 0.53, \tau_g = 0.65$.

Cases	Model	Specification	$\sigma_{LBM}(\times 10^{-6})$	$|u_{max}|$
1	I	With the stencil $\tau_2 = \tau_1$	–	–
2	I	Without the stencil $\tau_2 = \tau_1$	5.750	7.76×10^{-8}
3	II	$\tau_2 = \tau_1, \lambda = 0.1$	6.209	4.26×10^{-9}
4	II	$\tau_2 = 1.0, \lambda = 0.1$	6.137	3.93×10^{-9}
5	III	$\tau_2 = \tau_1, \lambda = 0.1$	6.217	3.50×10^{-9}
6	III	$\tau_2 = 1.0, \lambda = 0.1$	6.202	2.32×10^{-9}
7	IV$_a$	$\tau_2 = 1.0, \lambda = 0.1$	5.487	2.94×10^{-7}
8	IV$_a$	$\tau_2 = 1.0, \lambda = 1.0$	6.028	4.58×10^{-7}
9	IV$_a$	$\tau_2 = \tau_1, \lambda = 9$	6.112	5.00×10^{-7}
10	IV$_a$	$\tau_2 = 1.0, \lambda = 10$	6.097	6.57×10^{-7}
11	IV$_b$	$\tau_2 = \tau_1, \lambda = 0.1$	6.154	7.31×10^{-9}
12	IV$_b$	$\tau_2 = 1.0, \lambda = 0.1$	6.141	3.33×10^{-9}

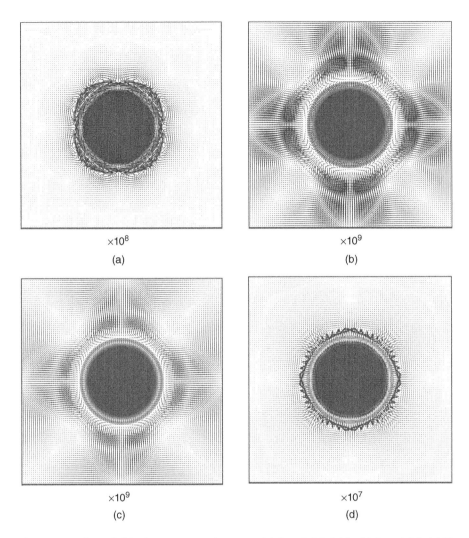

Figure 9.4 Velocity fields after 2,000,000 time steps. (a) Case 2 (Model I), (b) Case 4 (Model II), (c) Case 6 (Model III), and (d) Case 8 (Model IV$_a$). Values of **u** are magnified by $10^8, 10^9, 10^9$, and 10^7 times in (a), (b), (c), and (d), respectively.

see that using Model I without this stencil the velocity field is symmetric. The spurious current is illustrated in Figure 9.4(a). Note that this stencil was only proposed for Model I (Lee and Lin 2005). In some simulations of complex flow phenomena we also observed that this stencil is critical for Model I because without it certain simulations will blow up, for example the case of a droplet splashing on a thin film in Section 9.2.5. Here, we do not intend to discuss the fundamental mechanics of such a stencil. In the other models that we will discuss below, no such stencil is necessary (Amaya-Bower and Lee 2010; Lee and Liu 2010).

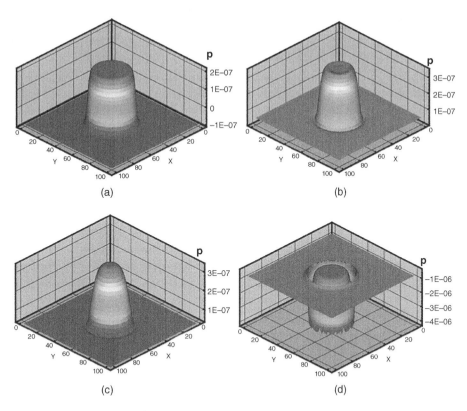

Figure 9.5 Pressure fields for (a) Case 2 (Model I), (b) Case 4 (Model II), (c) Case 6 (Model III), and (d) Case 8 (Model IV_a).

Next we discuss Model II. From Cases 3 and 4 in Table 9.1 we can see that using Model II, $\tau_2 = \tau_1$ or $\tau_2 = 1.0$, and $\lambda = 0.1$ all give converged results. The σ_{LBM} is consistent with the specified σ. The spurious current in Case 4 is illustrated in Figure 9.4 (b). The magnitude of spurious current is of $O(10^{-9})$ and the spurious currents are slightly away from the interface vicinity. The pressure field of Case 4 is illustrated in Figure 9.5(b).

For the value of the mobility (λ), Jacqmin (1999) discussed the range of λ in detail. Jacqmin (1999) suggested that an appropriate upper bound of λ is $O(W/L)$ and a lower bound is $O((W/L)^2)$, where L is the characteristic length of the flow. It is usually chosen as the diameter of the drop. Here $\lambda = 0.1$ is approximately inside the range of values suggested. To investigate the effect of λ we also carried out simulations of cases with $\lambda = 1.0$ and $\tau_2 = \tau_1$ or $\tau_2 = 1.0$. We found that the calculated surface tension did not converge for this parameter set. Hence, large λ appears to be difficult to simulate.

For Model III we simulated two cases, Cases 5 and 6, which have identical parameters to Cases 3 and 4. From Table 9.1 we can see that the converged σ_{LBM} is also consistent with the specified σ. The spurious current magnitudes are all of the same order as Cases 3 and 4. From Figure 9.4(c) we can see that the spurious current profile is similar to that of Case 4, Model II in Figure 9.4(b). Figure 9.5(b) and (c) illustrates that the pressure fields of Cases 4 and 6 are similar. The small difference is that the high-pressure plateau area in Case 6 is smaller than that of Case 4. Obviously, Models II and III are similar; comparing Eqs (9.45) and (9.71) shows that Models II and III yield almost identical macroscopic equations if the "artificial gradient", i.e. the term $\partial_\alpha(\partial_\rho p_2)\partial_\alpha\rho$, is omitted.

For Model IV_a we performed several simulations (Cases 7, 8, 9, and 10). From Table 9.1 we can see that using this model the magnitudes of the spurious currents are larger than they are in Models I, II, and III. However, the spurious current is still very small and only confined to the interface vicinity (Figure 9.4(d)). In applications it usually does not affect the flow simulations. Case 7 in Table 9.1 shows that the σ_{LBM} is 12% smaller than the specified one, while in Case 8 with $\lambda = 1$, the σ_{LBM} agrees well with the specified σ. The discrepancy may be due to the small λ in Case 7. Values of $\lambda = 0.1$ seemed appropriate in Cases 3, 4, 5, and 6, while higher values gave better results in Cases 8, 9, and 10. From Eqs (9.45), (9.71), and (9.75) we found that in Model IV the diffusion term in the CH equation is $(\tau_2 - 0.5)(\Delta t)\partial_\alpha(\lambda\partial_\alpha\mu)$, while in Models II and III the diffusion term in the CH equation is more complecated, e.g., including extra terms $-(\tau_2 - 0.5)(\Delta t)\partial_\alpha[u_\alpha\lambda(\partial_\gamma^2\mu)]$ in Model II. A possible explanation for the difference is that the terms like $-(\tau_2 - 0.5)(\Delta t)\partial_\alpha[u_\alpha\lambda(\partial_\gamma^2\mu)]$ may also play an important role in the diffusion effect in the CH equation.

We note in Cases 7 and 8 that $\tau_2 = 1$ is a constant. In Case 9 if we choose $\tau_2 = \tau_1$, since $\tau_2 \in (0.53, 0.65)$, it is smaller than that in Cases 7 and 8. From Eq. (9.75) we know that a larger λ is needed to achieve the same magnitude of diffusion effect in the CH equation in Cases 7 and 8. As an approximation we chose $\lambda = 9$ for Case 9 with $\tau_2 = \tau_1$. Here we can see that the σ_{LBM} agrees well with the specified one. However, we also observe a significant compressibility effect in Case 9. The density in most of the area outside the drop is found to be significantly larger than 0.001; for example, even far away from the interface vicinity the density is about 0.0013. In Case 10, with $\tau = 1$ and $\lambda = 10$, the diffusion effect in the CH equation should be much larger than in Cases 7 and 8. The density far away from the interface vicinity is found to be about $\rho = 0.0034$ after equilibrium is achieved. Hence, the larger λ is, the more prominent the diffusion effect in the CH equation.

In Cases 11 and 12, which apply Model IV_b, the spurious velocity and pressure fields not shown are similar to those of Models II and III.

Figure 9.5(d) illustrates the pressure field of Case 8 using Model IV_a. It is a typical pressure profile when using the model. We can see it is very different from the pressure fields obtained from Models I, II, and III. In Figure 9.5(d) there is a deep ravine in the interface vicinity. The depth of the ravine is much larger than the pressure difference between the inside and outside of the drop.

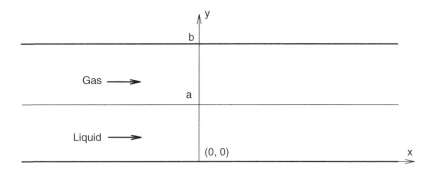

Figure 9.6   Immiscible parallel two-phase flow in a 2D channel.

This is not surprising because the N–S equations derived in Model IV_a are identical to those in the HCZ model but different from those of the other models. Further numerical tests confirmed this point. We simulated a case of density ratio 10 for the HCZ model (He et al. 1999) and found that the pressure distribution of the HCZ model and Model IV_a are similar. The pressure fields of Model IV_b are found to be very similar to those of Models I, II, and III. Hence, the recovered N–S equations determine the style of pressure field.

9.6.2   Layered two-phase flows in a channel

Immiscible two-phase flow between two parallel plates was simulated because it has an analytical solution (Gross et al. 2009; Huang and Lu 2011), which is given in Appendix A of this chapter. In the simulations, periodic boundary conditions were applied on the inlet/outlet boundaries (Figure 9.6). No-slip (bounce-back) boundary conditions were applied on the upper and lower plates. In the simulation, as illustrated in Figure 9.6, the gas flows in the upper region $a < y < b$ and the liquid flows in the lower region $0 < y < a$.

Cases of density ratio 50 and 200 are simulated, which are referred to as Cases (a) and (b), respectively. In the simulations, a 20×100 mesh is used with $a = 50$ lu and $b = 100$ lu. The relaxation times for the f_i are all $\tau_1 = 1.0$. Although the interface thickness is set to be 4 lu, the actually thickness will increase and depends on the density ratio. The larger the density ratio is, the thicker the interface becomes. In the simulations with Model I the special stencil is not used.

In Case (a), $\sigma = 10^{-4}$ mu/ts^2, $\tau_l = 0.6$, and $\tau_g = 1.0$. In Case (b) (density ratio 200), $\sigma = 10^{-4}$ mu/ts^2 and $\tau_l = \tau_g = 1.0$. In all of the cases, the body force acting on both liquid and gas is $G = 10^{-8}$ mu · lu/ts^2.

Typical comparisons between the LBM and two analytical solutions are illustrated in Figure 9.7. In Analytical Solution I, the interface thickness is assumed to be zero. In Analytical Solution II (see Appendix A), a finite interface thickness is considered and it is assumed that the velocity inside the interface can be obtained

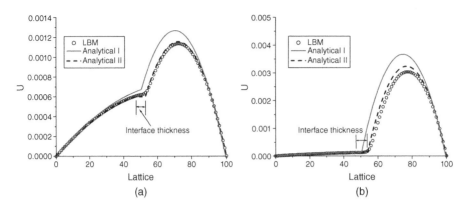

Figure 9.7   Velocity profile comparison between analytical solution and LBM results for immiscible parallel two-phase flow when using Model I. The body force is applied to both liquid and gas. (a) Density ratio 50. (b) Density ratio 200. In Analytical Solution I the interface thickness is assumed to be zero. In Analytical Solution II a finite interface thickness is considered. The thicknesses are measured from the density profile (not shown), and the interface thicknesses in Cases (a) and (b) are 6 lu and 8 lu, respectively.

by interpolation, which is determined by the velocity near the interface (refer to Appendix A for more details). Measured from the density profile (not shown), the interface thicknesses in Cases (a) and (b) are about 6 lu and 8 lu, respectively. There are large discrepancies between the LBM result and Analytical Solution I. The errors in Cases (a) and (b) are 12.3% and 27.9%, respectively. The definition of error is as follows:

$$error = \frac{\sum |u_{LBM} - u_a|}{\sum u_a}, \tag{9.81}$$

where u_{LBM} and u_a are the LBM result and the analytical solution respectively. The summation is performed for all of the lattice nodes (101 lattice nodes) in a vertical slice in the center of the channel. Comparing with Analytical Solution II, the LBM result is much more satisfactory. The corresponding errors in Cases (a) and (b) are 1.6% and 8.4%, respectively. Hence, the discrepancies are mainly caused by the difference between a finite interface thickness in simulations and zero thickness in the theoretical assumption.

We also simulate Cases (a) and (b) using more lattice nodes. Figure 9.8 shows the LBM results. From Figure 9.8 we can see that using more lattice nodes (fine mesh) the LBM results are much closer to the analytical solutions because the interface thickness is negligible in these cases. Now the errors in Cases (a) and (b) are 7.5% and 9.9%, respectively.

For the other models (Models II, III, and IV_b), compared to the analytical solutions, the velocity profile errors are similar to those in Model I. When the interface thickness is considered in the analytical solution for density ratio 50 the

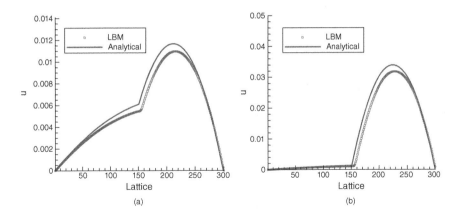

Figure 9.8 Velocity profile comparison between Analytical Solution I and LBM results using Model I for fine mesh. In LBM simulations 300 lu is used to represent the width of the channel and $G = 10^{-8}$ mu · lu/ts^2 is applied. (a) Density ratio 50 ($\rho_l = 1.0, \rho_g = 0.02, \tau_l = 0.6, \tau_g = 1.0$). (b) Density ratio 200 ($\rho_l = 1.0, \rho_g = 0.005, \tau_l = \tau_g = 1.0$). In the analytical solutions the interface thickness is assumed to be zero.

error is about 1.6%. When density ratio increases to 200 the error increases to about 8.4%. We also noted that in these simulations the results are not sensitive to the mobility value.

9.6.3 Galilean invariance

Theoretically speaking, the HCZ model is Galilean invariant. However, Models I, II, III, and IV are not Galilean invariant if $E_\alpha = \partial_\alpha p$. From Eq. (7.39) we can see that terms such as $\partial_\alpha(u_\alpha \partial_\beta p)$ would not cancel if $E_\alpha = \partial_\alpha p$. These terms may break the Galilean invariance law if the pressure gradient near the interface is large (Holdych et al. 1998).

Two cases of Couette flow are simulated to test Galilean invariance. The simulation setting is basically identical to that in Figure 9.6 except that there are no body forces applied to the fluids. The fluids are driven by shear stress induced by the moving plates. The computational domain size is 101×101. Case A is with the upper plate fixed and the lower plate moving at $u = 0.05$, and Case B is with the lower plate fixed and the upper plate moving with a velocity $u = -0.05$. Cases A and B are the same physical problem, only solved in reference frames moving at different constant velocities. The velocity profiles are shown in Figure 9.9. For comparison purpose, the velocity is normalized through $u^* = \frac{u_y - u_{\min}}{u_{\max} - u_{\min}}$. For the analytical solution, a finite interface thickness with 6 lu is also considered (refer to the above section). The velocity profiles of both cases are consistent with the analytical one. Although through this simple test it seems Model III is Galilean invariant, in the following a slightly more complicated test is also performed according to Holdych et al. (1998).

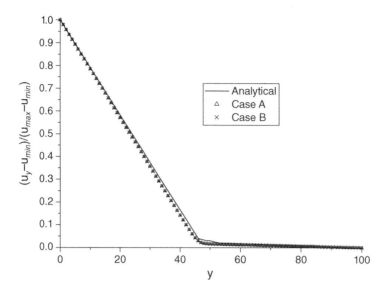

Figure 9.9   Couette flow velocity profiles for Model III. Case A: lower boundary moving with $u_w = 0.05$. Case B: upper boundary moving with $u_w = -0.05$. The density ratio is 50.

The cases of a moving circular droplet between two moving walls (Holdych et al. 1998) are tested. The computational domain is 90×90. A stationary circular droplet with a radius of 20 lu is initially put at the center of the domain. The interface thickness is $W = 4$ lu. Periodic boundary conditions are applied to the left and right boundaries. In Case (a) and Case (b) both the top and bottom plates move with a constant velocity $u_w = 0.0003$ lu/ts and $u_w = 0.005$ lu/ts, respectively. The moving boundary condition proposed by d'Humiéres et al. (2002) is applied, i.e., in the upper wall the unknown distribution functions $g_4, g_7,$ and g_8 are obtained through

$$g_4 = g_2 + 2w_2\rho_g \frac{\mathbf{e}_4 \cdot \mathbf{u}_w}{c_s^2}, \quad g_7 = g_5 + 2w_5\rho_g \frac{\mathbf{e}_7 \cdot \mathbf{u}_w}{c_s^2}, \quad g_8 = g_6 + 2w_6\rho_g \frac{\mathbf{e}_8 \cdot \mathbf{u}_w}{c_s^2}.$$
(9.82)

In Case (a) the densities of the liquid and gas are $\rho_l = 1$ and $\rho_g = 0.002$, respectively (density ratio 500). In Case (b) the densities of the liquid and gas are $\rho_l = 1$ and $\rho_g = 0.02$, respectively (density ratio 50). In the simulations $\tau_1 = \tau_2 = 1$. The surface tension $\sigma = 0.0001$.

Here we take Model III as an example. In the tests the mobility is set as $\lambda = 0.1$. The droplet shape and velocity vectors in the computational domain are shown in Figure 9.10. From the figure we can see that in Case (a) the velocities throughout the flow field approach the velocity of the plates, and are constant in the fully developed stage after the non-dimensional time $t^* = \frac{t u_w}{L} = 14.467$, where $L = 90$ lu. During the procedure the droplet is found to keep its circular shape. Hence, for smaller velocity, Galilean invariance seems satisfied (Holdych

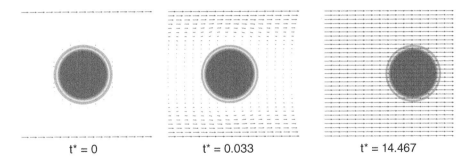

t* = 0	t* = 0.033	t* = 14.467

Figure 9.10   Density contours and velocity field for Model III. Case (a): density ratio 500 with velocities of the two plates $u_w = 0.0003$ lu/ts.

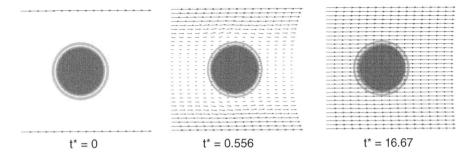

t* = 0	t* = 0.556	t* = 16.67

Figure 9.11   Density contours and velocity field for Model III. Case (b): density ratio 50 with velocities of the two plates $u_w = 0.005$ lu/ts.

et al. 1998). In the gas region the maximum gas density is about $\rho_g = 0.008$ at $t^* = 14.467$ (maximum gas density shift $\delta\rho_g = 0.006$), which is even larger than the initial $\rho_g = 0.002$.

For Case (b) with higher velocity (Figure 9.11) even the flow system reaches an equilibrium state ($t^* > 15$), and the flow velocity in the center area is less than velocities of the upper and lower walls (see Figure 9.11(c)). A possible explanation is that since conservation of the momentum (the N–S equations) is not satisfied, the drop does not advect correctly (Holdych et al. 1998). After the system reaches the equilibrium state there is also a density shift. However, for this smaller density ratio case the gas density shift is about $\delta\rho_g = -0.004$, which is not so significant compared to the initial gas density.

The droplet centroid velocities as a function of time are shown in Figure 9.12. In the figure, in Cases (a) and (b), the droplet centroid velocities all reach an equilibrium value after $t^* > 12$ and $t^* > 15$, respectively. In Case (a) the droplet centroid velocity reaches the velocities of the plates ($u^* = \frac{u}{u_w}$ equals to unity). However, in Case (b) the droplet centroid velocity $u^* = 0.79$.

As a conclusion, the density shift may increase with the density ratio. If plates' velocities or density ratio are small enough, terms such as $\partial_\alpha(u_\alpha \partial_\beta p)$ in the N–S

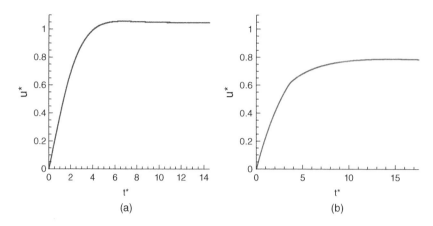

Figure 9.12   Droplet centriod velocity as function of time for Model III. (a) Density ratio 500. (b) Density ratio 50. The non-dimensional velocity is $u^* = \frac{u}{u_w}$.

equations, which break Galilean invariance, are of very small magnitude and can be omitted. It is quite reasonable because in the terms $\partial_\alpha(u_\alpha \partial_\beta p)$ a very small velocity eliminates the effect of large gradient near the interface. However, if the velocity is significantly large, Galilean invariance will be broken.

Here we also note the effect of the mobility. A case using Model III with $M = 0.0$ and density ratio 50 is simulated ($\rho_l = 1, \rho_g = 0.02, u_w = 0.0003$). It is found that the final droplet centroid velocity is $u^* = \frac{u}{u_w} \approx 0.76$, which means Galilean invariance is broken.

For the other models we observed similar results, that is, with proper choice of the mobility and very small plate velocities, Galilean invariance can be satisfied. The mobility choices are consistent with those in Table 9.1. Usually $\lambda = 0.1$ for Models II and III, and $\lambda = 1.0$ for Model IV$_a$ approximately guarantees Galilean invariance.

9.7 Conclusions

A series of multiphase LBMs based on the HCZ model are analyzed theoretically through Chapman–Enskog expansions. Models I, II, and III are found to be unable to recover the interface-tracking equation (the CH equation) exactly, although to some extent the CH equation is approximately recovered. Theoretically, Models II and III are found to be almost identical. We proposed a simple way to incorporate the diffusion term in the CH equation in Model IV. Model IV is simpler to implement and the CH equation is recovered exactly.

Numerical tests of a stationary droplet inside a box were performed. In the numerical tests we paid particular attention to the high-density ratio models (i.e., Models I, II, III, and IV). We found that without the special stencil suggested in Lee and Lin (2005) we were not able to get a converged result for Model I. Results obtained from Models II and III are similar in terms of the calculated surface tension, the magnitude of the spurious current, and the pressure field.

Through simulations of parallel two-phase flows in a channel, the accuracy of the models was evaluated. Compared with analytical solutions, the LBM models are able to give accurate results provided that the mesh size is large enough. Because of the terms $\partial_\alpha(u_\alpha\partial_\beta p)$ in the recovered N–S equations, the models are not Galilean invariant if the velocity is not small enough.

9.8 Appendix A: Analytical solutions for layered two-phase flow in a channel

Here two analytical solutions for the velocity profiles of layered two-phase flows in a channel are given. Because of the different setup, the basic analytical solution is slightly different from that in the appendix in Chapter 2. The flow is illustrated in Figure 9.6, where liquid and gas flow in the lower and upper parts of the channel, respectively. When the constant body forces $\mathcal{G}_1$ and $\mathcal{G}_2$ are applied to the liquid and gas, respectively, the fluid flow of each phase is governed by the following equations:

$$\rho_1 v_1 \nabla^2 u_1 = \mathcal{G}_1 \quad \text{and} \quad \rho_2 v_2 \nabla^2 u_2 = \mathcal{G}_2,$$

where $u_1(y)$ and $u_2(y)$ are the velocity of liquid and gas, respectively. If the interface thickness is assumed to be zero, the boundary conditions are

$$\begin{cases} u_1|_{y=a} = u_2|_{y=a}, & \rho_1 v_1(\partial_y u_1)|_{y=a} = \rho_2 v_2(\partial_y u_2)|_{y=a} \\ u_1|_{y=0} = 0, & u_2|_{y=b} = 0 \end{cases}.$$

Solving the above equations with the boundary conditions we obtained the analytical solutions

$$u_1 = A_1 y^2 + B_1 y + C_1 \quad \text{and} \quad u_2 = A_2 y^2 + B_2 y + C_2$$

where

$$A_1 = -\frac{\mathcal{G}_1}{2\rho_1 v_1}, \quad A_2 = -\frac{\mathcal{G}_2}{2\rho_2 v_2},$$

$$B_1 = \frac{1}{M}(2A_2 a + B_2) - 2A_1 a,$$

$$B_2 = \left[\left(-A_1 + \frac{2A_2}{M}\right)a^2 + A_2(b^2 - a^2)\right] / \left(a - \frac{a}{M} - b\right),$$

$$C_1 = 0, \quad C_2 = -A_2 b^2 - B_2 b,$$

and $M = \frac{\eta_1}{\eta_2}$ is the dynamic viscosity ratio. This is Analytical Solution I mentioned in Section 9.6.2.

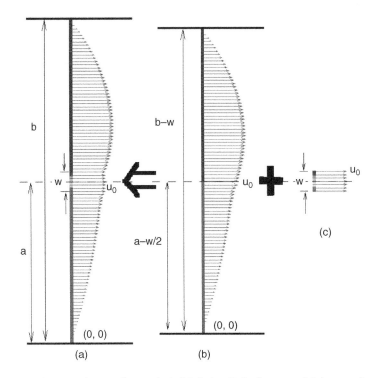

Figure 9.13 Schematic diagram for Analytical Solution II. Dark grey and light grey denote two different phases. (a) A layered two-phase flow with channel width b and interface thickness W. (b) Analytical Solution I for a channel width $(b - W)$ with interface at $y = a - \frac{W}{2}$. (c) The velocity inside the interface region is assumed to be u_0.

Analytical Solution II (see Section 9.6.2) can be constructed in the following way. Figure 9.13 shows the schematic diagram for Analytical Solution II. The dashed line represents the center of the interface. The dark grey and light grey regions denote two different phases. The original points are illustrated in Figure 9.13(a) and (b). Suppose in our LBM simulation the channel width is b and the interface thickness is W (see Figure 9.13(a)). The interface is located between $y \in (a - \frac{W}{2}, a + \frac{W}{2})$. First, Analytical Solution I for a channel width $(b - W)$ with zero-thickness interface at $y = a - \frac{W}{2}$ can be obtained (see Figure 9.13(b)). In Analytical Solution I the velocity in the zero-thickness interface is u_0. Second, as illustrated in Figure 9.13(c), the velocities inside the interface are supposed to be a constant, i.e., u_0. Third, we split the velocity profile in Figure 9.13(b) apart through the dashed line. The width of the split space is W, which is symmetric with respect to the dashed line. Finally, the velocity profile inside the interface (Figure 9.13(c)) is inserted into the space ($y \in (a - \frac{W}{2}, a + \frac{\delta}{2})$ in Figure 9.13(a)). As illustrated in Figure 9.13, Analytical Solution II (Figure 9.13(a)) consists of Analytical Solution I (Figure 9.13(b)) and the interface region (Figure 9.13(c)).

9.9   Appendix B: 2D code based on Amaya-Bower and Lee (2010)

```
! 1. This program is aimed to simulate a droplet or a bubble immersed
! in another fluid.
! 2. The fortran files: head.inc,  Main.for, Macro.for,
! StreamCollision.for, Initial.for,
!  Utility.for, Output.for, params.in should be put in an identical
! folder and compile
!  together.
! 3. The main parameters control the flow are in params.in.
! "params.in" begin with
!  the first line "101   101   !lx ly"

head.inc
C===============================================================
      integer lx,ly
      PARAMETER(lx=101,ly=101)

      common/fg/ ff(0:8,lx,ly), gp(0:8,lx,ly),
     &  pb(lx,ly), u_x(lx,ly), u_y(lx,ly),t_k(0:8)
      real*8 ff,   gp, pb, u_x, u_y, t_k

      common/rh/ rho(lx,ly), rh(lx,ly), rhp(lx,ly),
     &  up(lx,ly),vp(lx,ly), xc(0:8), yc(0:8), ex(0:8), ey(0:8)
      real*8 rho, rh, rhp, xc, yc, up, vp
      integer ex, ey

      common/cons/ cc, c_squ, con2, rho_h, rho_l, RR, thick, kappa
      real*8 cc, c_squ, con2, rho_h, rho_l, RR, thick, kappa

      common/aa/ beta, tau_h, tau_l, tau, rhoa, rhom, error
      real*8 beta,  tau_h, tau_l, tau, rhoa, rhom, error

      common/bb/ LmG, Eo, Mo, DD, UU, We, Mob, sigma
      real*8 LmG, Eo, Mo, DD, UU, We, Mob, sigma

      common/dd/ rhdx(lx,ly), rhdy(lx,ly), mudx(lx,ly), mudy(lx,ly)
     & ,rhpdx(lx,ly), rhpdy(lx,ly), brhpdx(lx,ly),brhpdy(lx,ly)
      real*8 rhdx, rhdy, mudx, mudy,rhpdx,rhpdy, brhpdx,brhpdy

      common/lap/ lap(lx,ly),  mu(lx,ly)
      real*8 lap,  mu

      common/tt/ t_max, Nwri, x_cen, y_cen
      integer t_max, Nwri, x_cen, y_cen

      common /ob/  obst(lx,ly)
      integer obst

      common/mu/ bmudx(lx,ly), bmudy(lx,ly), brhdx(lx,ly),
     & brhdy(lx,ly),
     & lapmu(lx,ly)
      real*8 bmudx, bmudy, brhdx, brhdy, lapmu
```

```
Main.for
C============================================
      program main
      implicit none
      include 'head.inc'
      integer k, x,y, time
      real*8 t_0, t_1, t_2, feq(0:8), feq2(0:8)
      integer*4 now(3)
!------------------------------------------
! Author: Haibo Huang, huanghb@ustc.edu.cn
!------------------------------------------
! Discrete velocity model (D2Q9),
! components e_ix, e_iy, where i=0,1,2,3,...8
        data xc/0.d0,1.d0,0.d0, -1.d0, 0.d0, 1.d0, -1.d0, -1.d0, 1.d0/,
     &       yc/0.d0,0.d0,1.d0, 0.d0, -1.d0, 1.d0, 1.d0, -1.d0, -1.d0/,
     &       ex/0, 1, 0, -1,  0, 1, -1, -1,  1/,
     &       ey/0, 0, 1,  0, -1, 1,  1, -1, -1/

      cc = 1.d0
      c_squ = cc *cc / 3.d0
! Weighting coefficient in the equilibrium distribution functions.
      t_0 =  4.d0 / 9.d0
      t_1 =  1.d0 / 9.d0
      t_2 =  1.d0 / 36.d0

      con2=(2.d0 * c_squ *c_squ)

      t_k(0) = t_0
      do 1 k =1,4
      t_k(k) = t_1
    1 continue
      do 2 k =5,8
      t_k(k) = t_2
    2 continue

      error= 1.d0
! Initialization
      call init()
C----------------------------------------------------
      open(39,file='out/residue.dat')

      do 100 time = 0, t_max

      if(mod(time,Nwri) .eq. 0) then
! Output the time
      write(*,*) time
      call itime(now)
      write(*,"(i2.2,' :',_i2.2,' :',_i2.2)") now

! Criterion for convergency (calculate velocity field error)
      if(time .gt. 100) call calc_error(time)
      do y= 1, ly
      do x= 1, lx
      up(x,y) = u_x(x,y)
      vp(x,y) = u_y(x,y)
```

```fortran
      enddo
      enddo

      if(error .lt. 1.0e-10) goto 101
      endif

C----------
! Collision step
      call collision()
! Streaming step
      call stream()
! Update macro-variables (density, velocity, and pressure)
      call getuv()

C----------------------------------------
      if(mod(time ,Nwri) .eq. 0) then

! Output Tecplot data file
      call write_results(u_x,u_y, time)

      endif
C----------------------------------------

  100 continue

  101 close(39)
      end


Macro.for
C=====================================
      subroutine getuv( )
      implicit none
      include "head.inc"
      integer x,y

      do 10 y = 1, ly
      do 10 x = 1, lx

      if(obst(x,y) .eq. 0) then
            rh(x,y) = ff(0,x,y) +ff(1,x,y) +ff(2,x,y)
     &                + ff(3,x,y) +ff(4,x,y) +ff(5,x,y)
     &                + ff(6,x,y) +ff(7,x,y) +ff(8,x,y)
     &                + 0.5d0* Mob *lapmu(x,y)

         rho(x,y) =rho_h* rh(x,y) + rho_l*(1.d0- rh(x,y) )

            u_x(x,y) = (gp(1,x,y) + gp(5,x,y) + gp(8,x,y)
     &          -(gp(3,x,y) + gp(6,x,y) + gp(7,x,y))) /rho(x,y)/c_squ

            u_y(x,y) = (gp(2,x,y) + gp(5,x,y) + gp(6,x,y)
     &          -(gp(4,x,y) + gp(7,x,y) + gp(8,x,y))) /rho(x,y)/c_squ

      endif
  10  continue
```

```
C------------------------
! Get density gradients (both central difference and biased
! difference)
      call Grad(rh, rhdx, rhdy, brhdx, brhdy, 1)

! Get Laplacian operator for 'rh'
      call laplace(rh,lap)

      do 40 y = 1, ly
      do 30 x = 1, lx
      mu(x,y) = 4.d0 * beta*(rh(x,y) -0.5d0)
     &          *(rh(x,y) -1.d0)*(rh(x,y)-0.d0)
     &          - kappa* lap(x,y)
  30 continue
  40 continue
! Laplacian operator
      call laplace(mu, lapmu)

! Get gradients for \mu  (both central difference and biased
! difference)
      call Grad(mu, mudx, mudy, bmudx, bmudy, 1)


C---------------
      do 80 y = 1, ly
      do 70 x = 1, lx
! Velocity components in x and y directions
      u_x(x,y) = u_x(x,y)  -0.5d0*rh(x,y)*mudx(x,y)/rho(x,y)
      u_y(x,y) = u_y(x,y)  -0.5d0*rh(x,y)*mudy(x,y)/rho(x,y)
! Pressure
      pb(x,y) = (gp(0,x,y) +gp(1,x,y) +gp(2,x,y)
     &           +gp(3,x,y) +gp(4,x,y) +gp(5,x,y)
     &           +gp(6,x,y) +gp(7,x,y) +gp(8,x,y) )
     &   + 0.5d0 *( u_x(x,y)* rhdx(x,y)
     &             + u_y(x,y)* rhdy(x,y) )*c_squ

      rhp(x,y) =rh(x,y)*c_squ -pb(x,y)
  70 continue
  80 continue

! Get gradients for \rhp
      call Grad(rhp, rhpdx, rhpdy, brhpdx, brhpdy, 1)

      end


StreamCollision.for
C=============================================================
      subroutine stream()
      implicit none
      include "head.inc"
      integer  k
      real*8 f_hlp(0:8,lx,ly)
      integer  x,y,x_e,x_w,y_n,y_s,l,n
! Streaming step
      do 10 y = 1, ly
         do 10 x = 1, lx
```

```
        y_n = mod(y,ly) + 1
        x_e = mod(x,lx) + 1

        y_s = ly - mod(ly + 1 - y, ly)
        x_w = lx - mod(lx + 1 - x, lx)

        f_hlp(1,x_e,y ) = ff(1,x,y)
        f_hlp(2,x  ,y_n) = ff(2,x,y)
        f_hlp(3,x_w,y ) = ff(3,x,y)
        f_hlp(4,x  ,y_s) = ff(4,x,y)
        f_hlp(5,x_e,y_n) = ff(5,x,y)
        f_hlp(6,x_w,y_n) = ff(6,x,y)
        f_hlp(7,x_w,y_s) = ff(7,x,y)
        f_hlp(8,x_e,y_s) = ff(8,x,y)

10 continue

   do 19 y = 1, ly
      do 20 x = 1, lx
      do 21 k = 1, 8
       ff(k,x,y) = f_hlp(k,x,y)
21 continue
20 continue
19 continue

   do 30 y = 1, ly
      do 30 x = 1, lx

        y_n = mod(y,ly) + 1
        x_e = mod(x,lx) + 1

        y_s = ly - mod(ly + 1 - y, ly)
        x_w = lx - mod(lx + 1 - x, lx)

        f_hlp(1,x_e,y ) = gp(1,x,y)
        f_hlp(2,x  ,y_n) = gp(2,x,y)
        f_hlp(3,x_w,y ) = gp(3,x,y)
        f_hlp(4,x  ,y_s) = gp(4,x,y)
        f_hlp(5,x_e,y_n) = gp(5,x,y)
        f_hlp(6,x_w,y_n) = gp(6,x,y)
        f_hlp(7,x_w,y_s) = gp(7,x,y)
        f_hlp(8,x_e,y_s) = gp(8,x,y)

30 continue

   do 39 y = 1, ly
      do 40 x = 1, lx
        do 31 k = 1, 8
        gp(k,x,y) = f_hlp(k,x,y)
31 continue
40 continue
39 continue

   end
c-------------------------------------------------
```

```
! Collision steps
      subroutine collision()
      implicit none
      include 'head.inc'
      integer x, y , k
      real*8 feq2(0:8), geq2(0:8), feq(0:8), ximud(0:8)
      real*8 xirhod(0:8), xirhodc(0:8), ximudc(0:8)
      real*8 xirhpd(0:8), xirhpdc(0:8)
      real*8 tmp1, tmp2, tmp3, tmp4, tmp5, tmp6

      do 10 y= 1, ly
      do 20 x =1, lx

      tau = tau_l + rh(x,y)*(tau_h - tau_l)

      call get_feq(u_x(x,y), u_y(x,y), feq)

      call xi_difference(x,y,rhp, xirhpd, xirhpdc)
      call xi_difference(x,y, mu, ximud, ximudc)
      call xi_difference(x,y, rho, xirhod, xirhodc)

      tmp1 = (rhdx(x,y)+brhdx(x,y))*0.5d0*LmG
      tmp2 = (rhdy(x,y)+brhdy(x,y))*0.5d0*LmG
      tmp3 = (mudx(x,y)+bmudx(x,y))*0.5d0*rh(x,y)
      tmp4 = (mudy(x,y)+bmudy(x,y))*0.5d0*rh(x,y)

      tmp5 = (rhpdx(x,y)+brhpdx(x,y))*0.5d0
      tmp6 = (rhpdy(x,y)+brhpdy(x,y))*0.5d0

      do 30 k=0, 8

        feq2(k)  = rh(x,y)*feq(k)  - 0.5d0 *feq(k)/c_squ
     &         *( xirhpdc(k)  -rh(x,y)*ximudc(k)
     &          +(-u_x(x,y) )*(rhpdx(x,y) - rh(x,y)*mudx(x,y))
     &          +(-u_y(x,y) )*(rhpdy(x,y) - rh(x,y)*mudy(x,y))
     &          + Mob*lapmu(x,y)*c_squ
     &          )

        geq2(k)  = pb(x,y)*t_k(k) + rho(x,y)*c_squ*(feq(k)-t_k(k) )
     &          - 0.5d0 *c_squ *(feq(k) -t_k(k) )
     &         *( xirhodc(k)
     &             -u_x(x,y)  *rhdx(x,y)*LmG
     &             -u_y(x,y)  *rhdy(x,y)*LmG
     &             )

     &          - 0.5d0 *feq(k)
     &         *(  -rh(x,y)*ximudc(k)
     &             -u_x(x,y) *( - rh(x,y)*mudx(x,y))
     &             -u_y(x,y) *( - rh(x,y)*mudy(x,y))
     &             )

c--------------------
      ff(k,x,y)  = ff(k,x,y)  -(ff(k,x,y)- feq2(k) )/tau
     &          + feq(k)/c_squ *(
     &          xirhpd(k)- rh(x,y)*ximud(k)
     &          +(-u_x(x,y) )*(tmp5   - tmp3   )
```

```fortran
     &             +(-u_y(x,y) )*(tmp6   - tmp4  )
     &             + Mob*lapmu(x,y)*c_squ
     &                )

        gp(k,x,y) = gp(k,x,y) -(gp(k,x,y)- geq2(k) )/tau
     &             + c_squ*(feq(k) -t_k(k) )
     &            *( xirhod(k)
     &               -u_x(x,y) *tmp1
     &               -u_y(x,y) *tmp2
     &               )

     &             + feq(k)
     &            *(  -rh(x,y)*ximud(k)
     &               -u_x(x,y) *( - tmp3)
     &               -u_y(x,y) *( - tmp4)
     &               )
 30   continue

 20   continue
 10   continue

      end


Initial.for
C========================================================

      subroutine  init()
      implicit none
      include 'head.inc'
      integer ix, iy, k, x, y
      real*8 a(lx,ly), b(lx,ly), feq(0:8)
      real*8 rho1, Re1
! Read main parameters from file params.in
      open(1,file='params.in')
      read(1,*)  ix, iy
      read(1,*)  t_max   ! Maximum time step
      read(1,*)  Nwri    ! Dump data per 'Nwri' steps
      read(1,*)  rho_h, rho_l    ! Maximum and minimum density
! Relaxation time for high-density and low-density fluid,
! respectively
      read(1,*)  tau_h, tau_l
      read(1,*)  thick   ! Thickness of the interface, several
! lattice units
      read(1,*)  sigma   ! Surface tension
      read(1,*)  RR      ! Droplet or bubble's radii
      read(1,*)  UU      ! Velocity of droplet or bubble
      read(1,*)  Mob     ! Mobility

      close(1)

      DD = 2.d0 *RR
      LmG = rho_h - rho_l
! If domain size ('ix' and 'iy' in the params.in) is different from
! that in head.inc,
! please check params.in again and rerun the program.
```

```fortran
      if(ix .ne. lx .or. iy .ne. ly) then
      write(*,*) "error in lx , ly input! "
      stop
      endif
! \beta and \kappa are main parameters in the free-energy expression
      beta    = 12.d0*sigma/(thick*( (rho_h - rho_l)**4 ))
      kappa   = 1.5d0*sigma*thick/( (rho_h - rho_l)**2 )

      Re1 = UU *DD*3.d0/(tau_h-0.5)
      We = UU*UU*DD*rho_h/sigma
! Write down the main parameters in the simulation (record in
! params.dat)
      open(2,file='./out/params.dat')
      write(2,'("lx=", i5, 3X, "ly=", i5)')  ix, iy
      write(2,'("t_max=", i10)')  t_max
      write(2,'("rho_h=", f12.5,4X, "rho_l=", f12.5)')  rho_h, rho_l
      write(2,'("tau_h=", f12.5,4X, "tau_l=", f12.5)')  tau_h, tau_l
      write(2,'("thickness=", f12.7)')  thick
      write(2,'("beta=", f12.6)')  beta
      write(2,'("Nwri=", i9)')  Nwri
      write(2,'("RR=", f12.5)')  RR
      write(2,'("sigma=",f12.8)') sigma
      write(2,'("kappa=",f12.8)') kappa
      write(2,'("UU=",f12.8)') UU
      write(2,'("Eo=",f12.8)') Eo
      write(2,'("Mo=",f12.8)') Mo
      write(2,'("We=",f12.8)') We
      write(2,'("Re1=",f12.8)') Re1
      write(2,'("Mob=",f12.8)') Mob
      close(2)
c--------------
      rhoa = 0.5d0*(rho_h+rho_l)
      rhom = 0.5d0*(rho_h-rho_l)
c--------------

      x_cen= (1+lx)/2
      y_cen= (1+ly)/2

      do 21 y = 1, ly
      do 31 x = 1, lx
      u_x(x,y) = 0.d0
      u_y(x,y) = 0.d0
        if( sqrt( dble(x-x_cen)**2+dble(y-y_cen)**2) .lt. RR ) then
          u_x(x,y) = UU
      endif

      rho(x,y) = rhoa - rhom
   &  *tanh( (dsqrt( dble(x-x_cen)**2
   &  + dble(y-y_cen)**2  ) - RR ) *2.d0/thick)

      rh(x,y) = (rho(x,y)-rho_l)/(rho_h-rho_l)
   31 continue
   21 continue

      do 25 y = 1, ly
       do 15 x = 1, lx
```

```
      obst (x,y)  = 0
      rhdx (x,y)  = 0.d0
      rhdy (x,y)  = 0.d0
      brhdx (x,y) = 0.d0
      brhdy (x,y) = 0.d0
      mudx (x,y)  = 0.d0
      mudy (x,y)  = 0.d0
      bmudx (x,y) = 0.d0
      bmudy (x,y) = 0.d0
      lap (x,y) = 0.d0
      pb (x,y)  = 0.d0
  15 continue
  25 continue
C-----------------------
! Laplacian operator
      call laplace (rh, lap)

      do 40 y = 1, ly
      do 30 x = 1, lx
      mu (x,y) = 4.d0 * beta* (rh (x,y) -0.5d0)
   &    * (rh (x,y) -1.d0)* (rh (x,y) -0.d0)
   &    - kappa* lap (x,y)
  30 continue
  40 continue
C-----------------------

      call Grad (rh, rhdx, rhdy, brhdx, brhdy, 1)
      call Grad (mu, mudx, mudy, bmudx, bmudy, 1)

C-----------------------
      do 80 y = 1, ly
      do 70 x = 1, lx
      call get_feq (u_x (x,y), u_y (x,y), feq)

      do k=0, 8
      ff (k,x,y) = rh (x,y)*feq (k) - 0.5d0 *feq (k)/c_squ
   &    * ( (xc (k) -u_x (x,y)  )* (rhdx (x,y) *c_squ - rh (x,y) *mudx (x,y) )
   &        + (yc (k) -u_y (x,y)  )* (rhdy (x,y) *c_squ - rh (x,y) *mudy (x,y) )
   &        )
      enddo

  70 continue
  80 continue

      end


Utility.for
C==============================================================
! Calculate the convergence of the numerical solution
      subroutine calc_error (time)
      implicit none
      include "head.inc"
      integer time,i,j,n_free
      real*8    a_vel,err_a,err_b
      err_a = 0.d0
```

```
      err_b = 0.d0
c
      do 30 j = 1, ly
      do 30 i = 1, lx
      if(obst(i,j) .eq. 0 ) then
      err_a = err_a + ( (u_x(i,j) - up(i,j))*(u_x(i,j) - up(i,j)) +
    &  (u_y(i,j) - vp(i,j))*(u_y(i,j) - vp(i,j))  )
        err_b = err_b + (u_x(i,j)*u_x(i,j) + u_y(i,j)*u_y(i,j))
      endif
  30 continue
        error = dsqrt(err_a/err_b)
      write(39,*) time, error
      return
      end
c----------------------
      subroutine get_feq(ux, uy, fequ)
      implicit none
      include 'head.inc'
      integer i
      real*8 ux,uy,fequ(0:8),u_n(0:8),u_squ,p1, tmp1

          u_squ = ux * ux + uy * uy
          tmp1  =  u_squ *1.5d0

        fequ(0) = t_k(0)* ( 1.d0 - tmp1 )

      do i = 1, 8
         u_n(i) =   ux *xc(i) + uy* yc(i)
         fequ(i) = t_k(i)*  (1.d0+ u_n(i) / c_squ
    &                  + u_n(i) * u_n(i) /con2
    &                  - tmp1 )
      enddo
      end
c--------------------------------------------------------
! Get gradient values for macro-variable 'p'
      subroutine Grad(p,dx,dy,bdx,bdy, id)
      implicit none
      include 'head.inc'
      integer  x,y,xp, yp, xp2, yp2, xn, xn2, yn, yn2, id
      real*8 p(lx,ly), dx(lx,ly), dy(lx,ly)
      real*8 bdx(lx,ly), bdy(lx,ly)

      do 10 y = 1, ly
      do 20 x = 1, lx
! Periodic boundary condition
      xp = x+1
      if(x .eq. lx)   xp = 1
      yp = y+1
      if(y .eq. ly)   yp = 1
      xn = x-1
      if(x .eq. 1)    xn = lx
      yn = y-1
      if(y .eq. 1)    yn = ly
      xp2 = x+2
      if(xp2 .gt. lx)  xp2 = xp2-lx
      xn2 = x-2
```

```
      if(xn2 .lt. 1)    xn2 = xn2+lx
      yp2 = y+2
      if(yp2 .gt. ly)   yp2 = yp2-ly
      yn2 = y-2
      if(yn2 .lt. 1)    yn2 = yn2+ly

      dx(x,y) = (p(xp,y)-p(xn,y))/3.d0  +(p(xp,yp)-p(xn,yn))/12.d0
   &        +   (p(xp,yn)-p(xn,yp))/12.d0

      dy(x,y) = (p(x,yp)-p(x,yn))/3.d0  +(p(xp,yp)-p(xn,yn))/12.d0
   &        +   (p(xn,yp)-p(xp,yn))/12.d0

! Biased difference
      if(id .eq. 1) then
      bdx(x,y) =(-p(xp2,y)+4.d0*p(xp,y)-(-p(xn2,y) +4.d0*p(xn,y)) )/6.d0
   &   +(-p(xp2,yp2)+4.d0*p(xp,yp) -(-p(xn2,yn2)+4.d0*p(xn,yn))
   &   +(-p(xp2,yn2)+4.d0*p(xp,yn))-(-p(xn2,yp2)+4.d0*p(xn,yp)))/24.d0

      bdy(x,y) = (-p(x,yp2)+4.d0*p(x,yp)-(-p(x,yn2) +4.d0*p(x,yn)))/6.d0
   &   +(-p(xp2,yp2)+4.d0*p(xp,yp) -(-p(xn2,yn2)+4.d0*p(xn,yn))
   &   +(-p(xn2,yp2)+4.d0*p(xn,yp))-(-p(xp2,yn2)+4.d0*p(xp,yn)) )/24.d0
      endif
 20 continue
 10 continue

      end

C---------------------------------------
! Get directional derivatives at position (x,y)
      subroutine xi_difference(x,y, p, xid, xidc)
      implicit none
      include 'head.inc'
      real*8 p(lx,ly), xid(0:8), xidc(0:8)
      integer x, y, xp, yp, xn, yn, xp2, yp2, xn2, yn2

      xp = x+1
      if(x .eq. lx)    xp = 1
      yp = y+1
      if(y .eq. ly)    yp = 1
      xn = x-1
      if(x .eq. 1)     xn = lx
      yn = y-1
      if(y .eq. 1)     yn = ly
      xp2 = x+2
      if(xp2 .gt. lx)  xp2 = xp2-lx
      xn2 = x-2
      if(xn2 .lt. 1)   xn2 = xn2+lx
      yp2 = y+2
      if(yp2 .gt. ly)  yp2 = yp2-ly
      yn2 = y-2
      if(yn2 .lt. 1)   yn2= yn2+ly
C----------------------
! Half of biased and central finite differences
      xid(0) = 0.d0
      xid(1) =0.25d0* (-p(xp2,y)+5.d0*p(xp,y) -3.d0*p(x,y) -p(xn,y))
      xid(2) =0.25d0* (-p(x,yp2)+5.d0*p(x,yp) -3.d0*p(x,y) -p(x,yn))
```

```
      xid(3) =0.25d0* (-p(xn2,y)+5.d0*p(xn,y) -3.d0*p(x,y) -p(xp,y))
      xid(4) =0.25d0* (-p(x,yn2)+5.d0*p(x,yn) -3.d0*p(x,y) -p(x,yp))
      xid(5) =0.25d0*(-p(xp2,yp2)+5.d0*p(xp,yp)-3.d0*p(x,y) -p(xn,yn))
      xid(6) =0.25d0*(-p(xn2,yp2)+5.d0*p(xn,yp)-3.d0*p(x,y) -p(xp,yn))
      xid(7) =0.25d0*(-p(xn2,yn2)+5.d0*p(xn,yn)-3.d0*p(x,y) -p(xp,yp))
      xid(8) =0.25d0*(-p(xp2,yn2)+5.d0*p(xp,yn)-3.d0*p(x,y) -p(xn,yp))

!  Central difference
      xidc(0) =0.d0
      xidc(1) =0.5d0* (p(xp,y)   -p(xn,y))
      xidc(2) =0.5d0* (p(x,yp)   -p(x,yn))
      xidc(3) =0.5d0* (p(xn,y)   -p(xp,y))
      xidc(4) =0.5d0* (p(x,yn)   -p(x,yp))
      xidc(5) =0.5d0*(p(xp,yp) -p(xn,yn))
      xidc(6) =0.5d0*(p(xn,yp) -p(xp,yn))
      xidc(7) =0.5d0*(p(xn,yn) -p(xp,yp))
      xidc(8) =0.5d0*(p(xp,yn) -p(xn,yp))

   10 continue

      end
c-------------------------------------------
! Laplacian operator
      subroutine laplace(p,nab)
      implicit none
      include 'head.inc'
      real*8 p(lx,ly)
      real*8 nab(lx,ly)
      integer  x,y,xp,xn,yp,yn

      do 10 y = 1, ly
      do 20 x = 1, lx
         yp = mod(y,ly) + 1
         xp = mod(x,lx) + 1
         yn = ly - mod(ly + 1 - y, ly)
         xn = lx - mod(lx + 1 - x, lx)

         nab(x,y) = (  p(xp,yp) +p(xn,yp) +p(xp,yn) +p(xn,yn)
     &        +4.d0* (p(xp,y)+ p(xn,y) +p(x,yp) +p(x,yn) )
     &          -20.d0* p(x,y)   )/6.d0
   20 continue
   10 continue

      return
      end

Output.for
c====================================================
! Write binary Tecplot file
      subroutine  write_results(upx,upy, n)
      implicit none
      include "head.inc"
      integer  x,y,i,j,n, k
      real*8  upx(lx,ly),upy(lx,ly)
      REAL*4 ZONEMARKER, EOHMARKER
      integer*2 ii, jj
      integer*1 obob
```

```fortran
      integer, parameter:: kmax=1
      character*40 Title,var
      character*40 V1,V2,V3,V4, V5, V6, V7, V8
      character*40 Zonename1
      character filename*46, B*7, C*4

      write(B,'(i7.7)') n
      filename='out/_'//B//'.plt'

        open(41,file=filename,  form='binary')
C-------------------------------------
      ZONEMARKER=             299.0
      EOHMARKER =             357.0
C-------------------------------------
      write(41) "#!TDV101"
      write(41) 1
      Title="LeeCF10"
      call dumpstring(Title)
! Number of variables (NumVar) in the datafile.
      write(41) 7

      V1='X'
      call dumpstring(V1)
      V2='Y'
      call dumpstring(V2)
      V3='u'
      call dumpstring(V3)
      V4='v'
      call dumpstring(V4)
      V5='rho'
      call dumpstring(V5)
      V7='obst'
      call dumpstring(V7)
      V8='pb'
      call dumpstring(V8)
! Zones---------------------
! Zone marker. Value = 299.0
      write(41) ZONEMARKER
      Zonename1="ZONE_001"
      call dumpstring(Zonename1)
! Zone Color
      write(41) -1
! ZoneType
      write(41) 0
! DataPacking 0=Block, 1=Point
      write(41) 1
! Specify Var Location.
! Don't specify=0, all data is located at the nodes. Specify=1
      write(41) 0
! Number of user defined face neighbor connections (value >= 0)
      write(41) 0
! IMax,JMax,KMax
      write(41) lx
      write(41) ly
      write(41) kmax
```

```
! Auxiliary name/value pair to follow=1   No more Auxiliary
! name/value pairs=1.
      write(41) 0
!  HEADER OVER-------
!  EOHMARKER, value=357.0
         write(41) EOHMARKER
! II. Data section++++++++++++++++++++++++
         write(41) Zonemarker
! Data format, 1=Float, 2=Double, 3=LongInt, 4=ShortInt, 5=Byte,
! 6=Bit
      write(41) 4
      write(41) 4
      write(41) 1
      write(41) 1
      write(41) 1
      write(41) 5
      write(41) 1
! Has variable sharing 0 = no, 1 = yes.
      write(41) 0
! Zone number to share connectivity list with (-1 = no sharing).
      write(41) -1
! Zone Data. Each variable is in data format as specified above.

      do k=1,kmax
      do j=1,ly
      do i=1,lx
      if(obst(i,j).ne. 0) then
      else
      endif
      ii = i
      jj = j
      obob = obst(i,j)

      write(41) ii
      write(41) jj
      write(41) real(upx(i,j))
      write(41) real(upy(i,j))
      write(41) real(rho(i,j))
      write(41) obob
      write(41) real(pb(i,j))
        end do
      end do
      end do
        close(41)

      end

c-------------------------------
      subroutine dumpstring(instring)
      character(40) instring
      integer len

      len=LEN_TRIM(instring)
      do ii=1,len
      I=ICHAR(instring(ii:ii))
      write(41) I
```

```
        end do
        write(41) 0
        return
        end


params.in
C=================================================
101    101    ! lx,ly
2000000       ! t_max
5000          ! Nwri
1.0   0.001   ! rho_h, rho_l
0.53  0.85    ! tau_h, tau_l
4. 0          ! interface thickness
0.00000625    ! sigma
20            ! RR radius of a drop
0.0d0         ! UU
0.1d0         ! Mob
```

CHAPTER 10

Axisymmetric high-density ratio two-phase LBMs (extension of the HCZ model)

10.1 Introduction

As illustrated in Chapter 9, a series of models (Amaya-Bower and Lee 2010; Lee and Lin 2005; Lee and Liu 2010) based on the HCZ model (He et al. 1999) have been further developed to handle higher-density ratio multiphase flows. Lee and Lin (2005) proposed the use of directional derivatives. These play an important role in simulations of cases with high-density ratios. After 2006, Lee and Fischer (2006), Lee and Liu (2010), and Amaya-Bower and Lee (2010) applied an isotropic discretization scheme in their models as well as the directional derivatives.

In this chapter the axisymmetric model of Mukherjee and Abraham (2007c), which is based on Lee and Lin (2005), is introduced briefly and applied to simulate a droplet splashing. An axisymmetric model based on Lee and Liu (2010) is introduced in Section 10.3 and is applied to head-on collisions of two droplets and bubble rise problems. However, it is found that the mass of the bubble may be not conserved. As a remedy, the mass conservation is improved through applying the mass correction technique described in Section 8.7.3.

10.2 The model based on Lee and Lin (2005)

From Chapter 8 we know that, compared to purely 2D cases, there are some source terms due to the axisymmetric effect on the right-hand side of the N–S equation (Huang et al. 2014a):

$$H_\alpha = \rho v \left\{ \frac{\partial_r u_\alpha}{r} - \frac{u_r}{r^2} \delta_{\alpha r} + \partial_\alpha \left(\frac{u_r}{r} \right) \right\} - \frac{\rho u_r u_\alpha}{r}. \tag{10.1}$$

Multiphase Lattice Boltzmann Methods: Theory and Application, First Edition.
Haibo Huang, Michael C. Sukop and Xi-Yun Lu.
© 2015 John Wiley & Sons, Ltd. Published 2015 by John Wiley & Sons, Ltd.
Companion Website: www.wiley.com/go/huang/boltzmann

In the following description, all the source terms in the N–S equation are integrated as $\mathbf{F} = \mathbf{H} + \mathbf{F}_s + \mathbf{G}$, where $\mathbf{F}_s$ is the surface tension. If the surface tension takes the form

$$\mathbf{F}_s = \kappa \rho \nabla (\nabla^2 \rho), \tag{10.2}$$

then in cylindrical coordinates, it can be written as

$$(F_s)_\alpha = \kappa \rho \partial_\alpha (\partial_\gamma^2 \rho) + \kappa \rho \partial_\alpha \left(\frac{1}{r} \partial_r \rho \right). \tag{10.3}$$

The last term represents the effect of cylindrical coordinates.

There are some typographical errors in Mukherjee and Abraham (2007c). From the above, it is seen that the extra terms due to axisymmetry on the right-hand side of both Eqs (22) and (24) in Mukherjee and Abraham (2007c) should be

$$H_\alpha = \eta \left\{ \frac{\partial_r u_\alpha}{r} - \frac{u_r}{r^2} \delta_{\alpha r} + \partial_\alpha \left(\frac{u_r}{r} \right) \right\} - \frac{\rho u_r u_\alpha}{r} + \kappa \rho \partial_\alpha \left(\frac{1}{r} \partial_r \rho \right) \tag{10.4}$$

instead of

$$\frac{\eta}{r} [\partial_r u_\alpha + \partial_i u_r] - \frac{\rho u_r u_\alpha}{r} + \kappa \rho \partial_\alpha \left(\frac{1}{r} \partial_r \rho \right). \tag{10.5}$$

Here we briefly introduce the model of Mukherjee and Abraham (2007c). In this model two distribution functions $\bar{g}_i$ and $\bar{f}_i$ are introduced to recover the axisymmetric incompressible multiphase N–S equations and a macro interface-tracking equation (a CH-like equation), respectively.

$$\bar{g}_i(\mathbf{x} + \mathbf{e}_i \Delta t, t + \Delta t) = \bar{g}_i(\mathbf{x}, t) - \frac{1}{\tau_1} (\bar{g}_i(\mathbf{x}, t) - \bar{g}_i^{eq}(\mathbf{x}, t)) + S_i(\mathbf{x}, t)\Delta t, \tag{10.6}$$

$$\bar{f}_i(\mathbf{x} + \mathbf{e}_i \Delta t, t + \Delta t) = \bar{f}_i(\mathbf{x}, t) - \frac{1}{\tau_2} (\bar{f}_i(\mathbf{x}, t) - \bar{f}_i^{eq}(\mathbf{x}, t)) + S_i'(\mathbf{x}, t)\Delta t, \tag{10.7}$$

where τ_1 is a relaxation time related to the kinematic viscosity as $\nu = c_s^2(\tau_1 - 0.5) \Delta t$. τ_2 is related to the mobility in the CH equation. $S_i(\mathbf{x}, t)$ and $S_i'(\mathbf{x}, t)$ are the source terms in Eqs (10.6) and (10.7), respectively. The equilibrium distribution function $\bar{g}_i^{eq}(\mathbf{x}, t)$ and $\bar{f}_i^{eq}(\mathbf{x}, t)$ can be calculated as described in the following section.

In He et al. (1999) and Lee and Lin (2005), there are two slightly different definitions for the equilibrium distribution function (EDF) for $\bar{g}_i$. Correspondingly, the macrovariables obtained from summation of the distribution function are slightly different. However, the difference did not affect the simulated result or the macroscopic equations they recovered.

10.2.1 The equilibrium distribution functions I

The equilibrium distribution functions in He et al. (1999) are

$$\bar{g}_i^{eq}(\mathbf{x}, t) = w_i \left[p + c_s^2 \rho \left(\frac{e_{i\alpha} u_\alpha}{c_s^2} + \frac{e_{i\alpha} u_\alpha e_{i\beta} u_\beta}{2c_s^4} - \frac{u_\alpha u_\alpha}{2c_s^2} \right) \right] \qquad (10.8)$$

and

$$\bar{f}_i^{eq}(\mathbf{x}, t) = w_i \rho \left[1 + \frac{e_{i\alpha} u_\alpha}{c_s^2} + \frac{e_{i\alpha} u_\alpha e_{i\beta} u_\beta}{2c_s^4} - \frac{u_\alpha u_\alpha}{2c_s^2} \right], \qquad (10.9)$$

respectively, where p and ρ are the hydrodynamic pressure and density of the fluid, respectively.

The macroscopic variables are given by

$$\rho = \sum \bar{f}_i - \frac{\rho}{2} \frac{u_r}{r}, \qquad (10.10)$$

$$p = \sum \bar{g}_i + \frac{\Delta t}{2} u_\beta E_\beta - \frac{1}{2} c_s^2 \rho \frac{u_r}{r} \Delta t, \qquad (10.11)$$

where

$$E_\beta = \partial_\beta (c_s^2 \rho), \qquad (10.12)$$

and

$$\rho u_\alpha c_s^2 = \sum e_{i\alpha} \bar{g}_i + \frac{\Delta t}{2} c_s^2 F_\alpha. \qquad (10.13)$$

For more details refer to Section 8.2.2.

10.2.2 The equilibrium distribution functions II

The equilibrium distribution function in Lee and Lin (2005) for $\bar{f}_i^{eq}$ is identical to Eq. (10.9) given in the above section. The equilibrium distribution function in Lee and Lin (2005) for $\bar{g}_i^{eq}$ is slightly different from the above Eq. (10.8):

$$\bar{g}_i^{eq}(\mathbf{x}, t) = w_i \left[\frac{p}{c_s^2} + \rho \left(\frac{e_{i\alpha} u_\alpha}{c_s^2} + \frac{e_{i\alpha} u_\alpha e_{i\beta} u_\beta}{2c_s^4} - \frac{u_\alpha u_\alpha}{2c_s^2} \right) \right]. \qquad (10.14)$$

The macroscopic variables p and u_α are given as

$$p = c_s^2 \sum \bar{g}_i + \frac{\Delta t}{2} u_\beta E_\beta - \frac{1}{2} c_s^2 \rho \frac{u_r}{r} \Delta t \qquad (10.15)$$

and

$$\rho u_\alpha = \sum e_{i\alpha} \bar{g}_i + \frac{\Delta t}{2} F_\alpha. \qquad (10.16)$$

In Eq. (61) of Mukherjee and Abraham (2007c) there is a typographical error that reads $E_\beta = -\partial_\beta(p - c_s^2 \rho)$, which is not consistent with $E_\beta = \partial_\beta(c_s^2 \rho)$ in Eq. (59) of Mukherjee and Abraham (2007c).

10.2.3   Source terms

The source terms appearing in Eqs (10.6) and (10.7) are (He et al. 1999):

$$S_i = \left(1 - \frac{1}{2\tau_1}\right) \left\{ (e_{i\alpha} - u_\alpha) F_\alpha \Gamma_i(\mathbf{u}) + (e_{i\alpha} - u_\alpha) E_\alpha [\Gamma_i(\mathbf{u}) - \Gamma_i(0)] - w_i c_s^2 \frac{\rho u_r}{r} \right\},$$
(10.17)

where $\Gamma_i(\mathbf{u}) = \bar{f}_i^{eq}/\rho$, and

$$S_i' = \left(1 - \frac{1}{2\tau_2}\right) \left\{ \frac{(e_{i\alpha} - u_\alpha) F_\alpha'}{c_s^2 \rho} \bar{f}_i^{eq} - w_i \frac{\rho u_r}{r} \right\},$$
(10.18)

where

$$F_\alpha' = \partial_\alpha (c_s^2 \rho) + H_\alpha.$$
(10.19)

In the Chapman–Engskog expansion derivation in Chapter 8 we can see that only when $E_\beta = -\partial_\beta (p - c_s^2 \rho)$ and $F_\alpha' = -\partial_\alpha (p - c_s^2 \rho)$ can the axisymmetric N–S equation be recovered correctly. Here Eqs (10.12) and (10.19) are different from those formulae. Hence, theoretically, this model of Mukherjee and Abraham (2007c) recovers momentum equations slightly different from the desired N–S equations.

10.2.4   Stress and potential form of intermolecular forcing terms

For pure 2D cases the intermolecular forcing term is (Lee and Lin 2005)

$$F_\alpha^i = \partial_\beta (\rho c_s^2 - P) \delta_{\alpha\beta} + \kappa \partial_\beta \left[\left(\frac{1}{2} \partial_\delta \rho \partial_\delta \rho + \rho \partial_\delta^2 \rho \right) \delta_{\alpha\beta} - \partial_\alpha \rho \partial_\beta \rho \right],$$
(10.20)

where P is the thermodynamic pressure. Lee and Lin (2005) suggested two forms for the above equation in the simulation: the stress form and the potential form.

In the 2D Lattice Boltzmann equation that recovers the momentum equations the stress form should be used (Lee and Lin 2005), i.e.,

$$\begin{aligned}
F_\alpha^i &= \partial_\beta (\rho c_s^2 - p) \delta_{\alpha\beta} + \kappa \partial_\beta (\partial_\delta \rho \partial_\delta \rho \delta_{\alpha\beta} - \partial_\alpha \rho \partial_\beta \rho) \\
&= \partial_\beta (\rho c_s^2 - p) \delta_{\alpha\beta} + \kappa \partial_\alpha (\partial_\delta \rho \partial_\delta \rho) - \kappa (\partial_\beta \rho \partial_\alpha \partial_\beta \rho + \partial_\alpha \rho \nabla^2 \rho),
\end{aligned}$$
(10.21)

in which the modified pressure (hydrodynamic pressure) is defined as

$$p = P - \kappa \rho \nabla^2 \rho + \frac{\kappa}{2} \partial_\delta \rho \partial_\delta \rho.$$
(10.22)

For the 2D Lattice boltzmann equation for the order parameter, which is able to recover the CH equation macroscopically, the intermolecular force is suggested to take the potential form (Lee and Lin 2005):

$$F_\alpha^i = \partial_\beta (\rho c_s^2 - P) \delta_{\alpha\beta} + \kappa \rho \partial_\alpha \nabla^2 \rho.$$
(10.23)

To understand the equivalence of the potential and stress forms, refer to Section 9.2.1.

For the axisymmetric cases in cylindrical coordinates (x, r), the stress form of the intermolecular forcing term should be similar to Eq. (10.21), with a slight difference due to axisymmetric effect First, Eq. (10.22) becomes (Mukherjee and Abraham 2007c)

$$p = P - \kappa \rho \left(\partial_\delta^2 \rho + \frac{\partial_r \rho}{r} \right) + \frac{\kappa}{2} \partial_\delta \rho \partial_\delta \rho \qquad (10.24)$$

because $\nabla^2 \rho$ in cylindrical coordinates is

$$\nabla^2 \rho = \partial_\delta^2 \rho + \frac{\partial_r \rho}{r}. \qquad (10.25)$$

This is Eq. (41) in Mukherjee and Abraham (2007c).

Because $\partial_\beta \rho \partial_\alpha \partial_\beta \rho + \partial_\alpha \rho \partial_\beta^2 \rho = \partial_\beta(\partial_\alpha \rho \partial_\beta \rho)$, in cylindrical coordinates, the stress form should be (Mukherjee and Abraham 2007c)

$$\begin{aligned}
F_\alpha^i &= \partial_\beta(\rho c_s^2 - p)\delta_{\alpha\beta} + \kappa \partial_\alpha(\partial_\delta \rho \partial_\delta \rho) - \kappa \left[\partial_\beta \rho \partial_\alpha \partial_\beta \rho + \partial_\alpha \rho \left(\partial_\beta^2 \rho + \frac{\partial_r \rho}{r} \right) \right] \\
&= \partial_\beta(\rho c_s^2 - p)\delta_{\alpha\beta} + \left\{ \kappa \partial_\alpha(\partial_\delta \rho \partial_\delta \rho) - \kappa \partial_\beta(\partial_\alpha \rho \partial_\beta \rho) - \kappa \frac{\partial_r \rho \partial_\alpha \rho}{r} \right\}. \qquad (10.26)
\end{aligned}$$

Hence, in the Mukherjee and Abraham (2007c) model, the above braced terms are surface tension, i.e.,

$$(F_s)_\alpha = \kappa \partial_\alpha(\partial_\beta \rho \partial_\beta \rho) - \kappa \partial_\beta(\partial_\alpha \rho \partial_\beta \rho) - \kappa \frac{\partial_r \rho \partial_\alpha \rho}{r}. \qquad (10.27)$$

Through redefinition of pressure (Eq. (10.24)), the potential form is suggested to be (Mukherjee and Abraham 2007c)

$$F_\alpha^i = \partial_\beta(\rho c_s^2 - P)\delta_{\alpha\beta} + \kappa \rho \partial_\alpha \left(\partial_\delta^2 \rho + \frac{\partial_r \rho}{r} \right). \qquad (10.28)$$

This formula can be obtained through substituting Eq. (10.24) into Eq. (10.26); refer to the similar procedure in Eq. (9.10).

There are some additional typographical errors in Mukherjee and Abraham (2007c), for example in Eqs (42) and (43) the additional term should be $-\kappa \frac{\partial_r \rho \partial_\alpha \rho}{r}$ instead of $\kappa \frac{\partial_r \rho \partial_\alpha \rho}{r}$.

10.2.5 Chapman–Enskog analysis

N–S equations

Applying the Taylor expansion to Eq. (10.6) and using the Chapman–Enskog expansion $\partial_t = \partial_{t_1} + \varepsilon \partial_{t_2} + \ldots$ and $\bar{g}_i = \bar{g}_i^{(0)} + \varepsilon \bar{g}_i^{(1)} + \varepsilon^2 \bar{g}_i^{(2)}$, where $\varepsilon = \Delta t$ and retaining terms to $O(\varepsilon^2)$, in different scales, we have

$$O(\varepsilon^0) : \ (\bar{g}_i^{(0)} - \bar{g}_i^{eq})/\tau_1 = 0, \qquad (10.29)$$

$$O(\varepsilon^1) : \ (\partial_{t_1} + e_{i\alpha} \partial_\alpha)\bar{g}_i^{(0)} + \frac{1}{\tau_1} \bar{g}_i^{(1)} - S_i = 0, \qquad (10.30)$$

$$O(\varepsilon^2) : \quad \partial_{t_2}\bar{g}_i^{(0)} + \left(1 - \frac{1}{2\tau_1}\right)(\partial_{t_1} + e_{i\alpha}\partial_\alpha)\bar{g}_i^{(1)} + \frac{1}{2}(\partial_{t_1} + e_{i\alpha}\partial_\alpha)S_i$$

$$+\frac{1}{\tau_1}\bar{g}_i^{(2)} = 0. \tag{10.31}$$

Since the detailed derivation has been given in Chapter 8, here only brief conclusions are given. From the zeroth moments of Eqs (10.30) and (10.31) we obtain

$$\partial_t p + c_s^2 \left(\rho\partial_\alpha u_\alpha + \frac{\rho u_r}{r}\right) = 0. \tag{10.32}$$

Hence, only when $\partial_t p = 0$ can the incompressibility condition $\partial_\alpha u_\alpha + \frac{\rho u_r}{r} = 0$ in cylindrical coordinates be satisfied.

Through a procedure similar to that in the appendix to Chapter 8 we have the recovered momentum equations,

$$\partial_t(\rho u_\beta) + \partial_\alpha(\rho u_\alpha u_\beta) + \rho\frac{u_r u_\alpha}{r} = -\partial_\beta p + v\partial_\alpha\{\rho[\partial_\beta u_\alpha + \partial_\alpha u_\beta]\}$$

$$+\rho v\left\{\frac{\partial_r u_\alpha}{r} - \frac{u_r}{r^2}\delta_{\alpha r} + \partial_\alpha\left(\frac{u_r}{r}\right)\right\} + (F_s)_\alpha + Q', \tag{10.33}$$

where $v = c_s^2(\tau_1 - 0.5)\Delta t$ and $(F_s)_\alpha$ is given by Eq. (10.27). Here

$$Q' = -(\tau_1 - 0.5)\partial_\alpha(u_\alpha\partial_\beta p + u_\beta\partial_\alpha p) \tag{10.34}$$

is the extra source term in the recovered momentum equations due to the definition of $E_\gamma = \partial_\gamma(c_s^2\rho)$. For more detail please refer to the appendix to Chapter 8. This equation is the recovered axisymmetric two-phase N–S equation.

CH equation

To obtain the equation that is used to track interfaces, we start from equations analogous to Eqs (10.29)–(10.31) for the distribution functions $\bar{f}_i$. From Eqs (10.10) and (10.29) we know that $\sum_i \bar{f}_i^{(0)} = \rho$ and $\sum_i \bar{f}_i^{(1)} = \frac{\rho}{2}\frac{u_r}{r}$. Summing both sides of Eq. (10.30) over i yields

$$\partial_{t_1}\rho + \partial_\alpha(\rho u_\alpha) = -\frac{\rho u_r}{r}. \tag{10.35}$$

Summing Eq. (10.31) over i yields

$$\partial_{t_2}\rho + \left(1 - \frac{1}{2\tau_2}\right)\partial_\alpha\left(\sum_i e_{i\alpha}\bar{f}_i^{(1)}\right) + \frac{1}{2}\partial_\alpha\left(\sum_i e_{i\alpha}S_i'\right) = 0. \tag{10.36}$$

In the simulations, since $\bar{f}_i e_{i\alpha}$ is not evaluated, the term $\sum_i e_{i\alpha}\bar{f}_i^{(1)}$ is unknown. If we assume this term can be omitted and substituting $\sum_i e_{i\alpha}S_i' = \left(1 - \frac{1}{2\tau_2}\right)F_\alpha'$ into Eq. (10.36), we have

$$\partial_{t_2}\rho + \frac{1}{2}\left(1 - \frac{1}{2\tau_2}\right)\partial_\alpha F_\alpha' = 0. \tag{10.37}$$

Substituting $F'_\alpha = \partial_\alpha(c_s^2 \rho) + H_\alpha$ into the above equation and combining Eqs (10.35) and (10.37) together gives

$$\partial_t \rho + \partial_\alpha(\rho u_\alpha) + \frac{\rho u_r}{r} = -\frac{1}{2}\left(1 - \frac{1}{2\tau_2}\right)\partial_\alpha(\partial_\alpha c_s^2 \rho + H_\alpha). \tag{10.38}$$

This is the macro interface-tracking equation, which is a CH-like equation in the HCZ model.

10.2.6 Implementation

The MRT collision model (Lallemand and Luo 2000; McCracken and Abraham 2005) is used in this model.

The model is based on Mukherjee and Abraham (2007c) and is implemented in three steps: the pre-streaming collision step, the streaming step, and the post-streaming step.

Pre-streaming collision step

The pre-streaming collision step for the mass and N–S component is

$$g_i(\mathbf{x}, t) = \bar{g}_i(\mathbf{x}, t) - \sum_j \bar{\mathbf{S}}_{1ij}(\bar{g}_j - \bar{g}_j^{eq})|_{(\mathbf{x},t)}$$

$$+ \frac{\Delta t}{2}\frac{(e_{i\alpha} - u_\alpha)\partial_\alpha \rho c_s^2}{c_s^2}[\Gamma_i(\mathbf{u}) - \Gamma_i(0)]|_{(\mathbf{x},t)}$$

$$+ \frac{\Delta t}{2}\frac{(e_{i\alpha} - u_\alpha)F_\alpha}{c_s^2}\Gamma_i(\mathbf{u})|_{(\mathbf{x},t)} - w_i \frac{\rho u_r}{r}\frac{\Delta t}{2}\Big|_{(\mathbf{x},t)}, \tag{10.39}$$

where $\bar{\mathbf{S}}_1 = \mathbf{M}^{-1}\hat{\mathbf{S}}_1\mathbf{M}$.

For the order parameter in the CH equation, the Lattice Boltzmann equation equation is

$$f_i(\mathbf{x}, t) = \bar{f}_i(\mathbf{x}, t) - \sum_j \bar{\mathbf{S}}_{2ij}(\bar{f}_j - \bar{f}_j^{eq})|_{(\mathbf{x},t)}$$

$$+ \frac{\Delta t}{2}\frac{(e_{i\alpha} - u_\alpha)}{c_s^2}\left[\partial_\alpha \rho c_s^2 - \rho\partial_\alpha\left(\frac{\partial E_f}{\partial \rho} - \kappa\partial_\delta^2\rho - \kappa\frac{\partial_r\rho}{r}\right) + H_\alpha\right]\Gamma_i(\mathbf{u})|_{(\mathbf{x},t)}$$

$$- w_i \frac{\rho u_r}{r}\frac{\Delta t}{2}\Big|_{(\mathbf{x},t)}, \tag{10.40}$$

where $\bar{\mathbf{S}}_2 = \mathbf{M}^{-1}\hat{\mathbf{S}}_2\mathbf{M}$, and

$$\frac{\partial E_f}{\partial \rho} = \mu_0(\rho) = 4\beta(\rho - \rho_g)(\rho - \rho_l)\left(\rho - \frac{\rho_l + \rho_g}{2}\right). \tag{10.41}$$

The matrix $\mathbf{M}$ is illustrated in Chapter 8. The diagonal collision matrix $\hat{\mathbf{S}}$ is given by Lallemand and Luo (2000) as $\hat{\mathbf{S}} \equiv diag(s_0, s_1, s_2, s_3, s_4, s_5, s_6, s_7, s_8)$. Usually, the parameters are chosen as $s_0 = s_3 = s_5 = 1.0$, $s_1 = 1.64$, $s_2 = 1.54$, $s_4 = s_6 = 1.7$, and $s_7 = s_8 = \frac{1}{\tau}$. In the collision matrices $\hat{\mathbf{S}}_1$ and $\hat{\mathbf{S}}_2$, s_0 to s_6 are identical to those illustrated in the above except $s_7 = s_8 = \frac{1}{\tau_1}$ in $\hat{\mathbf{S}}_1$ and $s_7 = s_8 = \frac{1}{\tau_2}$ in $\hat{\mathbf{S}}_2$.

Streaming step

The streaming steps for the N–S equation and order parameter are

$$g_i(\mathbf{x} + \mathbf{e}_i \Delta t, t + \Delta t) = g_i(\mathbf{x}, t), \tag{10.42}$$

$$f_i(\mathbf{x} + \mathbf{e}_i \Delta t, t + \Delta t) = f_i(\mathbf{x}, t). \tag{10.43}$$

Post-streaming step

Post-streaming, the following is applied to distribution function g_i

$$
\begin{aligned}
\bar{g}_i(\mathbf{x}, t) = &\; g_i(\mathbf{x}, t) \\
&+ \frac{\Delta t}{2} \frac{(e_{i\alpha} - u_\alpha)\partial_\alpha \rho c_s^2}{c_s^2}[\Gamma_i(\mathbf{u}) - \Gamma_i(0)]|_{(\mathbf{x},t)} \\
&+ \frac{\Delta t}{2} \frac{(e_{i\alpha} - u_\alpha)F_\alpha}{c_s^2}\Gamma_i(\mathbf{u})|_{(\mathbf{x},t)} - w_i \frac{\rho u_r}{r} \frac{\Delta t}{2}\Big|_{(\mathbf{x},t)},
\end{aligned}
\tag{10.44}
$$

and the order parameter

$$
\begin{aligned}
\bar{f}_i(\mathbf{x}, t) = &\; f_i(\mathbf{x}, t) \\
&+ \frac{\Delta t}{2} \frac{(e_{i\alpha} - u_\alpha)}{c_s^2}\left[\partial_\alpha \rho c_s^2 - \rho \partial_\alpha\left(\frac{\partial E_f}{\partial \rho} - \kappa \partial_\delta^2 \rho - \kappa \frac{\partial_r \rho}{r}\right) + H_\alpha\right]\Gamma_i(\mathbf{u})|_{(\mathbf{x},t)} \\
&- w_i \frac{\rho u_r}{r} \frac{\Delta t}{2}\Big|_{(\mathbf{x},t)}.
\end{aligned}
\tag{10.45}
$$

Although $\tau_1 = \tau_2$ can be adopted, it is not necessary to use uniform $\tau_1 = \tau_2$ in the Lattice Boltzmann equations. On the other hand, τ_1 can be linearly interpolated from τ_l and τ_g

$$\tau_1 = \tau_g + \frac{\rho - \rho_g}{\rho_l - \rho_g}(\tau_l - \tau_g), \tag{10.46}$$

where τ_l and τ_g are relaxation parameters for the liquid and gas, respectively.

There are some first and second derivatives in the forcing terms in the Lattice Boltzmann equations. The "directional derivative" ($e_{i\alpha}$ times a derivative), e.g., $e_{i\alpha}\Delta t \frac{d\zeta}{dx_\alpha}$, appear in both. In this model we pay additional attention to these directional derivatives. For more details refer to Section 9.2.4.

For the derivatives other than the directional derivatives, the isotropic discretization formulae for first and second derivatives can be used (refer to Sections 8.6.1 and 9.3.1).

10.2.7   Droplet splashing on a thin liquid film

The case of a droplet splashing on a thin film is used to validate the numerical method in Lee and Lin (2005). However, in 2D simulations it may have some discrepancies with 3D experimental data (Josserand and Zaleski 2003). Here the axisymmetric case is expected to be more consistent with the experimental data.

In this case the Weber number (We) and the Reynolds number (Re) are defined as

$$We = \frac{\rho_l U^2 D}{\sigma} \tag{10.47}$$

and

$$Re = \frac{UD}{\nu_l}. \tag{10.48}$$

In the simulation, the computational domain is 800×400. Here, the MRT collision is used and $s_0 = 1$, $s_1 = 0.8$, $s_2 = 1.54$, $s_3 = s_5 = 1$, $s_4 = s_6 = 1.5$, and $s_7 = s_8 = \frac{1}{\tau}$. Here s_1, s_4, and s_6 are different from the choices in Mukherjee and Abraham (2007c). These values are obtained from trials because if $s_1 = 1.64$, the simulation blows up due to numerical instability. In this model $\rho_l = 1$ and $\rho_g = 0.001$. The relaxation times for liquid and gas are $\tau_l = 0.006$ and $\tau_g = 0.15$, respectively. The kinematic viscosity ratio is $\frac{\nu_l}{\nu_g} = 0.04$ and the dynamic viscosity ratio is 40. The interface thickness is 5 lu and the surface tension is set to be $\sigma = 0.625 \times 10^{-6}$ mu/ts^2. The diameter of the drop is $D = 200$ lu and the initial downward velocity of the drop is $U = 0.005$ lu/ts. The corresponding $Re = 500$ and $We = 8000$. The thin film thickness is 20 lu and initially the drop almost contacts the film. The result is shown in Figure 10.1.

The 3D view of the splashing is shown in Figure 10.2. We can see that the droplet initially contacts with the thin film at $t^* = 0$. As the droplet impinges, there is a circumferential jet along the outer periphery of the contact surface. The radial finger grows with time and gradually forms a corona ($t^* = 1.6$).

The normalized spreading radius r, i.e., $\frac{r}{D}$ as function of $t^* = \frac{Ut}{D}$, is shown in Figure 10.3. It is seen that at longer time the numerical result obeys the power law (Josserand and Zaleski 2003) well.

In the simulation it is found that the choice of the parameters $s_0, s_1, \ldots$ in the MRT collision are important, e.g., $s_2 = 1.54$, $s_4 = s_6 = 1.9$ is not applicable due to numerical instability.

10.2.8   Head-on droplet collision

Using this axisymmetric model, here the droplet head-on collision was simulated, in which the separation distance between the droplet centers is parallel to the relative velocity **U**. The relevant non-dimensional parameters, such as Re, We, and Oh, are defined as in Section 8.6.

We simulated a head-on droplet collision with $Re = 1000$ and $We = 100$. The computational domain is 800×230 and the diameters of both droplets are $D = 100$ lu. Initial velocities of the left and right droplets are $U = 0.01$ and $U = -0.01$, respectively. In the simulation the surface tension $\sigma = 0.0004$ and the interface thickness $W = 5$ lu. The densities of the liquid and gas are $\rho_l = 1$ and $\rho_g = 0.001$, respectively. The other relevant parameters are $\tau_l = 0.006$ and $\tau_g = 0.15$. The expected outcomes of such collisions are reviewed in Section 8.6.

Here, the expected outcome for $We > 20$ is reflexive separation (Ashgriz and Poo 1990). However, it is found that in our simulation using the above model only permanent coalescence is observed. Hence, this axisymmetric model seems

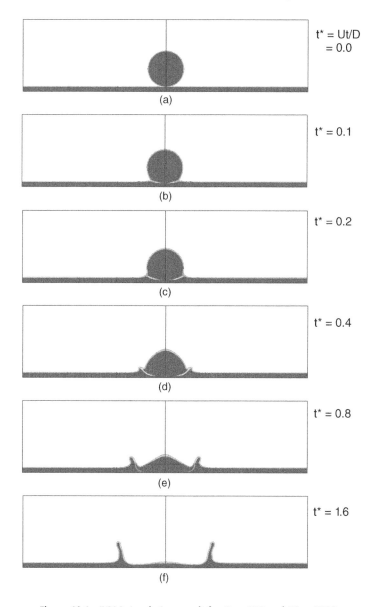

Figure 10.1   LBM simulation result for $Re = 500$ and $We = 8000$.

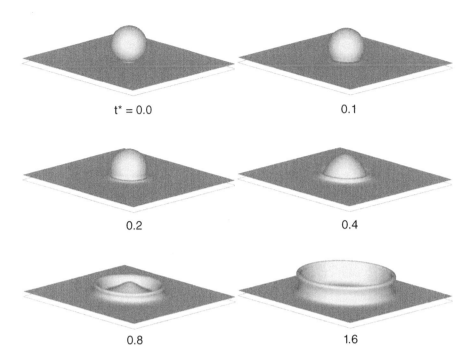

$t^* = 0.0$ 0.1

0.2   0.4

0.8   1.6

Figure 10.2   3D view of Figure 10.1 ($Re = 500$, $We = 8000$).

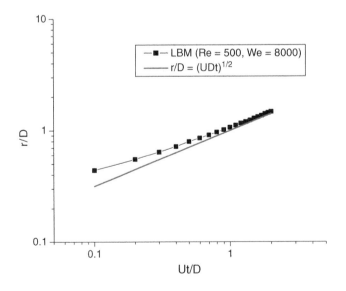

Figure 10.3   The measured spreading radius $\frac{r}{D}$ as a function of $t^* = \frac{Ut}{D}$ when $Re = 500$ and $We = 8000$. The straight line is the power law $r/D = \sqrt{UDt}$.

not perfect yet. A possible reason is that the surface tension in the model is not accurate (see Chapter 8).

10.3 Axisymmetric model based on Lee and Liu (2010)

Chiappini et al. (2009) applied the model of Lee and Liu (2010) to simulate the phenomenon of two-droplet coalescence. In the 2D simulations the simulation result is not consistent with experimental data (Chiappini et al. 2009). This may be due to the 2D effect.

Sun et al. (2013) extended the model to simulate axisymmetric flows, and their results seem consistent with the experimental data on the coalescence and separation of two-droplet head-on collisions. Since Amaya-Bower and Lee (2010) and Lee and Liu (2010) are very similar, we only introduce the axisymmetric model based on Lee and Liu (2010).

In the model of Lee and Liu (2010),

$$E_\alpha = c_s^2 \partial_\alpha \rho, \tag{10.49}$$

the surface tension takes the form

$$(F_s)_\alpha = -\phi \partial_\alpha \mu \tag{10.50}$$

and

$$F'_\alpha = c_s^2 \partial_\alpha \phi - \frac{\phi}{\rho} \partial_\alpha p - \frac{\phi^2}{\rho} \partial_\alpha \mu, \tag{10.51}$$

where

$$\mu = \partial_\phi E_f - \kappa \left(\frac{\partial^2 \phi}{\partial x^2} + \frac{\partial^2 \phi}{\partial r^2} + \frac{1}{r} \frac{\partial \phi}{\partial r} \right)$$

$$= 4\beta \phi (\phi - 1)(\phi - 0.5) - \kappa \left(\frac{\partial^2 \phi}{\partial x^2} + \frac{\partial^2 \phi}{\partial r^2} + \frac{1}{r} \frac{\partial \phi}{\partial r} \right). \tag{10.52}$$

In the above equations

$$\beta = \frac{12\sigma}{W}, \tag{10.53}$$

where W is the interfacial thickness and the parameter

$$\kappa = \frac{3}{2} \sigma W. \tag{10.54}$$

The index function is defined as

$$\phi = \sum \bar{f}_i - \frac{\phi u_r}{2r} \tag{10.55}$$

and hence, $\sum \bar{f}_i^{(1)} = \frac{\phi u_r}{2r}$ in this model. The other macro variables, pressures and velocities are obtained through Eqs (10.11) and (10.13), respectively.

We note that in the model there is an extra term

$$\frac{\lambda}{2}\nabla^2\mu\Gamma_i(\mathbf{u})|_{(\mathbf{x},t)} + \frac{\lambda}{2}\nabla^2\mu\Gamma_i(\mathbf{u})|_{(\mathbf{x}+e_i\Delta t,t)} \quad (10.56)$$

in the Lattice Boltzmann equation for $\bar{f}_i$. Through Taylor expansion with truncation error $O(\Delta t^2)$, this extra term can also be rewritten as

$$(2+\Delta t e_{i\alpha}\partial_\alpha)\frac{\lambda}{2}\nabla^2\mu\Gamma_i(\mathbf{u})|_{(\mathbf{x},t)}. \quad (10.57)$$

That means the source term S_i' in the Lattice Boltzmann equation for f_i is

$$S_i' = \left(1-\frac{1}{2\tau_2}\right)\left\{\frac{(e_{i\alpha}-u_\alpha)F_\alpha'}{c_s^2\rho}\bar{f}_i^{eq} - w_i\frac{\phi u_r}{r}\right\}$$

$$+(2+\Delta t e_{i\alpha}\partial_\alpha)\frac{\lambda}{2}\nabla^2\mu\Gamma_i(\mathbf{u})|_{(\mathbf{x},t)}. \quad (10.58)$$

In cylindrical coordinates,

$$\nabla^2\mu = \left(\partial_\gamma^2\mu + \frac{\partial_r\mu}{r}\right). \quad (10.59)$$

Hence, keeping the truncation error to $O(\Delta t)$ we have

$$\sum_i S_i' = -\left(1-\frac{1}{2\tau_2}\right)\frac{\phi u_r}{r} + \lambda\nabla^2\mu + \frac{\Delta t}{2}\partial_\alpha[u_\alpha(\lambda\nabla^2\mu)] \quad (10.60)$$

and

$$\sum_i S_i' e_{i\alpha} = \left(1-\frac{1}{2\tau_2}\right)F_\alpha' + u_\alpha\lambda\nabla^2\mu. \quad (10.61)$$

The corresponding Eqs (10.35) and (10.37) become

$$\partial_{t_1}\phi + \partial_\alpha(\phi u_\alpha) + \frac{\phi u_r}{r} = \lambda\nabla^2\mu + \frac{\Delta t}{2}\partial_\alpha[u_\alpha\lambda\nabla^2\mu] \quad (10.62)$$

and

$$\partial_{t_2}\phi = -\frac{1}{2}\partial_\alpha\left[\left(1-\frac{1}{2\tau_2}\right)F_\alpha' + u_\alpha\lambda\nabla^2\mu\right], \quad (10.63)$$

respectively.

Combining Eqs (10.62) and (10.63), the CH-like equation for this model is

$$\partial_t\phi + \partial_\alpha(\phi u_\alpha) + \frac{\phi u_r}{r} = \lambda\nabla^2\mu - \frac{\Delta t}{2}\left(1-\frac{1}{2\tau_2}\right)\partial_\alpha F_\alpha'. \quad (10.64)$$

Here again we can see that there are extra terms in Eq. (10.64) compared to the
CH equation. We note that in Lee and Liu (2010) the authors chose $\tau_2 = 1.0$ for
all cases because in this model $\tau_2 = \tau_2' + 0.5$, where $\tau_2' = 0.5$ is fixed in Lee and
Liu (2010). Even with this choice, the CH equation is still only approximately
recovered.

The recovered axisymmetric two-phase momentum equation is

$$\partial_t(\rho u_\beta) + \partial_\alpha(\rho u_\alpha u_\beta) + \rho\frac{u_r u_\alpha}{r} = -\partial_\beta p + v\partial_\alpha\{\rho[\partial_\beta u_\alpha + \partial_\alpha u_\beta]\}$$

$$+\rho v\left\{\frac{\partial_r u_\alpha}{r} - \frac{u_r}{r^2}\delta_{\alpha r} + \partial_\alpha\left(\frac{u_r}{r}\right)\right\} - \phi\partial_\alpha\mu + Q', \quad (10.65)$$

where $v = c_s^2(\tau_1 - 0.5)\Delta t$ and Q' is the extra source term (see Eq. (10.34)) in the recovered momentum equations due to the definition of $E_\gamma = \partial_\gamma(c_s^2\rho)$. For more detail refer to the appendix to the Appendix in Chapter 8.

10.3.1 Implementation

The equilibrium distribution functions are defined as

$$g_i^{eq} = \bar{g}_i^{eq}$$

$$-\frac{\Delta t}{2}\left\{(e_{i\alpha} - u_\alpha)\left[\partial_\alpha^{CD}\rho c_s^2(\Gamma_\alpha - \Gamma_\alpha(0)) - (\phi\partial_\alpha^{CD}\mu - F_\alpha)\Gamma_\alpha\right] - c_s^2 w_i\frac{\rho u_r}{r}\right\} \quad (10.66)$$

and

$$f_i^{eq} = \bar{f}_i^{eq}$$

$$-\frac{\Delta t}{2}\left\{(e_{i\alpha} - u_\alpha)\left[\partial_\alpha^{CD}\phi - \frac{\phi}{\rho c_s^2}\left(\partial_\alpha^{CD}p + \phi\partial_\alpha^{CD}\mu\right)\right]\Gamma_i(\mathbf{u}) - w_i\frac{\phi u_r}{r}\right\}, \quad (10.67)$$

where

$$\bar{g}_i^{eq}(\mathbf{x}, t) = w_i\left[p + c_s^2\rho\left(\frac{e_{i\alpha}u_\alpha}{c_s^2} + \frac{e_{i\alpha}u_\alpha e_{i\beta}u_\beta}{2c_s^4} - \frac{u_\alpha u_\alpha}{2c_s^2}\right)\right] \quad (10.68)$$

and

$$\bar{f}_i^{eq}(\mathbf{x}, t) = w_i\phi\left[1 + \frac{e_{i\alpha}u_\alpha}{c_s^2} + \frac{e_{i\alpha}u_\alpha e_{i\beta}u_\beta}{2c_s^4} - \frac{u_\alpha u_\alpha}{2c_s^2}\right]. \quad (10.69)$$

The evolution equations for $g_i(\mathbf{x}, t)$ and $f_i(\mathbf{x}, t)$ are

$$g_i(\mathbf{x} + \mathbf{e}_i\Delta t, t) = g_i(\mathbf{x}, t) - \frac{1}{\tau_1 + 0.5}(g_i - g_i^{eq})|_{(\mathbf{x},t)}$$

$$+\Delta t\left\{(e_{i\alpha} - u_\alpha)\left[\partial_\alpha^{MD}\rho c_s^2(\Gamma_i(\mathbf{u}) - \Gamma_i(0)) - (\phi\partial_\alpha^{MD}\mu - F_\alpha)\Gamma_i(\mathbf{u})\right] - c_s^2 w_i\frac{\rho u_r}{r}\right\}|_{(\mathbf{x},t)}$$

$$(10.70)$$

and

$$f_i(\mathbf{x} + \mathbf{e}_i\Delta t, t) = f_i(\mathbf{x}, t) - \frac{1}{\tau_2 + 0.5}(f_i - f_i^{eq})|_{(\mathbf{x},t)}$$

$$+\frac{\lambda}{2}\nabla^2\mu\Gamma_i(\mathbf{u})|_{(\mathbf{x},t)} + \frac{\lambda}{2}\nabla^2\mu\Gamma_i(\mathbf{u})|_{(\mathbf{x}+\mathbf{e}_i\Delta t,t)}$$

$$+\Delta t\left\{(e_{i\alpha} - u_\alpha)\left[\partial_\alpha^{MD}\phi - \frac{\phi}{\rho c_s^2}\left(\partial_\alpha^{MD}p + \phi\partial_\alpha^{MD}\mu\right)\right]\Gamma_i(\mathbf{u}) - w_i\frac{\phi u_r}{r}\right\}|_{(\mathbf{x},t)}.$$

$$(10.71)$$

Here τ_1 and τ_2 are slightly different from those in Eqs. (10.62) and (10.64). We can choose $\tau_1 = \tau_2$, which is determined from Eq.(10.46). The above derivatives are obtained through the finite difference scheme in Section 9.3.1.

After ϕ is known, the corresponding density of the fluid can be calculated from

$$\rho = \rho_g + \phi(\rho_l - \rho_g), \tag{10.72}$$

where ϕ changes from 0 (gas) to 1 (liquid).

10.3.2 Head-on droplet collision

In the experiment of Qian and Law (1997), there is a tetradecane and nitrogen two-phase flow system. In the system, gas density $\rho_g = 1.14 \times 10^{-3}$ g/cm^3, dynamic viscosity $\eta_g = 1.79 \times 10^{-4}$ g/(cm·s), liquid density $\rho_l = 0.758$ g/cm^3, and dynamic viscosity $\eta_l = 2.13 \times 10^{-2}$ g/(cm·s). The surface tension between the liquid and gas is $\sigma = 26$ g/s^2.

To match the density ratio in the experiments (Qian and Law 1997), we tried cases with $\rho_g = 0.0015$ in our simulations ($\frac{\rho_l}{\rho_g} = \frac{1.0}{0.0015} = \frac{0.758 \text{ g/cm}^3}{1.14\times10^{-3} \text{ g/cm}^3}$). However, numerical instability appears after the two droplets collide and contract into a disk. In the simulation to improve the numerical stability, the cut-off technique is used, i.e.,

$$\phi(\mathbf{x}) = \begin{cases} 0, & \text{if } \phi < 0, \\ 1, & \text{if } \phi > 1, \end{cases} \tag{10.73}$$

The artificial free-energy technique suggested in Lee and Liu (2010) is also found able to achieve better numerical stability. However, it is not better than the cut-off technique.

In our simulations the left and right boundary conditions are extrapolation boundaries. That means that both the distribution functions f_i and g_i and the macro variables are extrapolated from the inner fluid nodes. The lower boundary is the axisymmetric boundary condition (refer to Section 8.2.4).

In Figure 10.4 it is seen that initially the head-on collision results in contact and coalescence into a disk-like larger droplet and then oscillation. The coalesced droplet is expected to reach an equilibrium spherical state after enough time. In Figure 10.5 it is seen that when We is larger, the initially coalesced droplet will

Table 10.1   Parameters in experimental study (Qian and Law 1997) ($\rho_l = 0.758$ g/cm^3, $\eta_l = 0.0213$ g/(cm·s), $\rho_g = 0.00114$ g/cm^3, $\eta_g = 0.000179$ g/(cm·s)).

Case	We	Re	D (cm)	U (cm/s)	Oh
A	32.8	210.8	0.0318	93.0	0.61
B	37.2	228	0.0328	97.6	0.60
C	61.4	296.5	0.0336	124.0	0.59

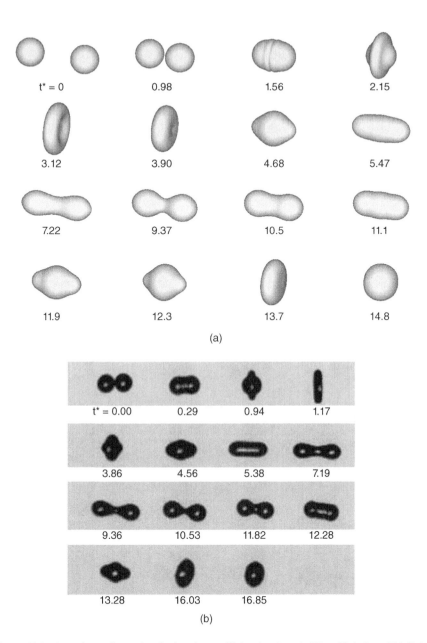

Figure 10.4 Snapshots of two-droplet head-on collision for Case A ($We = 32.8$, $Re = 210.8$) (a) compared with the experimental result and (b) in Qian and Law (1997). Source: Qian and Law 1997. Reproduced with permission of Cambridge University Press.

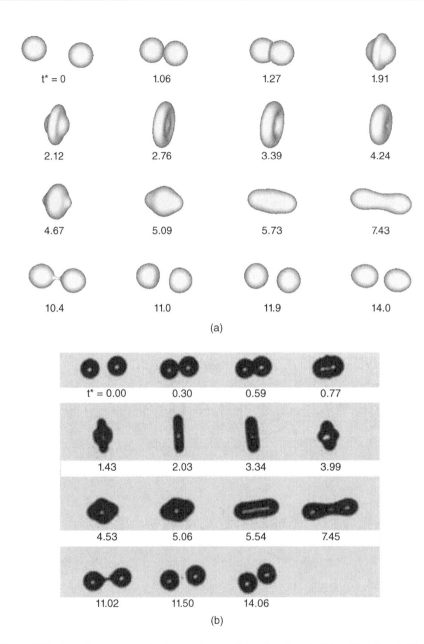

Figure 10.5 Snapshots of two-droplet head-on collision for Case B $(We = 37.2, Re = 228)$ (a) compared with the experimental result and (b) in Qian and Law (1997). Source: Qian and Law 1997. Reproduced with permission of Cambridge University Press.

form a disk-like drop. The drop contracts radially inward and pushes the liquid from its center. Eventually, the droplet will be elongated into a long cylinder with rounded ends and then break into two droplets. This is the reflexive separation. When We is larger, the reflexive separation becomes more complicated. For example, when $Re = 61.4$ (Figure 10.6), besides two primary drops, a small satellite droplet is generated.

In Figures 10.4, 10.5, and 10.6 it is seen that in terms of droplet evolution (coalescence and separation) our LBM results are all very consistent with the experimental result. However, there are discrepancies in time between the numerical and experimental results. In the experimental study Qian and Law (1997) mentioned that at $t^* = 0$ the distance between the two droplets is about $2D$. However, the two droplets touch earlier in the experiments than in our LBM simulations. We can estimate the time when the two droplets begin to touch. In all cases the expected time is $t^* = \frac{2R}{2U}\frac{U}{R} = 1$. However, in Figure 10.4(a) it is seen that before $t^* = 0.29$ the two droplets already contact one another. In Figures 10.5(b) and 10.6(b) the two droplets begin to contact at $t^* \approx 0.59$ and $t^* = 0.37$. Hence, the initial distance in the experiment must be much less than $2D$, although the value is claimed to be $2D$ in Qian and Law (1997). Hence, there will be a time discrepancy between the experiment and our LBM simulation.

In Table 10.2 the mobility is also shown. The mobility λ is found to affect the results. Actually, in Case A (Figure 10.4) gas was entrapped into the droplet but we are unable to see that in the 3D view. Figure 10.7 shows the corresponding 2D view. We can see that when the mobility $\lambda = 0.1$, a small bubble of gas was entrapped in the new droplet when the two drops begin to touch. After a long time, e.g., $t^* = 14.8$, the small bubble still remains in the large drop. If the mobility is large, e.g., $\lambda = 2.0$, from Figure 10.8 it is seen that the initially entrapped bubble was dissolved after $t^* = 5.47$. Hence, it seems that a larger mobility makes it easier for the gas to dissolve. Comparing Figures 10.7 and 10.8 we can see that evolution of the droplets' head-on collisions are similar. Hence, no matter if $\lambda = 0.1$ or $\lambda = 2$, overall the mobility does not affect the interface evolution. In the simulations of Cases B and C we also observed that with $\lambda = 2$ there is no gas entrapment or the gas is quickly dissolved. When $\lambda = 0.1$ the entrapped gas will remain in the coalesced droplet and can lead to a numerical instability when the large drop separates into two.

It is found that the density ratio also affects the evolution process. For example, when the parameters in Case B are adopted except that the densities $\rho_l = 1.0$ and $\rho_g = 0.005$ are used in the simulation, only pure coalescence without reflexive separation is observed (the result is not shown). This means that the density ratio 200 leads to a different result from that shown in the present Case B.

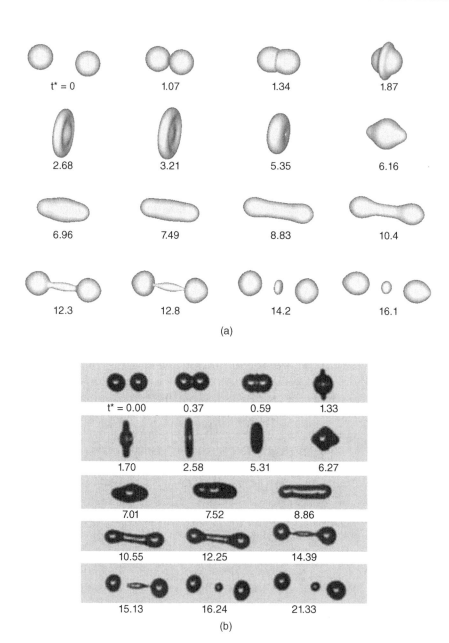

Figure 10.6 Snapshots of two-droplet head-on collision for Case C ($We = 61.4$, $Re = 296.5$) (a) compared with the experimental result and (b) in Qian and Law (1997). Source: Qian and Law 1997. Reproduced with permission of Cambridge University Press.

Table 10.2   Parameters in LBM simulations (in all these simulations $\sigma = 0.002$ mu/ts^2, $D = 120$ lu).

Case	We	Re	ρ_l	τ_l	ρ_g	τ_g	U	Oh	λ
A	32.9	210.8	1.0	0.04	0.003	0.226	0.0117	0.616	0.1
B	38.9	235	1.0	0.039	0.003	0.5	0.0127	0.60	2.0
C	61.9	296.5	1.0	0.039	0.003	0.5	0.0161	0.60	2.0

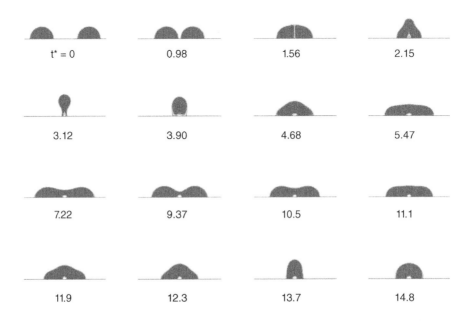

Figure 10.7   Gas entrapment in LBM simulation (Case A: $We = 32.8$, $Re = 210.8$, $\lambda = 0.1$). Dimensionless times indicated.

10.3.3   Bubble rise

In the experimental study of Bhaga and Weber (1981), the dynamic viscosity of the liquid is about 0.82 to 28 poise (g/(cm·s)). The density of liquid is approximately 1.33 g/cm^3. the surface tension is about 78 g/s^2. The air dynamic viscosity is 1.8×10^{-5} Pa · s = 1.8 g/(m·s) and $\rho = 1.29 \times 10^{-3}$ g/cm^3.

Case (d) in Figure 3 in Bhaga and Weber (1981) is simulated to test the model. Two cases with different meshes are simulated. The main parameters in the two cases are shown in Table 10.3. All parameters are in lattice units. The definitions of Mo, Eo, and how to determine the gravity force in the simulation can be found in Section 8.8.1.

For the setting of the flow, refer to Figure 8.1. For example, in Case I the computational domain is 1000×250, which means the dimensions along the x and r coordinates are 1000 lu and 250 lu, respectively. Hence, the tube diameter

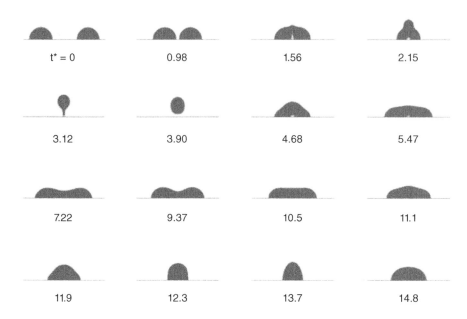

t* = 0	0.98	1.56	2.15
3.12	3.90	4.68	5.47
7.22	9.37	10.5	11.1
11.9	12.3	13.7	14.8

Figure 10.8   Entrapped gas dissolution in LBM simulation (Case A: $We = 32.8$, $Re = 210.8$, $\lambda = 2.0$). Dimensionless times indicated.

Table 10.3   Parameters in bubble rise simulations (cases with $\rho_l = 1$, $\rho_g = 0.01$, $Eo = 116$, $Mo = 5.51$, expected $Re = 13.3$, interface thickness 5 lu).

Case	Mesh	D	σ	$g(\times 10^{-6})$	ν	τ_l	τ_g	Expected U
I	1000×250	80	0.0002	3.625	0.0591	0.677	1.0	0.0098
II	1000×350	120	0.0002	1.61	0.0723	0.717	1.0	0.008

is 500 lu. Along the boundary $r = 250$ lu (top boundary) and the left and right boundaries (refer to Figure 8.1), the no-slip wall boundary condition, i.e., the bounce-back scheme, is applied.

Initially a spherical bubble is released from the low part of the tube with diameter $D = 80$ lu (the left part in Figure 8.1). Because of buoyant force, the bubble rises. Fifteen snapshots for the bubble rise in Case I are shown in Figure 10.9. In the figure, the contour of $\rho = \frac{\rho_l + \rho_g}{2}$ is shown. The distance between snapshots is arbitrary.

From Figure 10.9 it is seen that in terms of terminal bubble shape the simulated result has a large discrepancy with the experimental data (Bhaga and Weber 1981). For this case, the shape of the bubble is a spherical cap in the experiment (Bhaga and Weber 1981) (the top right picture in Figure 10.9). However, the simulated result finally becomes a toroidal bubble. A possible reason for this inconsistency is that the surface tension formula may significantly affect the bubble shape evolution (see Section 8.8.2).

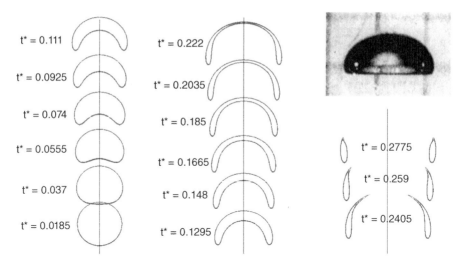

Figure 10.9 Bubble rise (Case I: $Eo = 116$, $Mo = 5.51$, $\lambda = 2.0$). The top right is the experimental snapshot for terminal bubble shape (Figure3(d) in Bhaga and Weber (1981)). The normalized time ($t^* = \frac{tv}{D^2} = 0.0185$) is shown beside each bubble. The vertical black lines represent the axisymmetric axes.

The bubble rise velocity as a function of time is shown in Figure 10.10(b). We can see that the terminal rising velocity is approximately 0.009 lu/ts, which seems consistent with the expected U (Table 10.3) that is calculated from the experimental Re. The integrated volume of the bubble as a function of time is shown in Figure 10.10(a). It is seen that the volume of the bubble decreases

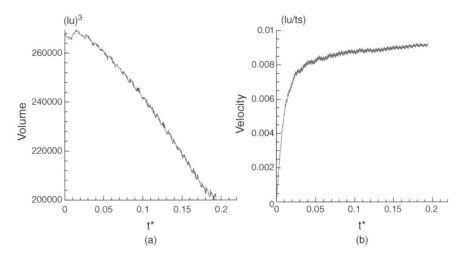

Figure 10.10   (a) Mass of the bubble as a function time and (b) bubble rise velocity as a function of time for Case I with $Eo = 116$, $Mo = 5.51$, and $\lambda = 2.0$.

continuously. Hence, the mass conservation of this model may not be so satisfactory. In the implementation, the volume of the bubble V is determined in the following way. For a computational node $\mathbf{x}$, if $\rho(\mathbf{x}) < \frac{\rho_l + \rho_g}{2}$ it is supposed to be occupied by gas and it is labeled as $\mathbf{x}_g$. The formula for calculation of the volume (integration scheme) is

$$V = \sum_{\mathbf{x}_g} 2\pi r \Delta r \Delta x, \tag{10.74}$$

where $\Delta x = \Delta r = 1$ lu and r is the radius from $\mathbf{x}_g$ to $r = 0$ in lattice units. Here V has units of lu^3.

In Case II a larger computational domain is used and the mass correction technique described in Section 8.7.3 in Chapter 8 is applied. The result is shown in Figure 10.11. It is seen that the result is much better than Figure 10.9 in terms of bubble shape. From Figure 10.12(a) we can see that the mass conservation becomes much better. The volume of the bubble decreases slightly before the

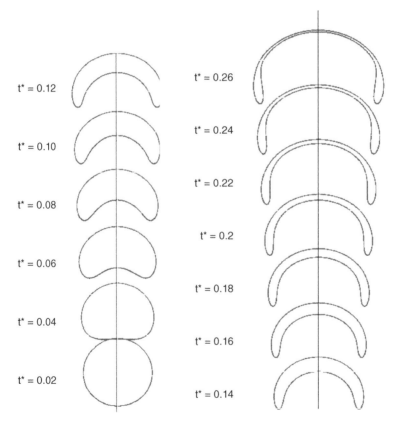

Figure 10.11 Bubble rise (Case II: $Eo = 116$, $Mo = 5.51$, $\lambda = 2.0$). The normalized time is shown beside each bubble. The vertical black lines represent the axisymmetric axes.

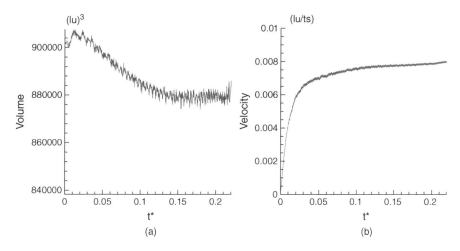

Figure 10.12   (a) Mass of the bubble as a function time. (b) Bubble rise velocity as a function of time in Case II with $Eo = 116$, $Mo = 5.51$, and $\lambda = 2.0$.

volumes of the gas and liquid reach an equilibrium state. After $t^* = 0.1$, the volume of the bubble does not decrease. Figure 10.12(b) shows that the velocity of the bubble almost reaches its equilibrium value of 0.076 lu/ts, which is consistent with the expected value. The shape of the bubble also eventually becomes toroidal after $t^* > 2.4$, which is different from the experiment data. As stated in Chapter 8, the calculation of the surface tension may affect the bubble shape. Hence, in this model the surface tension calculation may be not accurate.

Another case with a large initial bubble ($D = 120$, $Eo = 116$, $Mo = 5.51$) but without the mass correction technique is also simulated and the result shows that the volume of the bubble decreases to the same extend as that in Case I. It is found that even increasing the initial volume of the bubble and the computational domain, the bubble shape evolution is almost identical to Case I. The volume of the bubble decreases to the same extent as that in Case I. Hence, the mass correction technique appears to play a critical role in achieving better results.

We also checked if mobility affected mass conservation. Our numerical simulations show that the mobility λ almost does not affect the result.

We would like to discuss further the mass conservation issue in the diffuse interface two-phase model. Zheng et al. (2014) found that the diffuse interface LBM may produce a density shift, or bubble or droplet shrinkage. Density shift means that the fluid densities (density ratio) in the simulation procedure would be different from the initial ones.

Inspired by Yue et al. (2007), Zheng et al. (2014) studied the phenomenon through investigation of a stationary bubble or droplet immersed in another fluid. Assuming that the ratio between the interface thickness W and the initial bubble or droplet radius r_0 is small, they analytically show the existence of this density shift, bubble or droplet radius shrinkage, and critical bubble or droplet survival radius. Numerical results show that the curvature, contact angle, and initial droplet volume have an effect on this spontaneous shrinkage process. However, in their study the density ratio is only about 10 and the bubbles are stationary. For the high-density ratio and moving bubble cases, the mass conservation is seldom discussed in the LBM community. Further study is necessary.

As a minor conclusion, in this section the axisymmetric model based on Lee and Liu (2010) is able to simulate the droplets' collision correctly. However, for bubble rise the simulation is not so satisfactory in terms of terminal bubble shape and mass conservation. The mass conservation can be improved by the mass correction technique. For the bubble shape issue, it is conjectured that the surface tension calculation in the model may not be accurate enough.

References

Adamson A and Gast A 1997 *Physical Chemistry of Surfaces*. John Wiley & Sons, Inc.

Ahrenholz B, Toelke J, Lehmann P, Peters A, Kaestner A, Krafczyk M and Durner W 2008 Prediction of capillary hysteresis in a porous material using lattice-Boltzmann methods and comparison to experimental data and a morphological pore network model. *Advances in Water Resources* **31**(9), 1151–1173.

Aidun CK and Clausen JR 2010 Lattice–Boltzmann method for complex flows. *Annual Review of Fluid Mechanics* **42**, 439–472.

Amaya-Bower L and Lee T 2010 Single bubble rising dynamics for moderate Reynolds number using lattice Boltzmann method. *Computers & Fluids* **39**(7), 1191–1207.

Amir B, Janben C and Grilli S 2015 An efficient lattice Boltzmann multiphase model for 3D flows with large density ratios at high Reynolds numbers. *Computers and Mathematics with Applications* **68**(12), 1819–1843.

Angelopoulos A, Paunov V, Burganos V and Payatakes A 1998 Lattice Boltzmann simulation of nonideal vapor-liquid flow in porous media. *Physical Review E* **57**(3), 3237.

Ashgriz N and Poo J 1990 Coalescence and separation in binary collisions of liquid drops. *Journal of Fluid Mechanics* **221**, 183–204.

Atkins P 1978 *Physical Chemistry*. WH Freeman and Company.

Ben Salah Y, Tabe Y and Chikahisa T 2012 Two-phase flow simulation in a channel of a polymer electrolyte membrane fuel cell using the lattice Boltzmann method. *Journal of Power Sources* **199**, 85–93.

Benzi R, Biferale L, Sbragaglia M, Succi S and Toschi F 2006 Mesoscopic modeling of a two-phase flow in the presence of boundaries: The contact angle. *Physical Review E* **74**(2), 021509.

Benzi R, Chibbaro S and Succi S 2009a Mesoscopic lattice Boltzmann modeling of flowing soft systems. *Physical Review Letters* **102**(2), 026002.

Benzi R, Sbragaglia M, Succi S, Bernaschi M and Chibbaro S 2009b Mesoscopic lattice Boltzmann modeling of soft-glassy systems: theory and simulations. *Journal of Chemical Physics* **131**(10), 104903.

Bhaga D and Weber M 1981 Bubbles in viscous liquids: shapes, wakes and velocities. *Journal of Fluid Mechanics* **105**, 61–85.

Biferale L, Perlekar P, Sbragaglia M and Toschi F 2012 Convection in multiphase fluid flows using lattice Boltzmann methods. *Physical Review Letters* **108**(10), 104502.

Boltzmann L 1964/1995 *Lectures on gas theory (translated by SG Brush)*. Dover Publications New York.

Boyd J, Buick J, Cosgrove J and Stansell P 2005 Application of the lattice Boltzmann model to simulated stenosis growth in a two-dimensional carotid artery. *Physics in Medicine and Biology* **50**(20), 4783.

Briant AJ and Yeomans JM 2004 Lattice Boltzmann simulations of contact line motion. II. Binary fluids. *Physical Review E* **69**(3), 031603.

Briant AJ, Papatzacos P and Yeomans JM 2002 Lattice Boltzmann simulations of contact line motion in a liquid–gas system. *Philosophical Transactions of the Royal Society of London, Series A Mathematical, Physical and Engineering Sciences* **360**(1792), 485–495.

Multiphase Lattice Boltzmann Methods: Theory and Application, First Edition.
Haibo Huang, Michael C. Sukop and Xi-Yun Lu.
© 2015 John Wiley & Sons, Ltd. Published 2015 by John Wiley & Sons, Ltd.
Companion Website: www.wiley.com/go/huang/boltzmann

Briant AJ, Wagner AJ and Yeomans JM 2004 Lattice Boltzmann simulations of contact line motion. I. Liquid–gas systems. *Physical Review E* **69**(3), 031602.

Buick JM and Greated CA 2000 Gravity in a lattice Boltzmann model. *Physical Review E* **61**(5), 5307–5320.

Cahn J and Hilliard J 1958 Free energy of a nonuniform system. I. Interfacial free energy. *Journal of Chemistry and Physics* **28**, 258–267.

Carnahan NF and Starling KE 1969 Equation of state for nonattracting rigid spheres. *Journal of Chemical Physics* **51**(2), 635–636.

Chan T, Srivastava S, Marchand A, Andreotti B, Biferale L, Toschi F and Snoeijer J 2013 Hydrodynamics of air entrainment by moving contact lines. *Physics of Fluids* **25**(7), 074105.

Chang Q and Alexander JID 2006 Application of the lattice Boltzmann method to two-phase Rayleigh–Benard convection with a deformable interface. *Journal of Computational Physics* **212**(2), 473–489.

Chao JH, Mei RW, Singh R and Shyy W 2011 A filter-based, mass-conserving lattice Boltzmann method for immiscible multiphase flows. *International Journal for Numerical Methods in Fluids* **66**(5), 622–647.

Chen H, Boghosian BM, Coveney PV and Nekovee M 2000 A ternary lattice Boltzmann model for amphiphilic fluids. *Proceedings of the Royal Society of London, Series A: Mathematical, Physical and Engineering Sciences* **456**(2000), 2043–2057.

Chen L, Kang Q, Mu Y, He YL and Tao WQ 2014 A critical review of the pseudopotential multiphase lattice Boltzmann model: Methods and applications. *International Journal of Heat and Mass Transfer* **76**, 210–236.

Chen S and Doolen GD 1998 Lattice Boltzmann method for fluid flows. *Annual Review of Fluid Mechanics* **30**, 329–364.

Chen S, Martinez D and Mei R 1996 On boundary conditions in lattice Boltzmann methods. *Physics of Fluids* **8**(9), 2527–2536.

Cheng M, Hua J and Lou J 2010 Simulation of bubble-bubble interaction using a lattice Boltzmann method. *Computers & Fluids* **39**(2), 260–270.

Chiappini D, Bella G, Succi S and Ubertini S 2009 Applications of finite-difference lattice Boltzmann method to breakup and coalescence in multiphase flows. *Internationl Journal of Modern Physics C* **20**(11), 1803–1816.

Chiappini D, Bella G, Succi S, Toschi F and Ubertini S 2010 Improved lattice Boltzmann without parasitic currents for Rayleigh–Taylor instability. *Communications in Computational Physics* **7**(3), 423.

Chin J 2002 Lattice Boltzmann simulation of the flow of binary immiscible fluids with different viscosities using the Shan–Chen microscopic interaction model. *Philosophical Transactions of the Royal Society of London, Series A: Mathematical, Physical and Engineering Sciences* **360**(1792), 547–558.

Chin J and Coveney PV 2002 Lattice Boltzmann study of spinodal decomposition in two dimensions. *Physical Review E* **66**(1), 016303.

Clark TT 2003 A numerical study of the statistics of a two-dimensional Rayleigh–Taylor mixing layer. *Physics of Fluids* **15**(8), 2413–2423.

de Gennes P 1985 Wetting: statics and dynamics. *Review of Modern Physics* **57**(3), 827–863.

Denniston C, Orlandini E and Yeomans J 2001 Lattice Boltzmann simulations of liquid crystal hydrodynamics. *Physical Review E* **63**(5), 056702.

d'Humiéres D, Ginzburg I, Krafczyk M, Lallemand P and Luo LS 2002 Multiple-relaxation-time lattice Boltzmann models in three dimensions. *Philosophical Transactions of the Royal Society of London, Series A Mathematical, Physical and Engineering Sciences* **360**(1792), 437–451.

Ding H and Spelt PDM 2007 Inertial effects in droplet spreading: a comparison between diffuse-interface and level-set simulations. *Journal of Fluid Mechanics* **576**, 287–296.

Ding H and Spelt PDM 2008 Onset of motion of a three-dimensional droplet on a wall in shear flow at moderate Reynolds numbers. *Journal of Fluid Mechanics* **599**, 341–362.

Dong B, Yan Y, Li W and Song Y 2010 Lattice Boltzmann simulation of viscous fingering phenomenon of immiscible fluids displacement in a channel. *Computers & Fluids* **39**(5), 768–779.

Dünweg B and Ladd AJ 2009 Lattice Boltzmann simulations of soft matter systems. *Advanced Computer Simulation Approaches for Soft Matter Sciences III* **221**, 89–166.

Dupuis A and Yeomans JM 2004 Lattice Boltzmann modelling of droplets on chemically heterogeneous surfaces. *Future Generation Computer Systems* **20**(6), 993–1001.

Evans R 1979 The nature of the liquid–vapour interface and other topics in the statistical mechanics of non-uniform, classical fluids. *Advances in Physics* **28**(2), 143–200.

Fakhari A and Rahimian MH 2010 Phase-field modeling by the method of lattice Boltzmann equations. *Physical Review E* **81**(3), 036707.

Falcucci G, Ubertini S and Succi S 2010 Lattice Boltzmann simulations of phase-separating flows at large density ratios: the case of doubly-attractive pseudo-potentials. *Soft Matter* **6**(18), 4357–4365.

Falcucci G, Ubertini S, Bella G and Succi S 2013 Lattice Boltzmann simulation of cavitating flows. *Communications in Computational Physics* **13**(3), 685–695.

Fan L, Fang H and Lin Z 2001 Simulation of contact line dynamics in a two-dimensional capillary tube by the lattice Boltzmann model. *Physical Review E* **63**(5), 051603.

Finn R 2006 The contact angle in capillarity. *Physics of Fluids* **18**(4), 047102.

Frank X and Perré P 2012 Droplet spreading on a porous surface: A lattice Boltzmann study. *Physics of Fluids* **24**(4), 042101.

Frank X, Funfschilling D, Midoux N and Li HZ 2005 Bubbles in a viscous liquid: lattice Boltzmann simulation and experimental validation. *Journal of Fluid Mechanics* **546**, 113–122.

Frisch U, Hasslacher B and Pomeau Y 1986 Lattice-gas automata for the Navier–Stokes equation. *Physics Review Letter* **56**(14), 1505.

Ginzbourg I and Adler P 1995 Surface tension models with different viscosities. *Transport in Porous Media* **20**(1–2), 37–76.

Gong S and Cheng P 2012 A lattice Boltzmann method for simulation of liquid–vapor phase-change heat transfer. *International Journal of Heat and Mass Transfer* **55**(17), 4923–4927.

Gonnella G, Orlandini E and Yeomans J 1997 Spinodal decomposition to a lamellar phase: effects of hydrodynamic flow. *Physical Review Letters* **78**(9), 1695.

González-Segredo N, Nekovee M and Coveney PV 2003 Three-dimensional lattice-Boltzmann simulations of critical spinodal decomposition in binary immiscible fluids. *Physical Review E* **67**(4), 046304.

Gross M, Moradi N, Zikos G and Varnik F 2011 Shear stress in nonideal fluid lattice Boltzmann simulations. *Physical Review E* **83**(1), 017701.

Grubert D and Yeomans JM 1999 Mesoscale modeling of contact line dynamics. *Computer Physics Communications* **121**, 236–239.

Grunau D, Chen S and Eggert K 1993 A lattice Boltzmann model for multiphase fluid flows. *Physics of Fluids A: Fluid Dynamics (1989–1993)* **5**(10), 2557–2562.

Gu X, Gupta A and Kumar R 2009 Lattice Boltzmann simulation of surface impingement at high-density ratio. *Journal of Thermophysics and Heat Transfer* **23**(4), 773–785.

Gunstensen AK, Rothman DH, Zaleski S and Zanetti G 1991 Lattice Boltzmann model of immiscible fluids. *Physical Review A* **43**(8), 4320–4327.

Guo ZL and Shu C 2013 *Lattice Boltzmann method and its applications in engineering (Advances in computational fluid dynamics)*. World Scientific Publishing Company.

Guo ZL, Zheng CG and Shi BC 2002 Discrete lattice effects on the forcing term in the lattice Boltzmann method. *Physical Review E* **65**(4), 046308.

Gupta A and Kumar R 2008 Lattice Boltzmann simulation to study multiple bubble dynamics. *International Journal of Heat and Mass Transfer* **51**(21–22), 5192–5203.

Halliday I, Hollis A and Care C 2007 Lattice Boltzmann algorithm for continuum multicomponent flow. *Physical Review E* **76**(2), 026708.

Hao L and Cheng P 2009 Lattice Boltzmann simulations of liquid droplet dynamic behavior on a hydrophobic surface of a gas flow channel. *Journal of Power Sources* **190**(2), 435–446.

Hao L and Cheng P 2010 Lattice Boltzmann simulations of water transport in gas diffusion layer of a polymer electrolyte membrane fuel cell. *Journal of Power Sources* **195**(12), 3870–3881.

Harting J, Chin J, Venturoli M and Coveney PV 2005 Large-scale lattice Boltzmann simulations of complex fluids: advances through the advent of computational grids. *Philosophical Transactions of the Royal Society A: Mathematical, Physical and Engineering Sciences* **363**(1833), 1895–1915.

Hazewinkel M 1993 *Encyclopaedia of Mathematics*. Springer.

Házi G and Márkus A 2008 Modeling heat transfer in supercritical fluid using the lattice Boltzmann method. *Physical Review E* **77**(2), 026305.

Hazi G and Markus A 2009 On the bubble departure diameter and release frequency based on numerical simulation results. *International Journal of Heat and Mass Transfer* **52**(5), 1472–1480.

Házi G, Imre AR, Mayer G and Farkas I 2002 Lattice Boltzmann methods for two-phase flow modeling. *Annals of Nuclear Energy* **29**(12), 1421–1453.

He XY and Doolen GD 2002 Thermodynamic foundations of kinetic theory and lattice Boltzmann models for multiphase flows. *Journal of Statistical Physics* **107**(1–2), 309–328.

He XY and Luo LS 1997 Theory of the lattice Boltzmann method: From the Boltzmann equation to the lattice Boltzmann equation. *Physical Review E* **56**(6), 6811–6817.

He XY, Chen SY and Zhang RY 1999 A lattice Boltzmann scheme for incompressible multiphase flow and its application in simulation of Rayleigh–Taylor instability. *Journal of Computational Physics* **152**(2), 642–663.

He XY, Shan XW and Doolen GD 1998 Discrete Boltzmann equation model for nonideal gases. *Physical Review E* **57**(1), R13–R16.

He XY, Zou QS, Luo LS and Dembo M 1997 Analytic solutions of simple flows and analysis of nonslip boundary conditions for the lattice Boltzmann bgk model. *Journal of Statistical Physics* **87**(1–2), 115–136.

Higuera F, Succi S and Benzi R 1989 Lattice gas dynamics with enhanced collisions. *EPL (Europhysics Letters)* **9**(4), 345.

Hilpert M 2007 Capillarity-induced resonance of blobs in porous media: Analytical solutions, lattice–Boltzmann modeling, and blob mobilization. *Journal of Colloid and Interface Science* **309**(2), 493–504.

Hilpert M 2011 Determination of dimensional flow fields in hydrogeological settings via the MRT lattice-Boltzmann method. *Water Resources Research*.

Holdych D, Georgiadis J and Buckius R 2001 Migration of a van der Waals bubble: lattice Boltzmann formulation. *Physics of Fluids* **13**(4), 817–825.

Holdych DJ, Rovas D, Georgiadis JG and Buckius RO 1998 An improved hydrodynamics formulation for multiphase flow lattice-Boltzmann models. *International Journal of Modern Physics C* **9**(8), 1393–1404.

Hou SL, Shan XW, Zou QS, Doolen GD and Soll WE 1997 Evaluation of two lattice Boltzmann models for multiphase flows. *Journal of Computational Physics* **138**(2), 695–713.

Hua J and Lou J 2007 Numerical simulation of bubble rising in viscous liquid. *Journal of Computational Physics* **222**(2), 769–795.

Huang HB, Huang JJ and Lu XY 2014a A mass-conserving axisymmetric multiphase lattice Boltzmann method and its application in simulation of bubble rising. *Journal of Computational Physics* **269**, 386–402.

Huang HB, Huang JJ and Lu XY 2014b Study of immiscible displacements in porous media using a color-gradient-based multiphase lattice Boltzmann method. *Computers & Fluids* **93**, 164–172.

Huang HB, Krafczyk M and Lu X 2011a Forcing term in single-phase and Shan-Chen-type multiphase lattice Boltzmann models. *Physical Review E* **84**(4), 046710.

Huang HB, Thorne DT, Schaap MG and Sukop MC 2007 Proposed approximation for contact angles in Shan-and-Chen-type multicomponent multiphase lattice Boltzmann models. *Physical Review E* **76**(6), 066701.

Huang HB and Lu XY 2009 Relative permeabilities and coupling effects in steady-state gas-liquid flow in porous media: A lattice Boltzmann study. *Physics of Fluids* **21**(9), 092104.

Huang HB, Huang JJ, Lu XY and Sukop MC 2013 On simulations of high-density ratio flows using color-gradient multiphase lattice Boltzmann models. *International Journal of Modern Physics C* **24**(4), 1350021.

Huang HB, Wang L and Lu XY 2011b Evaluation of three lattice Boltzmann models for multiphase flows in porous media. *Computers & Mathematics with Applications* **61**(12), 3606–3617.

Huang JJ, Shu C and Chew YT 2009 Mobility-dependent bifurcations in capillarity-driven two-phase fluid systems by using a lattice Boltzmann phase-field model. *International Journal for Numerical Methods in Fluids* **60**(2), 203–225.

Hyväluoma J and Harting J 2008 Slip flow over structured surfaces with entrapped microbubbles. *Physical Review Letters* **100**(24), 246001.

Hyväluoma J, Koponen A, Raiskinmäki P and Timonen J 2007 Droplets on inclined rough surfaces. *European Physical Journal E: Soft Matter and Biological Physics* **23**(3), 289–293.

Hyväluoma J, Raiskinmäki P, Jäsberg A, Koponen A, Kataja M and Timonen J 2006 Simulation of liquid penetration in paper. *Physical Review E* **73**(3), 036705.

Inamuro T, Konishi N and Ogino F 2000 A Galilean invariant model of the lattice Boltzmann method for multiphase fluid flows using free-energy approach. *Computer Physics Communications* **129**(1), 32–45.

Inamuro T, Ogata T, Tajima S and Konishi N 2004 A lattice Boltzmann method for incompressible two-phase flows with large density differences. *Journal of Computational Physics* **198**(2), 628–644.

Iwahara D, Shinto H, Miyahara M and Higashitani K 2003 Liquid drops on homogeneous and chemically heterogeneous surfaces: A two-dimensional lattice Boltzmann study. *Langmuir* **19**(21), 9086–9093.

Jacqmin 1999 Calculation of two-phase Navier-Stokes flows using phase-field modeling. *Journal of Computational Physics* **155**, 96–127.

Joshi AS and Sun Y 2009 Multiphase lattice Boltzmann method for particle suspensions. *Physical Review E* **79**(6), 066703.

Joshi AS and Sun Y 2010 Wetting dynamics and particle deposition for an evaporating colloidal drop: A lattice Boltzmann study. *Physical Review E* **82**(4), 041401.

Josserand C and Zaleski S 2003 Droplet splashing on a thin liquid film. *Physics of Fluids* **15**(6), 1650–1657.

Junk M and Yong WA 2009 Weighted L(2)-stability of the lattice Boltzmann method. *SIAM Journal on Numerical Analysis* **47**(3), 1651–1665.

Junk M, Klar A and Luo LS 2005 Asymptotic analysis of the lattice Boltzmann equation. *Journal of Computational Physics* **210**(2), 676–704.

Kalarakis A, Burganos V and Payatakes A 2002 Galilean-invariant lattice-Boltzmann simulation of liquid–vapor interface dynamics. *Physical Review E* **65**(5), 056702.

Kang Q, Zhang D and Chen S 2004 Immiscible displacement in a channel: simulations of fingering in two dimensions. *Advances in Water Resources* **27**(1), 13–22.

Kang QJ, Zhang DX and Chen SY 2002 Displacement of a two-dimensional immiscible droplet in a channel. *Physics of Fluids* **14**(9), 3203–3214.

Kang QJ, Zhang DX and Chen SY 2005 Displacement of a three-dimensional immiscible droplet in a duct. *Journal of Fluid Mechanics* **545**, 41–66.

Kendon VM, Cates ME, Pagonabarraga I, Desplat JC and Bladon P 2001 Inertial effects in three-dimensional spinodal decomposition of a symmetric binary fluid mixture: a lattice Boltzmann study. *Journal of Fluid Mechanics* **440**, 147–203.

Kikkinides E, Yiotis A, Kainourgiakis M and Stubos A 2008 Thermodynamic consistency of liquid–gas lattice Boltzmann methods: Interfacial property issues. *Physical Review E* **78**(3), 036702.

Kim J 2005 A continuous surface tension force formulation for diffuse-interface models. *Journal of Computational Physics* **204**(2), 784–804.

Koido T, Furusawa T and Moriyama K 2008 An approach to modeling two-phase transport in the gas diffusion layer of a proton exchange membrane fuel cell. *Journal of Power Sources* **175**(1), 127–136.

Kuksenok O, Yeomans J and Balazs AC 2002 Using patterned substrates to promote mixing in microchannels. *Physical Review E* **65**(3), 031502.

Kupershtokh AL, Medvedev DA and Karpov DI 2009 On equations of state in a lattice Boltzmann method. *Computers & Mathematics with Applications* **58**(5), 965–974.

Kusumaatmaja H and Yeomans JM 2007 Controlling drop size and polydispersity using chemically patterned surfaces. *Langmuir* **23**(2), 956–959.

Kusumaatmaja H and Yeomans JM 2009 Anisotropic hysteresis on ratcheted superhydrophobic surfaces. *Soft Matter* **5**(14), 2704–2707.

Kusumaatmaja H, Leopoldes J, Dupuis A and Yeomans J 2006 Drop dynamics on chemically patterned surfaces. *EPL (Europhysics Letters)* **73**(5), 740.

Kusumaatmaja H, Vrancken R, Bastiaansen C and Yeomans J 2008 Anisotropic drop morphologies on corrugated surfaces. *Langmuir* **24**(14), 7299–7308.

Ladd AJC and Verberg R 2001 Lattice-Boltzmann simulations of particle-fluid suspensions. *Journal of Statistical Physics* **104**(5–6), 1191–1251.

Lafrance P 1975 Nonlinear breakup of a laminar liquid jet. *Physics of Fluids* **18**(4), 428–432.

Lallemand P and Luo LS 2000 Theory of the lattice Boltzmann method: Dispersion, dissipation, isotropy, Galilean invariance, and stability. *Physical Review E* **61**(6), 6546–6562.

Lallemand P and Luo LS 2003 Theory of the lattice Boltzmann method: Acoustic and thermal properties in two and three dimensions. *Physical Review E* **68**(3), 036706.

Lallemand P, Luo LS and Peng Y 2007 A lattice Boltzmann front-tracking method for interface dynamics with surface tension in two dimensions. *Journal of Computational Physics* **226**(2), 1367–1384.

Lamb H 1932 *Hydrodynamics*. Cambridge University Press, London.

Lamura A, Gonnella G and Yeomans J 1999 A lattice Boltzmann model of ternary fluid mixtures. *EPL (Europhysics Letters)* **45**(3), 314.

Langaas K and Papatzacos P 2001 Numerical investigations of the steady state relative permeability of a simplified porous medium. *Transport in Porous Media* **45**(2), 241–266.

Langaas K and Yeomans JM 2000 Lattice Boltzmann simulation of a binary fluid with different phase viscosities and its application to fingering in two dimensions. *European Physical Journal B* **15**(1), 133–141.

Latva-Kokko M and Rothman D 2005a Static contact angle in lattice Boltzmann models of immiscible fluids. *Physical Review E* **72**(4), 046701.

Latva-Kokko M and Rothman DH 2005b Diffusion properties of gradient-based lattice Boltzmann models of immiscible fluids. *Physical Review E* **71**(5), 056702.

Latva-Kokko M and Rothman DH 2007 Scaling of dynamic contact angles in a lattice-Boltzmann model. *Physical Review Letters* **98**(25), 254503.

Leclaire S, Reggio M and J-Y T 2012 Numerical evaluation of two recoloring operators for an immiscible two-phase flow lattice Boltzmann model. *Applied Mathematical Modelling* **36**, 2237–2252.

Leclaire S, Reggio M and Trepanier JY 2011 Isotropic color gradient for simulating very high-density ratios with a two-phase flow lattice Boltzmann model. *Computers & Fluids* **48**(1), 98–112.

Lee T and Fischer PF 2006 Eliminating parasitic currents in the lattice Boltzmann equation method for nonideal gases. *Physical Review E* **74**(4), 046709.

Lee T and Lin CL 2005 A stable discretization of the lattice Boltzmann equation for simulation of incompressible two-phase flows at high density ratio. *Journal of Computational Physics* **206**(1), 16–47.

Lee T and Liu L 2010 Lattice Boltzmann simulations of micron-scale drop impact on dry surfaces. *Journal of Computational Physics* **229**(20), 8045–8063.

Lenormand R, Touboul E and Zarcone C 1988 Numerical models and experiments on immiscible displacements in porous media. *Journal of Fluid Mechanics* **189**, 165–187.

Leopoldes J, Dupuis A, Bucknall D and Yeomans J 2003 Jetting micron-scale droplets onto chemically heterogeneous surfaces. *Langmuir* **19**(23), 9818–9822.

Li H, Pan C and Miller C 2005 Pore-scale investigation of viscous coupling effects for two-phase flow in porous media. *Physical Review E* **72**(2), 026705.

Li Q and Wagner A 2007 Symmetric free-energy-based multicomponent lattice Boltzmann method. *Physical Review E* **76**(3), 036701.

Lishchuk SV, Care CM and Halliday I 2003 Lattice Boltzmann algorithm for surface tension with greatly reduced microcurrents. *Physical Review E* **67**(3), 036701.

Liu H and Zhang Y 2011 Droplet formation in microfluidic cross-junctions. *Physics of Fluids* **23**(8), 082101.

Liu H, Valocchi A and Kang Q 2012 Three-dimensional lattice Boltzmann model for immiscible two-phase flow simulations. *Physical Review E* **85**(4), 046309.

Liu HH, Valocchi AJ, Kang QJ and Werth C 2013 Pore-scale simulations of gas displacing liquid in a homogeneous pore network using the lattice Boltzmann method. *Transport in Porous Media* **99**(3), 555–580.

Luo LS 1998 Unified theory of lattice Boltzmann models for nonideal gases. *Physical Review Letters* **81**(8), 1618–1621.

Martys NS and Chen HD 1996 Simulation of multicomponent fluids in complex three-dimensional geometries by the lattice Boltzmann method. *Physical Review E* **53**(1), 743–750.

Martys NS and Douglas JF 2001 Critical properties and phase separation in lattice Boltzmann fluid mixtures. *Physical Review E* **63**(3), 031205.

McCracken ME and Abraham J 2005 Multiple-relaxation-time lattice-Boltzmann model for multiphase flow. *Physical Review E* **71**(3), 036701.

McNamara GR and Zanetti G 1988 Use of the Boltzmann equation to simulate lattice-gas automata. *Physical Review Letters* **61**(20), 2332.

Miller C and Scriven L 1968 Oscillations of a fluid droplet immersed in another fluid. *Journal of Fluid Mechanics* **32**, 417.

Mukherjee S and Abraham J 2007a Crown behavior in drop impact on wet walls. *Physics of Fluids* **19**(5), 052103.

Mukherjee S and Abraham J 2007b Investigations of drop impact on dry walls with a lattice-Boltzmann model. *Journal of Colloid and Interface Science* **312**(2), 341–354.

Mukherjee S and Abraham J 2007c Lattice Boltzmann simulations of two-phase flow with high density ratio in axially symmetric geometry. *Physical Review E* **75**(2), 026701.

Nekovee M, Coveney PV, Chen H and Boghosian BM 2000 Lattice-Boltzmann model for interacting amphiphilic fluids. *Physical Review E* **62**(6), 8282.

Ngachin M, Galdamez RG, Gokaltun S and Sukop MC 2015 Lattice Boltzmann simulation of rising bubble dynamics using an effective buoyancy method. *International Journal of Modern Physics C* **26**(3), 1550031.

Niu XD, Munekata T, Hyodo SA and Suga K 2007 An investigation of water-gas transport processes in the gas-diffusion-layer of a PEM fuel cell by a multiphase multiple-relaxation-time lattice Boltzmann model. *Journal of Power Sources* **172**(2), 542–552.

Nourgaliev R, Dinh TN and Sehgal B 2002 On lattice Boltzmann modeling of phase transition in an isothermal non-ideal fluid. *Nuclear Engineering and Design* **211**(2), 153–171.

Nourgaliev RR, Dinh TN, Theofanous T and Joseph D 2003 The lattice Boltzmann equation method: theoretical interpretation, numerics and implications. *International Journal of Multiphase Flow* **29**(1), 117–169.

Palmer BJ and Rector DR 2000 Lattice-Boltzmann algorithm for simulating thermal two-phase flow. *Physical Review E* **61**(5), 5295.

Pan C, Hilpert M and Miller CT 2004 Lattice-Boltzmann simulation of two-phase flow in porous media. *Water Resources Research* **40**(1), W01501.

Park J and Li X 2008 Multi-phase micro-scale flow simulation in the electrodes of a PEM fuel cell by lattice Boltzmann method. *Journal of Power Sources* **178**(1), 248–257.

Pooley C, Kusumaatmaja H and Yeomans J 2008 Contact line dynamics in binary lattice Boltzmann simulations. *Physical Review E* **78**(5), 056709.

Pooley CM, Kusumaatmaja H and Yeomans JM 2009 Modelling capillary filling dynamics using lattice Boltzmann simulations. *European Physical Journal Special Topics* **171**(1), 63–71.

Porter ML, Schaap MG and Wildenschild D 2009 Lattice-Boltzmann simulations of the capillary pressure–saturation–interfacial area relationship for porous media. *Advances in Water Resources* **32**(11), 1632–1640.

Premnath KN and Abraham J 2005a Lattice Boltzmann model for axisymmetric multiphase flows. *Physical Review E* **71**(5), 056706.

Premnath KN and Abraham J 2005b Simulations of binary drop collisions with a multiple-relaxation-time lattice-Boltzmann model. *Physics of Fluids* **17**(12), 122105.

Premnath KN and Abraham J 2007 Three-dimensional multi-relaxation time (MRT) lattice-Boltzmann models for multiphase flow. *Journal of Computational Physics* **224**(2), 539–559.

Qian J and Law C 1997 Regimes of coalescence and separation in droplet collision. *Journal of Fluid Mechanics* **331**, 59–80.

Qian Y, d'Humières D and Lallemand P 1992 Lattice BGK models for Navier–Stokes equation. *EPL (Europhysics Letters)* **17**(6), 479.

Raabe D 2004 Overview of the lattice Boltzmann method for nano-and microscale fluid dynamics in materials science and engineering. *Modelling and Simulation in Materials Science and Engineering* **12**(6), R13.

Raiskinmäki P, Koponen A, Merikoski J and Timonen J 2000 Spreading dynamics of three-dimensional droplets by the lattice-Boltzmann method. *Computational Materials Science* **18**(1), 7–12.

Rannou G 2008 *Lattice-Boltzmann method and immiscible two-phase flow.* Master's thesis Georgia Institute of Technology.

Rayleigh L 1878 On the instability of jets. *Proceedings of the London Mathematical Society* **10**, 4–13.

Reis T and Phillips TN 2007 Lattice Boltzmann model for simulating immiscible two-phase flows. *Journal of Physics A: Mathematical and Theoretical* **40**(14), 4033–4053.

Rothman D and Keller J 1988 Immiscible cellular-automaton fluids. *Journal of Statistical Physics* **52**(3/4), 1119–1127.

Rothman D and Zaleski S 1997 *Lattice-gas cellular automata: Simple models of complex hydrodynamics.* Cambridge.

Rowlinson J and Widom B 1982 *Molecular Theory of Capillarity.* Clarendon Press, Oxford.

Rutland D and Jameson G 1971 A nonlinear effect in the capillary instability of liquid jets. *Journal of Fluid Mechanics* **46**(2), 267–271.

Sankaranarayanan K, Kevrekidis I, Sundaresan S, Lu J and Tryggvason G 2003 A comparative study of lattice Boltzmann and front-tracking finite-difference methods for bubble simulations. *International Journal of Multiphase Flow* **29**(1), 109–116.

Sankaranarayanan K, Shan X, Kevrekidis I and Sundaresan S 2002 Analysis of drag and virtual mass forces in bubbly suspensions using an implicit formulation of the lattice Boltzmann method. *Journal of Fluid Mechanics* **452**, 61–96.

Sbragaglia M, Benzi R, Biferale L, Succi S and Toschi F 2006 Surface roughness-hydrophobicity coupling in microchannel and nanochannel flows. *Physical Review Letters* **97**(20), 204503.

Sbragaglia M, Benzi R, Biferale L, Succi S, Sugiyama K and Toschi F 2007 Generalized lattice Boltzmann method with multirange pseudopotential. *Physical Review E* **75**(2), 026702.

Scardovelli R and Zaleski S 1999 Direct numerical simulation of free-surface and interfacial flow. *Annual Review of Fluid Mechanics* **31**, 567–603.

Schaap MG, Porter ML, Christensen BSB and Wildenschild D 2007 Comparison of pressure-saturation characteristics derived from computed tomography and lattice Boltzmann simulations. *Water Resources Research* **43**(12), W12S06.

Sehgal B, Nourgaliev R and Dinh T 1999 Numerical simulation of droplet deformation and break-up by lattice-Boltzmann method. *Progress in Nuclear Energy* **34**(4), 471–488.

Shan X 2006 Analysis and reduction of the spurious current in a class of multiphase lattice Boltzmann models. *Physical Review E* **73**(4), 047701.

Shan X 2008 Pressure tensor calculation in a class of nonideal gas lattice Boltzmann models. *Physical Review E* **77**(6), 066702.

Shan X and Chen H 1994 Simulation of nonideal gases and liquid-gas phase transitions by the lattice Boltzmann equation. *Physical Review E* **49**(4), 2941–2948.

Shan X and Doolen G 1996 Diffusion in a multicomponent lattice Boltzmann equation model. *Physical Review E* **54**(4), 3614.

Shan XW and Chen HD 1993 Lattice Boltzmann model for simulating flows with multiple phases and components. *Physical Review E* **47**(3), 1815–1819.

Shan XW and Doolen G 1995 Multicomponent lattice-Boltzmann model with interparticle interaction. *Journal of Statistical Physics* **81**(1-2), 379–393.

Shan XW, Yuan XF and Chen HD 2006 Kinetic theory representation of hydrodynamics: a way beyond the Navier–Stokes equation. *Journal of Fluid Mechanics* **550**, 413–441.

Shao J, Shu C, Huang H and Chew Y 2014 Free-energy-based lattice Boltzmann model for the simulation of multiphase flows with density contrast. *Physical Review E* **89**(3), 033309.

Sinha PK, Mukherjee PP and Wang CY 2007 Impact of GDL structure and wettability on water management in polymer electrolyte fuel cells. *Journal of Materials Chemistry* **17**(30), 3089–3103.

Sofonea V and Mecke K 1999 Morphological characterization of spinodal decomposition kinetics. *European Physical Journal B – Condensed Matter and Complex Systems* **8**(1), 99–112.

Sofonea V and Sekerka RF 2001 BGK models for diffusion in isothermal binary fluid systems. *Physica A: Statistical Mechanics and its Applications* **299**(3), 494–520.

Sofonea V, Lamura A, Gonnella G and Cristea A 2004 Finite-difference lattice Boltzmann model with flux limiters for liquid–vapor systems. *Physical Review E* **70**(4), 046702.

Son G 2001 A numerical method for bubble motion with phase change. *Numerical Heat Transfer, Part B: Fundamentals* **39**(5), 509–523.

Spaid MA and Phelan Jr FR 1998 Modeling void formation dynamics in fibrous porous media with the lattice Boltzmann method. *Composites Part A: Applied Science and Manufacturing* **29**(7), 749–755.

Srivastava S, Perlekar P, Boonkkamp JHMT, Verma N and Toschi F 2013 Axisymmetric multiphase lattice Boltzmann method. *Physical Review E* **88**(1), 013309.

Stratford K, Adhikari R, Pagonabarraga I and Desplat JC 2005 Lattice Boltzmann for binary fluids with suspended colloids. *Journal of Statistical Physics* **121**(1–2), 163–178.

Succi S 2001 *The Lattice Boltzmann Equation for Fluid Dynamics and Beyond*. Oxford University Press.

Succi S, Karlin IV and Chen H 2002 Colloquium: Role of the H theorem in lattice Boltzmann hydrodynamic simulations. *Reviews of Modern Physics* **74**(4), 1203.

Suekane T, Soukawa S, Iwatani S, Tsushima S and Hirai S 2005 Behavior of supercritical CO_2 injected into porous media containing water. *Energy* **30**(11), 2370–2382.

Sukop M and Thorne D 2006 *Lattice Boltzmann Modeling: An Introduction for Geoscientists and Engineers*. Springer.

Sukop MC and Or D 2003 Invasion percolation of single component, multiphase fluids with lattice Boltzmann models. *Physica B: Condensed Matter* **338**(1), 298–303.

Sukop MC and Or D 2004 Lattice Boltzmann method for modeling liquid–vapor interface configurations in porous media. *Water Resources Research* **40**(1), W01509.

Sukop MC and Or D 2005 Lattice Boltzmann method for homogeneous and heterogeneous cavitation. *Physical Review E* **71**(4), 046703.

Sun C, Migliorini C and Munn LL 2003 Red blood cells initiate leukocyte rolling in postcapillary expansions: a lattice Boltzmann analysis. *Biophysical Journal* **85**(1), 208–222.

Sun K, Jia M and Wang TY 2013 Numerical investigation of head-on droplet collision with lattice Boltzmann method. *International Journal of Heat and Mass Transfer* **58**(1–2), 260–275.

Suppa D, Kuksenok O, Balazs AC and Yeomans J 2002 Phase separation of a binary fluid in the presence of immobile particles: A lattice Boltzmann approach. *Journal of Chemical Physics* **116**(14), 6305–6310.

Swift MR, Orlandini E, Osborn WR and Yeomans JM 1996 Lattice Boltzmann simulations of liquid-gas and binary fluid systems. *Physical Review E* **54**(5), 5041–5052.

Swift MR, Osborn WR and Yeomans JM 1995 Lattice Boltzmann simulation of nonideal fluids. *Physical Review Letters* **75**(5), 830–833.

Takada N, Misawa M, Tomiyama A and Fujiwara S 2000 Numerical simulation of two-and three-dimensional two-phase fluid motion by lattice Boltzmann method. *Computer Physics Communications* **129**(1), 233–246.

Takada N, Misawa M, Tomiyama A and Hosokawa S 2001 Simulation of bubble motion under gravity by lattice Boltzmann method. *Journal of Nuclear Science and Technology* **38**(5), 330–341.

Teng S, Chen Y and Ohashi H 2000 Lattice Boltzmann simulation of multiphase fluid flows through the total variation diminishing with artificial compression scheme. *International Journal of Heat and Fluid Flow* **21**(1), 112–121.

Tiribocchi A, Stella N, Gonnella G and Lamura A 2009 Hybrid lattice Boltzmann model for binary fluid mixtures. *Physical Review E* **80**(2), 026701.

Tölke J 2001 *Gitter-Boltzmann-Verfahren zur Simulation von Zweiphasenströmungen*, PhD thesis Lehrstuhl Bauinformatik, Munich.

Tölke J, Freudiger S and Krafczyk M 2006 An adaptive scheme for LBE multiphase flow simulations on hierarchical grids. *Computers & Fluids* **35**, 820–830.

Tölke J, Krafczyk X, Schulz M and Rank E 2002 Lattice Boltzmann simulations of binary fluid flow through porous media. *Philosophical Transactions A: Mathematical, Physical and Engineering Sciences* **360**(1792), 535–545.

Van der Graaf S, Nisisako T, Schroen C, Van Der Sman R and Boom R 2006 Lattice Boltzmann simulations of droplet formation in a t-shaped microchannel. *Langmuir* **22**(9), 4144–4152.

Van der Sman R and Van der Graaf S 2006 Diffuse interface model of surfactant adsorption onto flat and droplet interfaces. *Rheologica Acta* **46**(1), 3–11.

Van der Sman R and Van der Graaf S 2008 Emulsion droplet deformation and breakup with lattice Boltzmann model. *Computer Physics Communications* **178**(7), 492–504.

Varnik F, Truman P, Wu B, Uhlmann P, Raabe D and Stamm M 2008 Wetting gradient induced separation of emulsions: A combined experimental and lattice Boltzmann computer simulation study. *Physics of Fluids (1994–present)* **20**(7), 072104.

Vogel HJ, Tölke J, Schulz V, Krafczyk M and Roth K 2005 Comparison of a lattice-Boltzmann model, a full-morphology model, and a pore network model for determining capillary pressure–saturation relationships. *Vadose Zone Journal* **4**(2), 380–388.

Vrancken RJ, Kusumaatmaja H, Hermans K, Prenen AM, Pierre-Louis O, Bastiaansen CW and Broer DJ 2009 Fully reversible transition from Wenzel to Cassie–Baxter states on corrugated superhydrophobic surfaces. *Langmuir* **26**(5), 3335–3341.

Wagner A and Yeomans J 1999 Phase separation under shear in two-dimensional binary fluids. *Physical Review E* **59**(4), 4366.

Wagner AJ 2003 The origin of spurious velocities in lattice Boltzmann. *International Journal of Modern Physics B* **17**(1–2), 193–196.

Wagner AJ 2006 Thermodynamic consistency of liquid–gas lattice Boltzmann simulations. *Physical Review E* **74**(5), 056703.

Wagner AJ and Yeomans J 1998 Breakdown of scale invariance in the coarsening of phase-separating binary fluids. *Physical Review Letters* **80**(7), 1429.

Wang L, Huang HB and Lu XY 2013 Scheme for contact angle and its hysteresis in a multiphase lattice Boltzmann method. *Physical Review E* **87**(1), 013301.

Wolf-Gladrow D 2000 *Lattice-Gas Cellular Automata and Lattice Boltzmann Models*. Springer.

Wolfram S 1983 Statistical mechanics of cellular automata. *Reviews of Modern Physics* **55**(3), 601–644.

Wolfram S 2002 *A new kind of science* vol. 5. Wolfram Media Champaign.

Wu L, Tsutahara M, Kim LS and Ha M 2008 Three-dimensional lattice Boltzmann simulations of droplet formation in a cross-junction microchannel. *International Journal of Multiphase Flow* **34**(9), 852–864.

Xi H and Duncan C 1999 Lattice Boltzmann simulations of three-dimensional single droplet deformation and breakup under simple shear flow. *Physical Review E* **59**(3), 3022.

Xu A 2005 Finite-difference lattice-Boltzmann methods for binary fluids. *Physical Review E* **71**(6), 066706.

Xu A, Gonnella G and Lamura A 2003 Phase-separating binary fluids under oscillatory shear. *Physical Review E* **67**(5), 056105.

Xu A, Gonnella G and Lamura A 2004 Phase separation of incompressible binary fluids with lattice Boltzmann methods. *Physica A: Statistical Mechanics and its Applications* **331**(1), 10–22.

Yan Y and Zu Y 2007 A lattice Boltzmann method for incompressible two-phase flows on partial wetting surface with large density ratio. *Journal of Computational Physics* **227**(1), 763–775.

Yang Z, Dinh TN, Nourgaliev R and Sehgal B 2001 Numerical investigation of bubble growth and detachment by the lattice-Boltzmann method. *International Journal of Heat and Mass Transfer* **44**(1), 195–206.

Yang Z, Palm B and Sehgal B 2002 Numerical simulation of bubbly two-phase flow in a narrow channel. *International Journal of Heat and Mass Rransfer* **45**(3), 631–639.

Yiotis AG, Psihogios J, Kainourgiakis ME, Papaioannou A and Stubos AK 2007 A lattice Boltzmann study of viscous coupling effects in immiscible two-phase flow in porous media. *Colloids and Surfaces A: Physicochemical and Engineering Aspects* **300**(1–2), 35–49.

Young T 1805 An essay on the cohesion of fluids. *Philosophical Transactions of the Royal Society of London* **95**, 65–87.

Yu D, Mei R, Luo LS and Shyy W 2003 Viscous flow computations with the method of lattice Boltzmann equation. *Progress in Aerospace Sciences* **39**(5), 329–367.

Yu Z and Fan LS 2010 Multirelaxation-time interaction-potential-based lattice Boltzmann model for two-phase flow. *Physical Review E* **82**(4), 046708.

Yu Z, Hemminger O and Fan LS 2007 Experiment and lattice Boltzmann simulation of two-phase gas–liquid flows in microchannels. *Chemical Engineering Science* **62**(24), 7172–7183.

Yuan P and Schaefer L 2006 Equations of state in a lattice Boltzmann model. *Physics of Fluids* **18**(4), 042101.

Yue P, Zhou C and Feng JJ 2007 Spontaneous shrinkage of drops and mass conservation in phase-field simulations. *Journal of Computational Physics* **223**(1), 1–9.

Zhang J 2011 Lattice Boltzmann method for microfluidics: models and applications. *Microfluidics and Nanofluidics* **10**(1), 1–28.

Zhang J and Kwok DY 2004 Lattice Boltzmann study on the contact angle and contact line dynamics of liquid–vapor interfaces. *Langmuir* **20**(19), 8137–8141.

Zhang J and Kwok DY 2005 A 2D lattice Boltzmann study on electrohydrodynamic drop deformation with the leaky dielectric theory. *Journal of Computational Physics* **206**(1), 150–161.

Zhang J, Li B and Kwok DY 2004 Mean-field free-energy approach to the lattice Boltzmann method for liquid–vapor and solid–fluid interfaces. *Physical Review E* **69**(3), 032602.

Zhang R and Chen H 2003 Lattice Boltzmann method for simulations of liquid–vapor thermal flows. *Physical Review E* **67**(6), 066711.

Zhang RY, He XY and Chen SY 2000 Interface and surface tension in incompressible lattice Boltzmann multiphase model. *Computer Physics Communications* **129**(1–3), 121–130.

Zheng HW, Shu C and Chew YT 2006 A lattice Boltzmann model for multiphase flows with large density ratio. *Journal of Computational Physics* **218**(1), 353–371.

Zheng L, Lee T, Guo Z and Rumschitzki D 2014 Shrinkage of bubbles and drops in the lattice Boltzmann equation method for nonideal gases. *Physical Review E* **89**(3), 033302.

Zhu L, Tretheway D, Petzold L and Meinhart C 2005 Simulation of fluid slip at 3D hydrophobic microchannel walls by the lattice Boltzmann method. *Journal of Computational Physics* **202**(1), 181–195.

Zou QS and He XY 1997 On pressure and velocity boundary conditions for the lattice Boltzmann BGK model. *Physics of Fluids* **9**(6), 1591–1598.

Index

adhesion force, 35
adhesion parameter, 76
advancing contact angle, 216
analytical solution, 58, 108, 311, 317
apparent contact angle, 217
artificial gradient, 302
Atwood number, 207, 208
axisymmetric boundary condition, 258, 263
axisymmetric model, 334, 345
axisymmetric multiphase LBM, 255
axisymmetric two-phase N-S equation, 339

BD (biased difference), 298, 302
BGK (Bhatnagar-Gross-Krook), 2, 3, 72, 274
biased derivative, 303
body force, 42
Bond number, 13, 38
bounce-back, 35, 39, 103, 155, 158, 206, 211, 258, 299, 311, 354
boundary condition, 33, 35, 38, 39, 47, 59, 83, 87, 89, 91, 95, 112, 117, 151, 154, 175, 206, 207, 211, 215, 258, 260, 299, 314, 317, 348, 354
breakthrough, 118
bubble rise, 279, 281, 353
bubble wake, 286
bulk free-energy density, 137

C-S EOS, 34, 36
Cahn-Hilliard equation, 173, 196, 253, 255, 291
capillary filling, 111, 112
capillary fingering, 118
capillary number, 12, 118
capillary pressure, 84, 111
capillary rise, 36, 213
Carnahan-Starling (C-S) EOS, 26, 27, 198, 253
Cartesian mesh, 19, 258, 277
Cellular Automata, 2
central difference (CD), 298, 302
central difference scheme, 156
Central Processing Unit (CPU), 6
CFD (computational fluid dynamics), 6
CFL (Courant-Friedrichs-Lewy) number, 12
chain rule, 146
Chapman-Enskog, 199
Chapman-Enskog expansion, 99, 143, 170, 288, 338

chemical potential, 303
coalescence, 351
coexistence, 150
coflow, 117
cohesion force, 73
color-gradient model, 7
compressibility, 77, 93, 207, 208
compressibility effect, 310
contact angle, 34, 78, 80, 82, 114, 154, 210
contact angle hysteresis, 215
contact line, 216
contact point, 216
continuity equation, 100, 143, 254
corresponding states, 26
critical temperature, 22
curvature-induced pressure, 151
cylindrical coordinates, 335, 338

D2Q7, 4
D2Q9, 4
D2Q9 velocity model, 140
D3Q15, 4, 167
D3Q19, 4
Darcy velocity, 13
density derivatives, 143
density shift, 357
diffuse-interface method, 54
directional derivative, 298, 341
discrete velocity, 3, 255
discrete velocity models, 4
displacement, 83–85, 87, 91
dissolved density, 76, 78
distribution function, 3
drainage, 111
drop impacting, 253
droplet collision, 175, 265
droplet splashing, 299, 342
dummy index, 15
dummy variable, 15

EDF (Equilibrium Distribution Function), 335
effective mass, 73
Einstein summation, 14
ellipsoidal droplet, 219, 259
EOS (Equation of State), 19, 21, 152

Multiphase Lattice Boltzmann Methods: Theory and Application, First Edition.
Haibo Huang, Michael C. Sukop and Xi-Yun Lu.
© 2015 John Wiley & Sons, Ltd. Published 2015 by John Wiley & Sons, Ltd.
Companion Website: www.wiley.com/go/huang/boltzmann

equilibrium distribution function, 3, 138, 168, 255, 302, 336, 347

equilibrium momentum tensor, 147

equilibrium state, 20

equilibrium stress tensor, 148, 149

equilibrium velocity, 20

Euler equation, 74, 101, 147

exact difference method, 41, 44

Eötvös number, 281

FE (free energy), 8

filter technique, 279

first-order Euler scheme, 278

first-order partial derivative, 155

flow pattern, 120

fluid-fluid cohesion, 72, 73

fluid-solid adhesion, 72, 73

forcing term, 40, 41, 43, 44

fourth-order difference, 270

free-energy, 137

free-energy-based multiphase LBM, 136

front-tracking method, 6

Galilean invariance, 146, 201, 313, 314

Geometric scheme, 215

ghost cells, 215

gravity, 38, 207

gravity acceleration, 214

grid-independence study, 12

H theorem, 6

HCZ (He-Chen-Zhang) model, 9, 196

head-on collision, 176, 265, 352

head-on droplet collision, 342, 348

high density ratio, 48

high-density-ratio model, 298

high-density-ratio two-phase flows, 167, 292

horizontal interface, 151

hydrodynamic pressure, 198, 255, 336

hysteresis, 215

hysteresis effect, 217

hysteresis window, 216, 218

ideal wall, 218

ideal-gas equation of state, 6

implementation, 347

implicit method, 301

Inamuro model, 167

incompressibility, 174, 282

incompressible flow, 174

index function, 196, 198, 255, 256, 301

inter-particle force, 20, 21, 55, 57

Inter-particle Force Model A, 31, 32

Inter-particle Force Model B, 31, 33

interface-tracking, 196

interface-tracking equation, 202

interfacial tension, 34, 95, 98, 99, 110, 168

intermolecular forcing term, 293, 338

intrinsic permeability, 118

invariant form, 26

isotropic discretization, 301, 334, 341

isotropy, 110

kinematic viscosity, 3, 72, 102, 197, 281, 299

kinetic energy, 176

Kronecker delta, 15

l'Hôpital's rule, 258

Lamb's natural resonance frequency, 220

Laplace law, 77, 83, 84, 95, 110, 155, 203, 258, 282, 306, 307

Laplacian, 267, 270

Lattice Boltzmann Equation, 72, 143

lattice Gas model, 94

lattice pinning, 94

lattice speed, 5

lattice tensor, 15

lattice unit, 11

layered two-phase flow, 204, 311, 317

Lee-Lin model, 10, 280

level set method, 6

main fluid density, 78

mass conservation, 260, 289, 356

mass correction, 278

mass correction technique, 334, 357

Maxwell Construction, 24, 49, 150, 175, 295

mechanical stability condition, 49, 50

mixed direction derivative, 303

mobility, 177, 197, 253, 292, 351

molecular interactions, 196

momentum conservation, 289

momentum equation, 171

momentum flux tensor, 101

Morton number, 281

MRT (Multiple-Relaxation-Time), 92, 113, 274

MRT collision model, 257

multi-component, 81

Multi-Component MultiPhase (MCMP), 7, 71

multirange pseudopotential, 55

natural boundary condition, 154

Navier-Stokes (N-S), 143

no-slip wall, 35, 103, 117, 258, 354

non-equilibrium bounce-back, 89

non-equilibrium distribution function, 172, 177

non-Galilean invariance, 148, 149

non-ideal surfaces, 216

non-wetting fluid, 118

numerical instability, 207

numerical stability, 342, 348

oblate ellipsoidal, 284

oblate ellipsoidal cap, 284

Ohnesorge number, 266

oil-water flow, 12

order parameter, 168

oscillation, 219, 259

parallel flow, 39, 58, 85, 103
perturbation operator, 94
phase separation, 153, 202
Poisson equation, 167, 174
porous medium, 91
post-stream, 294
post-streaming collision, 297
post-streaming step, 341
potential form, 294, 337
pre-stream, 294
pre-streaming collision, 297, 340
pressure boundary condition, 84, 88
pressure tensor, 21, 29, 33, 137, 173

random number, 153
Rayleigh–Taylor, 196
Rayleigh-Benard convection, 196
Rayleigh-Taylor instability, 196, 205, 209
receding contact angle, 216
recoloring step, 94, 98
Redlich-Kwong (R-K), 26, 27
relative permeabilities, 39, 40, 59
relative velocity, 176
relaxation time, 3, 12, 72, 97, 174, 197, 198, 255,
    259, 263, 282, 342
Reynolds number, 175, 207, 265, 299, 342
RK model, 95
Rothman-Keller (RK), 7

satellite droplets, 266
satellite drops, 263
SC (Shan-Chen) model, 7, 18
second-order derivative, 155
secondary wake circulations, 286
Shan-Chen (SC) EOS, 29
shrinkage, 357
Single Component MultiPhase (SCMP), 7, 18
sixth-order central difference, 270
slip boundary condition, 258
slip wall, 299
sound speed, 3, 72
source term, 3, 337, 339

spike, 208
spurious currents, 8, 19, 94, 115, 151, 259,
    309
stable displacement, 118
stress form, 294, 337
supercritical state, 22, 28
surface tension, 32, 33, 50, 52, 202, 277, 335
surface tension model, 302

tangent plane, 216
Taylor expansion, 31, 56, 74, 99, 346
temporal derivative, 171
thermodynamic coexistence, 56
thermodynamic consistency, 29, 42, 48, 49, 57
thermodynamic instability, 151
thermodynamic pressure, 197, 253
time step-independence study, 12
toroidal shape, 176
typographical error, 148, 256, 276, 297, 303, 335,
    336, 338

upwind finite difference scheme, 278

van der Waals, 22, 25, 27
van der Waals EOS, 155, 169
van der Waals fluid, 138
viscosity term, 102
viscous dissipation, 219, 259
viscous fingering, 118
viscous stress tensor, 168, 172, 177, 293
viscous tensor, 172
volume-of-fluid, 6

wall density, 35, 73
Weber number, 175, 208, 265, 299, 342
weighting coefficients, 3, 142, 197
wettability, 218
wetting fluid, 118
wetting potential, 154

Young-Laplace equation, 36, 79

Printed and bound by CPI Group (UK) Ltd, Croydon, CR0 4YY
17/11/2021
03092587-0001